D1250714

Mastering the Market

The State and the Grain Trade in Northern France, 1700–1860

Judith Miller's *Mastering the Market* is the first book to bring a long-term perspective to the historical study of French strategies to control the grain trade, rather than focusing only on the eighteenth century or the Revolutionary era, as previous scholars have done.

The grain trade – a crucial sector of the French economy – caused enormous concern throughout the eighteenth and early nineteenth centuries. Bread was the staple of French diets, so harvest shortfalls frequently triggered periods of unrest. The royal government had no effective means of drawing foodstuffs into restless cities during these times.

Professor Miller shows how successive regimes developed strategies to dominate the baking trades, influence prices along vital supply lines, and amass emergency stocks of grain that could meet months-long demand. She shows that as free-trade ideologies developed, French administrators at both the local and national levels sought to reconcile these ideologies with the perceived need to control the market.

Miller draws on an extensive range of archival material from Parisian and Departmental sources to show – surprisingly – that free trade in grain was rendered workable only through the French state's intervention.

Judith A. Miller is Associate Professor of History at Emory University. She is a recipient of the Society for French Historical Studies' William H. Koren Prize for the best article written by a North American, and the Economic History Association's Alexander Gerschenkron Dissertation Prize. Miller has received fellowships from the National Endowment for Humanities and the Bourses Chateaubriand (French government).

Mastering the Market

The State and the Grain Trade in Northern France, 1700–1860

Judith A. Miller
Emory University

PUBLISHED BY THE PRESS SYNDICATE OF THE UNIVERSITY OF CAMBRIDGE
The Pitt Building, Trumpington Street, Cambridge CB2 1RP, United Kingdom

CAMBRIDGE UNIVERSITY PRESS
The Edinburgh Building, Cambridge CB2 2RU, UK http://www.cup.cam.ac.uk
40 West 20th Street, New York, NY 10011-4211, USA http://www.cup.org
10 Stamford Road, Oakleigh, Melbourne 3166, Australia

First published 1999

Printed in the United States of America

Typeset in Sabon 10/12 pt, in Quark XPress™ [BTS]

Library of Congress Cataloging-in-Publication Data is available.

A catalog record for this book is available from the British Library

ISBN 0 512 62129 1 hardback

For my family, with love.

Contents

Figures and Tables

Figures

Tables

Abbreviations

AESC	*Annales. Economies, sociétés, civilisations*
AHR	*American Historical Review*
AHRF	*Annales historiques de la Révolution française*
EHR	*Economic History Review*
FHS	*French Historical Studies*
JEH	*Journal of Economic History*
JIH	*Journal of Interdisciplinary History*
JMH	*Journal of Modern History*
JSH	*Journal of Social History*
RHMC	*Revue d'histoire moderne et contemporaine*

Old Regime Weights and Measures for Wheat

Livre, poids de marc (Paris)	1.079 lbs
Livre, poids de vicomté (Rouen)	1.122 lbs
Livre, poids du roy (Honfleur, Harfleur)	1.165 lbs
Boisseau (Rouen)	22.75 liters
Boisseau (Paris)	13.01 liters
Boisseau (Le Havre)	39.1 liters
Mine (Rouen)	91 liters
Somme (Rouen)	330–340 lbs
Setier (Rouen)	182 liters
Setier (Paris)	156.1 liters
Muid (Rouen)	21.84 hectoliters (app. 3,747 lbs)
Muid (Paris)	18.73 hectoliters (app. 3,214 lbs)

"Mémoire sur l'unité des poids et mesures dans la généralité de Rouen . . . ," n.d., ADSM C 2120, p. 64; Ronald Edward Zupko, *French Weights and Measure Before the Revolution: A Dictionary of Provincial and Local Units* (Bloomington, 1978); Horace Doursther, *Dictionnaire universel des poids et mesures . . .* (n.p., 1850, reprint ed., Amsterdam, 1945); Arundel de Condé, "Mesures de capacité dans les foires et marchés, *Circuits commerciaux: Foires et marchés en Normandie* (Rouen, [1975]), pp. 229–30; J.H. Alexander, *Universal Dictionary of Weights and Measures* (Baltimore, 1850), pp. 55–6. A hectoliter of grain generally weighed between 76 and 80 kilograms.

Acknowledgments

Ray Bradbury has written that if you consulted only your intellect, you would never begin a friendship, nor would you would ever fall in love. Listening to your intellect alone is "nonsense." "You've got to jump off cliffs all the time and build your wings on the way down," he counseled. Perhaps he knew something about the less exalted voyage that begins with a dissertation and ends with a book. Certainly no graduate student can fully realize where a vague idea that a research paper generated might lead some years hence. This former graduate student did not. Fortunately, however, the path from research seminar to this book was one that brought more friendships than cliffs. That community of friends has helped build whatever wings were needed, and for such a gift, I am immensely grateful.

A number of fellowships have made it possible to think of Paris and Rouen as second homes. The Bourses Chateaubriand, Duke University, and Fulbright fellowships supported my dissertation research. Later, funding by the National Endownment for the Humanities, the Economic History Association, the American Council of Learned Societies, the University of Tennessee at Knoxville, and Emory University permitted me time at archives in France and the Library of Congress and as a visiting fellow at the Center for European Studies at Harvard.

Once in France, I had the great fortune to find a community of scholars with whom I have enjoyed intense discussions, archival forays, French Historical Studies banquets, and numerous *kirs* at the end of the day at the Petit Berry, a site that has had as deep an influence on the field of history as the salles Soubise and Clisson.

I am deeply indebted to the personnel of the many archives I have visited, especially Viviènne Miguet, formerly of the Archives de la Seine-Maritime, and Eric Dufour at the Archives Nationales. Caroline Ford, Liana Vardi, Françoise Bléchet, Philip Hoffman, Jacques Bottin, Colin Lucas, Bertrand Roehner, Philippe Minard, Jean-Pierre Hirsch, Claude and Simone Mazauric, and the late Harry Miskimin are but a few of the scholars and friends who have made the profession a place of both ideas and humanity.

Peter Jones, William Doyle, Daryl Hafter, Donald Sutherland, Clay Ramsay, Michael Sonenscher, Cynthia Bouton, Howard Brown, and especially Timothy Le Goff, read all or part of the manuscript, and in some cases, patiently saw it through several revisions.

The anonymous readers further gave very welcome advice, and editor Frank Smith of Cambridge University Press moved the final version along with lightning speed.

Ronald Cohen, my manuscript editor, dispensed equal parts of sage counsel and patient encouragement, and made the final stages of the process a pleasure.

This project took shape in graduate school at Duke University while I was working with William Reddy, whose questions and insights pushed the dissertation along. At the University of Tennessee, Knoxville, departmental colleagues John Morrow, John Bohstedt, James Farr, and Palmira Brummet were among those who gave encouragement. Later, at Emory University, a number of colleagues and friends, among them Irene Browne, Beth Reingold, Niall Slater, David Carr, Matthew Payne, Stephen White, Elizabeth Fox-Genovese, J. Russell Major, Rondo Cameron, Walter Adamson, Richard Hansen, Nan McMurry, Wendy Lord, Magda Walter, and especially William Beik, Kathryn Amdur, and Randall Packard, provided advice and inspiration. Eric Nitschke of the Reference Department of Woodruff Library should be singled out for his enthusiasm and assistance, along with the staff of the Interlibrary Loan Office. Research assistants, especially Rebecca Wendelken, Matt Dunne, Jeff Young, Lisa Greenwald, and Jeremy Hayhoe, then at York University, provided essential help along the line.

This project took several more years than expected to complete. In fact, that it is completed at all is something of a miracle. In 1990, I began to fall ill. Two years later, I was diagnosed with leukemia. In 1993, I received a bone marrow transplant from my brother, Robert Miller. An extraordinary medical team – Robert Geller, John Wingard, Kent Holland, Ira Horowitz, Mary Potter, and Ingrid Evans, among many others – saw me back to health.

My family took turns flying and driving to Atlanta to care for me for many months. Without their unfailing love and support, I could not have survived. This book is dedicated to them.

There are also three friends whom I met years ago at Duke whose companionship has been as necessary and sustaining as the wings Bradbury described – Melinda Reagor Flannery, Steven Wilf, and Charles Maier.

I am not sure that a book can in any way repay the energy and sacrifices of these medical teams, friends, and family members. I hope that its completion is at least a tribute to their faith that wings could be built and that I would fly, even if the cliffs were ragged and steep.

INTRODUCTION

Two Crises: 1709 and 1853

This book is about the reformulation of the French state during a critical century and a half of wars, famines, and revolutions. It takes as its theme the persistent problem of *subsistances* – the maintenance of the grain supply. The problem of the grain supply was a perpetual concern of townsfolk in early modern France. At the slightest rumor of failure, crowds in cities and towns would assail farmers as they brought sacks of grain to market, break into bakeries, and harass any visible representatives of public authority. In the countryside, too, such rumors halted the flow of grain, as rural consumers dissuaded producers, carters, or boatmen from moving cereals to urban markets, or even confiscated grain from barns or granaries. When supplies appeared to be short, officials grew edgy all over the country. The king, his ministers, his intendants, and greater and lesser law courts would struggle to keep order and to ensure that a minimum of grain and bread got to the consumers.

There were many reasons for such behavior. In most years, the country could produce enough cereals to feed itself. But periodically there were significant shortages. These occurred often enough – once every fifteen or twenty years on average – and the consequences were so disastrous that the memory of unsavory deeds done at those moments by desperate consumers and greedy suppliers left their mark on people of almost all ranks of society. The result was an abiding conviction among most people that the food supply was too important to be left to individuals. When the great jurist and political philosopher Montesquieu wrote in 1748 that the state "owes all its citizens regular means of subsistence," he was expressing what all but a tiny minority of French people took to be a truism.[1]

The tale of the shift from this regime of fear, bumbling, and disorder to

[1] Charles de Secondat, Baron de la Brède et de Montesquieu, *The Spirit of the Laws*, in *Selected Political Writings*, trans. and ed., Melvin Richter (Indianapolis and Cambridge, 1976), p. 228.

the free flow of supplies in a liberal economy has often been told. The story has been cast most often in the mode of political mobilization – the coming of a free market in grain provoking confusion and misgivings among Old Regime rulers and ruled alike. Both consumers and elites lost their faith in the institutions of a monarchical regime that had abandoned its responsibilities to the poor. Those who starved blamed the Crown and its new policies; those whose wealth sheltered them from the worst lamented the turmoil and suffering the policies had wrought. When the monarchy vacillated and backed away from free trade in 1770, in 1778, and again in 1788, it confirmed the growing suspicion that neither the king nor his ministers knew what to do. The coming of the French Revolution could be found in the stumbling policies of the grain trade alone – political disarray, popular resistance to capitalism, and the deepening sense that no true solutions would emanate from the chambers of Versailles.

This model of political mobilization characterizes most historical work on the organization of the cereal trade in the last twenty-five years. It doubtless has its uses, both as a historical explanation and for bolstering contemporary ideology. But there is another thread in the story that is seldom mentioned – the recreation of the state's power in the face of anarchy. For on reflection, what is striking about the period is less the disruptive effect of food shortages on the French state in the Old Regime, the Revolution, and after than the forces that restored order. Its actors are neither rioters nor sansculottes, although they hover in the wings. Instead, the story concerns the determined officials who mobilized resources to maintain order. They knew that provisioning could not be abandoned to the vagaries of the market; they knew equally well that regulations drove merchants away. Through their efforts, the state stepped in to smooth pathways to cities, supplementing and sheltering commercial networks while subsidizing merchants, bakers, and buyers. In the midst of wars, shortages, and revolutions, the state consolidated and defined its powers.

This book is intended for those who have caught the urgency of the insistent calls for total free trade by the liberal economists of the 1750s and 1760s such as Quesnay and Turgot, and have realized the contradiction between free trade and the sansculottes' cries for price ceilings in the Revolution, but who have not paused to consider the meaning of carefully calculated bread price schedules or export duties tied to Parisian flour prices. It is also for those who can measure economic performance but would prefer to keep the government at some distance from their models.

The evolving relationship between the state and economy can be found in daily tasks – the incessant struggles over how to set and enforce bread prices; private meetings with landlords who refused to send grain to market; scribes filling out endless forms, sometimes misadding their columns; secret government imports quietly sent to market; and mayors perched nervously on wagons, wrangling with crowds over the ownership of a single sack of grain. The broad picture that emerges is one of increasingly skillful state intervention in the economy through many decades of unrest, hunger, and war.

The book takes issue with the way our scholarly energies have been directed toward the question of what destroyed the state, whether during the Old Regime, the constitutional phase of the Revolution, the Terror, or the Directory. Within that historiography, provisioning crises hold an esteemed and dignified position. Whether or not Marie Antoinette ever dismissed the hungry with her famous admonition that they eat cake, the connection between shortages and violence is indisputable. Pushing such inquiries further, several generations of historians have used food riots to illuminate the poor's hostility to nascent capitalism. Their studies reveal the increasing politicization of provisioning in the waning years of the eighteenth century. An embittered working class was caught between low wages and rising food prices, their livelihoods further eroded by the liberal economic policies that effectively stripped them of the fragile security of guilds and paternalist policies.[2] In this framework, Marxist historians saw the nucleus of class warfare.[3] The equation seemed simple. Angry consumers in France not only seized grain wagons, they also marched a king to the guillotine and burst into the Convention demanding "bread or death."

Against this backdrop of working-class agitation, revisionist historians have postulated an alternative source for the attacks directed at the monar-

[2] E.P. Thompson's article gave the fullest articulation of the idea of food rioters' resistance to capitalism, "The Moral Economy of the English Crowd in the Eighteenth Century," *Past and Present* 50 (1971): 76–136. See also Cynthia Bouton, *The Flour War: Gender, Class, and Community in Late Ancien Régime French Society* (University Park, PA, 1993); Olwen Hufton, "Social Conflict and Grain Supply in Eighteenth-Century France," *JIH* 14 (1983): 303–331; Georges Rudé, *The Crowd in History: A Study of Popular Disturbance in France and England 1738–1840* (New York, 1964); Guy Lemarchand, "Les troubles de subsistances dans la généralité de Rouen," *AHRF* 35 (October–December 1963): 401–27; C.-E. Labrousse, *Esquisse du mouvement des prix et des revenus en France au XVIIIe siècle*, 2 vols., *Collection scientifique d'économie politique* 3 (Paris, 1933). Some of these points about Old Regime worklife have been problematized recently. See Michael Sonenscher, *Work and Wages: Natural Law, Politics and the Eighteenth-Century French Trades* (Cambridge, 1989); David R. Weir, "Les crises économiques et les origines de la Révolution française," *AESC* 46 (1991): 917–47.

[3] Georges Lefebvre, *The French Revolution*, trans. Elizabeth Moss Evanson (London and New York, 1962–4); Albert Soboul, *The Sans-culottes: The Popular Movement and Revolutionary Government, 1793–1794*, trans. Remy Inglis Hall (Princeton, 1980); Claude Mazauric, *Babeuf et la conspiration pour l'égalité* (Paris, 1962); R.B. Rose, "18th-Century Price-Riots, the French Revolution and the Jacobin Maximum," *International Review of Social History*, 4 (1959): 438–9; and "The French Revolution and the Grain Supply," *Bulletin of the John Rylands Library* 39 (1956–1957): 171–87.

chy. Their literature follows the pitched battles that destroyed absolutism ideologically. The origins of this scholarly tack are complex, but in short, this critique of the Marxist model scrutinized the actions and beliefs of the classes that *theoretically* drove the Revolution. Very quickly, the revisionists' research revealed that none of the historical actors stuck to the expected script. The Revolution was not led by a capitalist bourgeoisie intent on establishing free trade, but by lawyers and notables; section militants were not suitably proletarian, but perhaps opportunistic shopkeepers and neighborhood demagogues; landed elites, while shaken by the Revolution, consolidated their power in the early nineteenth century.[4] Having thus dashed the Marxist model, the revisionists developed political and cultural explanations for the monarchy's demise. Directing attention to ideological conflicts and the emergence of a public sphere in eighteenth-century France, these scholars charted absolutism's utter disintegration in the decades before 1789.[5] This literature now locates the impending breakdown of the Old Regime in the struggles between the Crown and its sovereign courts. Those contests escalated after the 1750s, and called each side's claims to legitimacy pointedly into question. By 1789, these battles had spilled out into a wider public. Impassioned journalists, lawyers, and novelists and their steadily expanding readership brought their indictments right up to the gates of Versailles. The king himself was corrupt and despotic and had betrayed the nation.

The post-revisionist model implies first that the monarchy was to blame for its own demise, or at least its own ideological destruction, and, second, that the Old Regime state was not only illegitimate, but ineffective. Perhaps as early as the mid-eighteenth century, these historians have argued, the Crown's cause was lost. It had ceded ground to its enemies by fighting the war on their terms. Each time the Crown invoked the public as its judge, it undermined even its most convincing assertions of authority. Every royal attempt to shore up absolutism, whether by disbanding parlements, by declaring partial bankruptcies, or by changing the rules of venal offices, gave credence to allegations that monarchy and despotism were one and the same. Such was the absolutist state of pamphlets, playwrights, and garrulous lawyers. It was grasping, confused and, ultimately, powerless and

[4] See especially William Doyle, *The Origins of the French Revolution*, 2nd ed. (Oxford, 1988).

[5] Jürgen Habermas, *The Structural Transformation of the Public Sphere: An Inquiry into a Category of Bourgeois Society*, trans. Thomas Burger (Cambridge, Mass., 1989); *The French Revolution and the Creation of Modern Political Culture*, Keith Baker, Colin Lucas, and François Furet, eds., 4 vols. (Oxford, 1987–94); Furet, *Interpreting the French Revolution*, trans. Elborg Forster (Cambridge and Paris, 1985); Lynn Hunt, *Politics, Culture and Class in the French Revolution* (Berkeley, 1984); Sarah Maza, *Private Lives and Public Affairs: The Causes Célèbres of Prerevolutionary France* (Berkeley, 1993).

illegitimate. Long before the opening processions of the Estates General, the corridors of power were vacant.

Such courtroom fireworks and public invective no doubt created a stir at the time, and did indeed signal a real crisis of authority. But I argue that such accounts present only one vision of the state. There was another state to be found outside the opera houses, salons, and other venues of recent scholarship. It had other countenances beyond the grizzled ones presented in *causes célèbres* newsheets and Grub Street bestsellers. It was there on the ground in the late eighteenth century, intermittently emerged in skeletal form during the Revolution, and finally achieved coordination and effectiveness in the early nineteenth century.[6] It labored in the bureaus of intendancies, in town halls, and in subdelegates' cramped chambers. Venturing into bakeries and marketplaces, it distributed profits and losses.

This state was not rudderless. Instead, it was conscientious, determined to navigate the twists and turns of policy changes and harvest shortfalls. The windows of countless petty officials looked out on market squares and bakeries, and these men knew that, day in and day out, people had to eat and that their charge was to feed them. Each time the central government changed course, lifting or reimposing the controls on the grain trade, for instance, they took note, adapted, and returned to their task. At times, these market officials, subdelegates, mayors, and police lieutenants were helpless before desperate crowds. The means they had to combat harvest shortfalls, muddy roads, and grain owners' intransigence were often insufficient. By the end of the eighteenth century, these officials also had the failed promises of both liberal economic ideologies and the harsh controls of the Revolution to explain. Nonetheless, they did not give up.

While one could see in the food riots of the Old Regime and the Revolution one further sign of the ideological bankruptcy of the state, that would be to misunderstand fundamentally the process at hand. The state was not tottering – waiting to fall – nor was it lost in either 1789 or 1794. Those stormy moments were only waystations in a longer process of growing competence and authority. It is a mistake to see the Old Regime's policial and economic policies as simply an endless series of miscalculations. This book tells a different story. It outlines the process through which the state harnessed the grain trade and turned it to meet the needs of cities over the course of nearly two centuries. This was a vital and beneficial undertaking if there was to be any hope for political stability.

[6] See P.M. Jones's recent synthesis, which suggests that the monarchy and many of its ministers evinced enthusiasm for reform throughout the late Old Regime. *Reform and Revolution in France: The Politics of Transition, 1774–1791* (Cambridge, 1996), especially pp. 88–9.

A Longer Time Frame: 1700–1860

In order to make such a case, the problem of *subsistances* must be placed in a long time frame, showing how, in a century and a half, administrators moved from confusion to assurance. The extension of this study into the nineteenth century points to a second glaring problem with historiography – namely, the prevailing formulation of the period's chronology. To date, much of the work on the Revolution has centered on either its origins or its radicalization under the Jacobins. Even the few historians who have continued to examine the economy have generally accepted those boundaries and have stopped short in either 1789 or 1794.[7] There remains, however, the crucial issue of how the Revolution was resolved. How did the denouement of the Revolution shape the first half of the nineteenth century? Of Napoleonic battles and sanguinary exploits, we know much. Of the resolute struggles of early nineteenth-century administrators to create a viable state, we know very little.[8] By extending the framework beyond 1789 or 1794, the Revolution can be understood in part as but one episode in the gradual evolution of the French state and economy.

This story of the state's mastery of urban provisioning trades begins in the early eighteenth century. At that time, two needs defined policy – cities clamoring for affordable bread, and a royal treasury desperate for tax revenues. By the mid-eighteenth century, the Crown had been persuaded to embrace the emerging liberal ideology of free trade. It looked toward increasing crop yields as a source of steady receipts for its coffers. Pushed by Physiocratic reformers, the monarchy lifted controls on the grain trade

[7] Of course, an important problem here is that the Revolution interrupted many of the data series that would be helpful in carrying explorations across these political divides. See the proceedings of several conferences from the Revolutionary bicentennial. Comité des travaux historiques et scientifiques, *La Révolution française et le monde rural, Actes du colloque tenu en Sorbonne les 23, 24 et 25 octobre 1987* (Paris, 1989); *La Révolution française et le développement du capitalisme, Actes du colloque de Lille, 19–21 novembre 1987, Revue du Nord*, special issue, no. 5, 1989, Gérard Gayot and Jean-Pierre Hirsch, eds.; *La pensée économique pendant la Révolution française, Actes du colloque international de Vizille (6–8 septembre 1989)*, G. Faccarello and Ph. Steiner, eds. (Grenoble, 1990); Ministère de l'économie, des finances et du budget, *Etat, finances et économie pendant la Révolution française, colloque tenu à Bercy les 12, 13, 14 octobre 1989 . . .* (Paris, 1991).

[8] One of the few exceptions is Isser Woloch's broad study of the new civic order and institutions, although he does not cover the economy. Isser Woloch, *The New Regime: Transformations of the French Civic Order, 1789–1820s* (New York and London, 1994). A call by two of the foremost experts on the period for "an administrative history of the economy" has gone largely unanswered. Guy Thuillier and Jean Tulard, "Conclusion," in *Administration et contrôle de l'économie, 1800–1914*, Michel Bruguière and Jean Clinquart, et al., eds., *Publications de l'Ecole Pratique des Hautes Etudes*, V, *Hautes Etudes médiévales et modernes 55* (Geneva, 1985), pp. 161–7.

three times between 1763 and 1789, and periodically allowed grain owners, anxious for profits, to export their grain. During these decades, however, agricultural production was sometimes insufficient to feed the kingdom, and each experiment gave way to high prices and riots. Growing restive and increasingly politicized, cities erupted during shortages, and the liberal policies brought no immediate agricultural breakthroughs.

Well into the nineteenth century, the pound or so of bread per day that most French consumers required remained a major component of their diet. As milling and commercial networks shifted, bread, not grain, became the commodity of contention. Thus, still greater animosities were generated whenever bakers were unable or unwilling to stock their shelves. Except in the most rural areas, most consumers bought bread, and so generally were more dependent on the provisioning trades than their Old Regime ancestors. Toiling many hours at looms and spinning wheels, they exchanged a few copper coins for loaves of wheaten bread, or asked the baker to extend them another week of credit. They frequented the grain market only in times of shortage, when they hoped to save a few centimes by purchasing their own grain and having it milled and baked. If the baker failed them, then, they had no alternatives that did not add time and effort to their already long days.[9] Given their increased reliance on bakeries and bread, it was critically important to ease the journey from field to mill and bakery.

The problem was not so much one of wholly inadequate harvests, but rather of sporadic shortages, whether regional or continent-wide. In the eighteenth century, poor harvests brought high prices in 1709–10, 1713, 1738–40, 1747, 1757, 1767–77, 1788–1790, 1792–3, and 1795–6. The deepest crises occurred in 1709, 1740, 1768, 1771, 1789, and 1795. The irregular occurrence of shortages lent a permanent feeling of uneasiness to French consumers. They could not predict when the next crop failures would happen, only that scarcity would inevitably strike. The seriousness of such failures undermined confidence and public order, and could not be ignored. Northern France and the Paris Basin witnessed steady increases in both production and productivity, although these were spaced unevenly, and shortages in one area could place heavy burdens on others.[10] Once authorities and merchants realized a harvest had been insufficient, delays

[9] Reply to the Ministry of Agriculture's 1853 survey, "Fabrication du Pain," 6 December 1853, Archives départementales de l'Eure (AD Eure) 6M 762.

[10] Philip T. Hoffman, *Growth in a Traditional Society: The French Countryside, 1450–1815* (Princeton, 1996). There is still intense debate over the timing and progress of agricultural change in France, although there is a growing consensus that areas near cities witnessed the earliest increases in productivity. See Michel Morineau, *Les faux-semblants d'un démarrage économique: agriculture et démographie en France au XVIIIe siècle*, *Cahiers des Annales* 30 (Paris, 1971); Jean-Marc Moriceau, "Au rendez-vous de la 'Révolution agricole' dans la France du XVIIIe siècle: A propos de la grande culture," *Annales HSS* 49

proved destabilizing. The weeks it took for merchants to send purchase orders to Amsterdam, Danzig, Liverpool, or Virginia, and the months that followed before the ships pulled into port, could spell disaster. Reliable crop yields, steady imports, and smooth transportation networks did not come on the scene until after 1860. Then, steam-powered farm equipment, railroads, and improved farming techniques ensured more stable harvests and their more even distribution. In addition, the wheat fields and railroads of the American Midwest after the Civil War provided essential emergency supplies.

Until these supplies were certain, shortages provoked outbursts in front of bakeries and handwringing in prefectural halls. When failures occurred, public authorities could not simply shrug their shoulders and counsel their *administrés* to wait for a better year. The situation had to be dealt with right then, supplies secured and distributed, and concerns confronted and allayed. The earnest attempts by local and national authorities to address these problems stirred tensions and misgivings throughout the late Old Regime, the Revolution, and into the nineteenth century when they finally yielded success.

Controller General Desmarets and the Shortages of 1709

Two episodes – the shortages of 1709 and 1853 – serve as the opening and closing scenes of this story. The contrast between the two reveals the confusion and impotence of the state in 1709 and its marked confidence and effectiveness in 1853. In 1709, the royal government's efforts to feed cities were scattershot, contradictory and, for the most part, counterproductive. In 1853, Paris escaped unscathed. This was no accident. Instead, it signalled the state's abilities to shape supply lines and ensure that grain was on hand well in advance of any shortages. Nearly two centuries of provisioning experiments had generated effective means to protect cities from the ravages of shortages.

A brief analysis of these two shortages helps to illustrate the nature of the changes underway. In 1709, no supplies were readily available. The Crown engaged in a long and nearly fruitless battle to dislodge grain from warehouses across the kingdom so that cities would not starve.[11] This

(January–February 1994): 27–63; George Grantham, "Agricultural Supply During the Industrial Revolution," *JEH* 49 (March 1989): 43–72.

[11] This account is drawn from the correspondence between the Controller General (CG) and the intendants regarding subsistence crises late in the reign of Louis XIV, found in Archives nationales (AN) G7 15–16. I am grateful to Elizabeth Nachison for calling it to my attention. See also *Correspondance des Contrôleurs Généraux avec les intendants des provinces . . .*, A.M. Boislisle and P. de Brotonne, eds., 3 vols. (Paris, 1897); Marcel Lachiver, *Les*

undertaking, often ad hoc and poorly coordinated, failed to offset the harvest shortfalls and the demands of war. At first, Controller General Desmarets was inclined to see if market forces themselves would solve the shortage. He hoped that high prices might induce owners to part with their wares. To avoid frightening the owners, officials were to use "less authority and more tact." Owners were to be exhorted – not ordered – to speed their sacks of grain to market.[12] To stretch scarce supplies, many cities prohibited the production of the finest white loaves. When that amalgam of efforts failed, the Crown, along with the Parlement of Paris, employed an escalating series of heavy-handed measures – grain censuses and confiscations, orders to provinces to release supplies for cities, and capital punishment for rioters and hoarders alike.[13]

Provincial magistrates and even some intendants fought those initiatives every step of the way by blocking rivers and roads and interfering with the Crown's attempt to prosecute the guilty.[14] Attacking the problems one by one, Louis XIV and Desmarets rebuffed these attempts to hold onto local supplies, and insisted that grain be allowed to move freely toward cities.[15] Many local authorities, along with members of the Parisian police, urged Louis to set price ceilings on grain. While the king wished to keep prices in

années de misère: La famine au temps du grand roi, 1680–1720 (Paris, 1991); and W. Gregory Monahan, *The Year of Sorrows: The Great Famine of 1709 in Lyon* (Columbus, Ohio, 1993).

12 This was in line with late seventeenth-century policies. "Police pour le blé et le pain," 1665, Bibliothèque nationale (BN) Manuscrits français (Mf) 8127.

13 There was some confusion about the jurisdiction and the chronology of a series of royal and parlementary measures, especially the Royal Declaration of 27 April 1709 ordering searches. This led to conflicts between intendants, parlements, and police officials. "Déclaration portant qu'il sera procédé à la visite des magasins . . . ," 27 April 1709, *Recueil général des anciennes lois françaises*, F.-A. Isambert, et al., eds., 29 vols. (Paris, 1822–33), 20: 539. See, for example, CG to M de Bernière, Procureur général du Parlement de Rouen, 28 June 1709; CG to Mssrs de Bonville, Pinon, de Sagonne, Trudaine, de Basville, Le Bret, Angervilliers and Turgot [de Tours], 25 March 1709; and CG to Mssrs d'Ormesson, Turgot de Tours, and de Haroüys, 22 April 1709, AN G7 15.

14 Among the parlements and intendancies enacting measures to obstruct the royal will were those of Burgundy, Paris, Bordeaux, Toulouse, Orléans, and Rouen. CG to M de Haroüys, 16 May 1709; to M de Basille, 9 May 1709; to de Courson, 8 May 1709; to M de Sagonne, 28 June 1709; to d'Ormesson, 11 June and 2 July 1709, AN G7 15; CG to Bouchus, Premier Président of the Burgundy Parlement, 15 December 1708; to Pinon, 29 August 1708; to Bouville, 13 May 1709, AN G7 16; Boislisle, *Correspondance*, 3: pièce 559; Monahan, *Year of Sorrows*, pp. 56–7; Abbot Payson Usher, *The History of the Grain Trade in France* (Cambridge, Mass., 1913), p. 321.

15 Trudaine to CG, 18 August [1708] CG to Mssrs d'Ormesson, Turgot de Tours, and de Haroüys, 22 April 1709; CG to M de Bernaye, 30 April 1709, AN G7 15, CG to Turgot d'Auvergne, 9 May [1709]; CG to M de Bouville, 13 May 1709, and CG to M de Colembercq, 15 May 1709, AN G7 16.

check, he worried that such limits would only drive merchants from the trade and make matters worse. It was acknowledged widely that similar controls clamped on barley by the Parlement of Paris had caused that grain to disappear from markets.[16] Yet pressure for more sweeping price controls continued unabated, and the monarch opened a brief kingdom-wide discussion with intendants and parlements on the merits of such measures. In the end, he decided against ceilings, but since he refused to make any formal pronouncement, the matter remained unsettled late into the summer of 1709.[17] Persuasion, threats, and controls aimed at merchants, bakers, local officials, and unruly crowds proved unreliable. Calm only appeared with the harvest at the end of the summer.

While the measures of 1709 were undeniably chaotic and ineffective, the crisis nonetheless articulated the main lines of later policies – price ceilings on grain were rejected in general, and the unimpeded transport of cereals to cities was to be protected. Because there were no stores of grain to be mobilized during shortages, the best Desmarets could do was to try to provision the cities after the crisis had been announced. To that end, he invoked laws protecting "*la liberté du commerce*." This was not the same as the *libre commerce* that would become so dear to the physiocrats of the mid-century, but was instead the *libre circulation* of Old Regime administrators intent on directing shipments to cities. It meant primarily the free flow of grain across local boundaries and secondarily and only occasionally some form of circumscribed freedom for grain owners to sell how, where, and for how much they wished. The vision of grain supplies moving easily across the kingdom, unopposed by provincial authorities or desperate crowds, occupied royal and urban reveries. Balancing the needs of the rural regions against those of the cities, Versailles and Paris placed their weight on the side of the latter.[18]

[16] The Paris Parlement probably had royal approval to impose the price controls, although the Crown specified that the parlement was acting under its own authority and that such matters fell within the traditional rights of those bodies. D'Ormesson to CG, 10 April 1709; CG to D'Ormesson, 16 April 1709; M de Bernaye to CG, 7 April 1709, and CG's reply [n.d.]; CG to the Procureurs généraux of the parlements, 6 May 1709; and CG, "Lettre circulaire," 8 July 1709, AN G7 15.

[17] CG, "Lettre circulaire," 8 July 1709; "Memoire," 28 July 1709; CG to Roujault, 8 June 1709, CG to Pinon, n.d. [ca. 7 June 1709], AN G7 15.

[18] In response to local attempts to curtail shipments leaving Burgundy in 1708, the Controller General ordered Intendant de Haroüys "to reestablish trade and the *liberté du transport* of wheat going to the Lyonnais" and later, to "let shipments for Lyons pass freely . . ." In Sens, Desmarets cautioned that "it would be an evil to prevent the *liberté* of that transport." Another less frequent expression, the *liberté des marchés*, generally meant a calm market environment, free of upheaval and riots, in which buyers and sellers bargain without interference. See the many letters that use these terms in AN G7 15 and 16.

Bakers and the Supplies of 1853

In contrast to the 1709 measures, which cut a wide swath across all sectors of the grain trade and its supervision, there was but one focus for the 1853 strategies – urban, and especially Parisian, bread prices.[19] After monitoring rising grain prices across the continent for nearly two years, the Minister of the Interior and the Paris Prefect of Police concluded in September 1853 that the prices had reached dangerous levels sure to incite disorder. Without hesitation, the two men placed a firm cap on bread prices in Paris, and encouraged other cities to do the same. For almost a year, the price of one kilogram of bread in Paris remained a mere 40 centimes, even though bakers lost as much as 16 centimes per kilogram. A second ceiling of 50 centimes was imposed in 1855 when flour prices again rose. There was brief confusion until the capital's suburbs had coordinated their bread prices with those of Paris, but once that was worked out, all went as planned. Between 1852 and 1855, there were reports of disturbances elsewhere, generally in cities outside of northern France or along transportation lines, but Paris and the cities of northern France remained relatively undisturbed.[20]

As for the capital's bakers, however, one might conclude that they either went bankrupt, to jail, or both, broken by the cost of keeping the city fed. And given the centrality of bread to urban order, one also might expect that bloodshed followed close behind empty bakery shelves. The rigid price ceilings of 1853–5 could not be sustained, and sooner or later the system would crack. Yet that was not how the tale ended. A few bakers did end up in court over shortweight loaves and other forms of fraud, but by and large, both the city and its bakers survived the crisis.

Their unexpected survival resulted not from good fortune, but rather from several decades of concerted efforts to reorganize the grain trade so that it could meet growing urban needs. This was no mean task. A host of bureaucrats in ministries, departments, and municipal halls spent the better part of the nineteenth century finding ways to combat shortages and the weaknesses of commercial networks. Under their watchful eyes, urban tran-

[19] The Parisian information is from the correspondence of the Paris Prefecture of Police, AN F11 2799–2802. See, in particular, the price tables for the 1850s, "Taxe périodique du Pain à Paris," AN F11 2801; and related correspondence, including Minister of the Interior (MI) to Prefects, "Confidentielle," 2 July 1853; Prefect, Department of the Seine, Circular, "Prix du Pain" to Mayors, 22 September 1853, AN F11 2799; and Prefect of Police to the Ministers of the Interior, Agriculture and Commerce, 10 August 1852, AN F11 2802. For price data, see Bertrand Roehner, et al., *Un siècle de commerce du blé en France, 1825–1913: Les fluctuations du champs des prix* (Paris, 1991), pp. 154, 223.

[20] See the court records relating to riots in AN BB24 448–56.

quility was sustained by emergency warehouses, massive expenditures, and when necessary, troops willing to fire on crowds. During the 1852–5 shortages, for instance, the Paris prefect of police ordered bakers to drain their private stocks of several months of grain.[21] Such granaries had been developed in the Year X (1801–2) and steadily increased in number for several decades, holding ready backups to cover any shortfalls.[22] To help carry the bakers through the shortage, the city of Paris set up an office, the *Caisse de service de la boulangerie de Paris*, which served as an intermediary for all the purchases that the city's bakers made. The *Caisse* opened lines of credit for the bakers and absorbed some of the losses, so that the bakers could continue to buy high-priced flour.[23] This was not the first time the government had subsidized bakers. For many years, government officials had expended considerable energy to determine ways to fund and distribute such outlays. Although some of their plans had failed miserably and had stirred greater hostility and agitation among buyers, they steadily gained coherence and effectiveness in the years after the Revolution.

Given that these were the centuries in which historians trace the slow adoption of liberal economic policies, we must ask how the more regulated provisioning policies of the nineteenth century developed under the aegis of the state. Clearly, the controls on the nineteenth-century baking trades were heavier than those of the Old Regime. Prefects and mayors set bread prices, using tables that had been sanctioned by ministers. In times of shortage, they forced bakers to accept losses and to exhaust their emergency supplies. The state's domination of the trade was more thorough, aimed more precisely, and more effective in the nineteenth century than it ever had been during the period before the Revolution. How was it that Paris and other northern cities had ready flour reserves and an army of bakers to send into battle in 1853, while Desmarets and his assistants could cobble together only the most insufficient stocks while attempting to confront widespread confusion, insubordination, and violence? Only through a full

[21] At least 80,000 metric quintals of flour had been taken from the bakers' required stocks of 128,000 metric quintals. Perhaps as many as 40,000 metric quintals were withdrawn in August 1854 alone. Over the next year, the Prefect of Police oversaw the reconstitution of those supplies. Prefect of Police to the Minister of Agriculture, 28 August and 16 November 1854, AN F11 2801.

[22] *Tableau des boulangers de Paris, pour l'éxercise de l'an 1831* (Paris, 1831), BN 8° Z Le Senne 14259. These records include laws and price tables from 1801 to 1831.

[23] The best quality white bread, *pain de luxe*, was exempt from the price ceilings. Prefect, Department of the Seine, "Prix du Pain," Circular to Mayors, 22 September 1853, AN F11 2799. Later, subsidies from the *Caisse* were extended to bakers in the department of the Seine, and the *Caisse* itself remained in service until the baking trades were freed by the Imperial Decree of 22 June 1863. "Decret . . . qui institue une caisse de service pour la boulangerie de Paris," 27 December 1853, *Bulletin de Lois* (Paris, 1854). See the legislation of 22 June and 31 August 1863 in that year's *Bulletin de Lois*.

understanding of the state's more rigorous regulation of provisioning trades may we grasp the complexities of political and economic development of this period.

Sources and Questions: The Interdisciplinary Conversation

The search for this understanding led to bakeries, market stalls, bankruptcy files, and administrative chambers over the course of nearly two centuries. The administrative correspondence reveals lengthy discussions of the appropriate role for the government. Here, intendants, prefects, and their subordinates offer frank reflections on the reasons commercial networks had failed and for the success of the state's efforts. These were, after all, the men charged with maintaining calm in the face of famine.[24] Because those famines often pushed starving consumers beyond control, police and judicial records revealed the tensions surrounding shortages.[25] I have not drawn the same conclusions from these reports as many other scholars have, and, as the reader will see, I have serious reservations about many of their conclusions. In general, the historians who have consulted these records have used them to delineate the poor's sharp rejection of free trade in the late eighteenth century and to argue that they adhered to a "moral economy" in which need and equity were to determine access to food stuffs.[26] While a fuller discussion of the intersection between market practices and free trade must wait for the next chapter, I will note here that an extensive study of administrative and court records makes it difficult to sustain the argument that the authorities, at least before 1789, renounced their

[24] These included those of the Joly de Fleury collection in the Bibliothèque nationale, which houses correspondence concerning the Paris Parlement's activities in provisioning and of the numerous ministries devoted to trade, such as AN F11 (Subsistances), AN F12 (Trade and Manufacturing), AN F10 (Agriculture), and AN F4 (Finances). Correspondence with Old Regime intendants is in AN G. For the Revolution, see AN C and D. In departmental collections, see the series C (Intendancies), L (Revolution), and M (Post-1789 Administration), primarily in the departments of the Seine-Maritime, Seine, Eure, Somme, Calvados, and Yvelines. They were supplemented by municipal collections in Rouen, Caen, and Le Havre.

[25] Among them were the records of the Châtelet Court of Paris (AN Y), the Paris Parlement (AN X and U), the Ministry of Justice (AN BB), and the Ministry of Police (AN F7). Departmental collections included the materials under the series rubrics B (Parlement and Police), L, and some parts of the series M.

[26] Thompson, "The Moral Economy of the English Crowd." Complementary studies include Louise Tilly, "The Food Riot as a Form of Political Conflict in France," *JIH* 1 (1971): 23–57; Hufton, "Social Conflict and Grain Supply." For one response to Thompson's model, see John Bohstedt, *Riots and Community Politics in England and Wales, 1790–1810* (Cambridge, Mass., 1983).

responsibility for the marketplace. The nature of their control changed; the fact of their control did not. The state did not abolish regulations. It did, however, take a long time to figure out which ones would best serve urban needs and the state's desire for political stability. To uncover that story, I have added a broad range of guild and bankruptcy records that give a fuller sense of how the government's decisions shaped the activities of the merchants and bakers involved in provisioning trades.[27] The municipal police records entitled "*la taxe du pain*," or bread price controls, were exceptionally revealing about the increased weight and effectiveness of regulation on bakers from the Old Regime deep into the nineteenth century.[28] Of particular interest are the records of the committees that restructured the Paris baking trades and redefined their relationship to the grain and flour markets in the first half of the nineteenth century.[29] It is important to note that many officials were well aware of the inherent contradictions between free-trade policies and the controls they believed the provisioning trades required. Their correspondence and actions reveal their conscious efforts to bring the two into harmony. All together, the records yield a rich and complicated story of economic activities and of the state's attempts to guide them.

This study of the evolving relationship between the French state and the grain trade is situated between two potentially conflicting disciplines – economics and history. Given the intensification of disciplinary tensions in recent decades, it is no easy task to find a meaningful way to address them both. This books explains how and why the French state tamed the grain trade, an undertaking that came at a pivotal moment in the nation's history. I hope this story will be of interest to those in both disciplines. There are possible points of contact between the two field's scholarly endeavors – both attempt to identify change and assign causality to forces or actors, although each would differ as to what constitutes either cause or effect.[30] With regard to France, a common set of questions would ask to what extent the economic regulations gave way to free trade, whether new practices, ideolo-

[27] These included the departmental series E (Guilds) and parts of BP (Police des arts et métiers).

[28] Local records on the regulation of bread prices are generally in the municipal collections FF and HH series. There were frequent conflicts over how to set those prices, and in some cases, parlements intervened. The records of the Paris Parlement's adjudication are in AN U; for other parlements, see the relevant departmental series B.

[29] These are the records of the Caisse syndicale des boulangers in AN F11* 3049–68. See also AN F 11* 2799–1802.

[30] See, for example, the illuminating essays in *Economics and the Historian*, Tom G. Rawski, et al., eds. (Berkeley, 1996); *Markets in History: Economic Studies of the Past*, David W. Galenson, ed. (Cambridge, 1989); and *Capitalism in Context: Essays on Economic Development and Cultural Change in Honor of R.M. Hartwell*, John A. James and Mark Thomas, eds. (Chicago, 1994).

gies, institutions, and policies produced or hindered economic growth, and with what costs or benefits. Historians or economists would emphasize one or another of those questions, but nonetheless they offer a starting point for an interdisciplinary dialogue.

The general framework for any exploration of the French economy traces the decline of the "mercantilist system" of Colbert and the emergence of capitalism in the nineteenth century.[31] The historiography of the grain trade runs along those lines, marking turning points in the royal and Revolutionary legislation that lifted controls on merchants and markets in 1763–4, 1774, 1787, and, permanently, in 1797. According to Kaplan, the 1763 edict was "among the most daring and revolutionary reforms attempted in France before 1789," representing a "drastic departure from the past," although he questions the extent to which the Crown's policy reforms actually changed marketplace activities.[32] Such a narrative places the reforms of the grain trade within the received chronology of the slow liberalization of the economy and of the resistance and anguish of the poor who were dealt the harshest blows each time prices rose. In the end, free trade prevailed – or so the present literature would lead one to believe. While historians of the nineteenth century can offer examples of shortages and accompanying unrest throughout their period, these crises declined in significance, and were supplanted finally by the labor movement.[33]

Historians have long understood that secure urban food supplies underlay political stability; we still have little idea how such resources were garnered. What explanations have been offered for the diminishing incidence and effect of food riots? One set of answers straightforwardly points to eco-

[31] Two classic outlines are Joseph A. Schumpeter, *History of Economic Analysis*, ed. Elizabeth Boody Schumpeter (New York, 1954); and *Histoire économique et sociale de la France*, 4 vols., Fernand Braudel and Ernest Labrousse, eds. (Paris, 1970–1982). The nature and appropriateness of the term "mercantilism" has been a topic of much historical discussion. The discussion turns on the issues of whether any coherent system can be discerned in seventeenth-century policies, and whether those policies were as oriented toward the heavy protectionism and regulation that the concept has come to imply. See Eli Heckscher, *Mercantilism*, 2 vols., rev. 2nd ed., E.F. Soderland, ed., trans. Mendel Shapiro (London, 1955); Charles Woolsey Cole, *French Mercantilism, 1683–1700* (New York, 1943); and *Colbert and a Century of French Mercantilism*, 2 vols. (Hamden, Conn., 1964); *Revisions in Mercantilism*, D.C. Coleman, ed. (London, 1969).

[32] "By the liberal decrees," Kaplan writes, "Louis XV estranged himself unequivocally from the past and dissociated his reign from the policies which it had inherited and faithfully executed for almost a half a century." Steven L. Kaplan, *Bread, Politics and Political Economy in the Reign of Louis XV*, 2 vols. (The Hague, 1976), 1: xxvi–xxvii, xxix, 72–86, 96. See also his *Provisioning Paris: Merchants and Millers in the Grain and Flour Trade during the Eighteenth Century* (Ithaca and London, 1984); and *The Bakers of Paris and the Bread Question, 1700–1775* (Durham, N.C. and London, 1996).

[33] Charles Tilly, "Food Supply and Public Order in Modern Europe," in *The Formation of National States in Western Europe*, Charles Tilly, ed. (Princeton, 1975), pp. 380–455.

nomic "modernization" as the key. Rising grain yields, improved transportation, changing diets, and more stable incomes brought security.[34] Nevertheless, each of these is an insufficient answer until the late nineteenth century. Transportation lines were uneven until at least the 1850s, and much later in areas outside of northern France. The greatest expansion of the railroads came during the last years of the July Monarchy and the Second Empire.[35] Although the production of grain and other foodstuffs grew more or less sufficiently to match a slow but steadily expanding population, the gains were never enough to guarantee the country a safe margin against shortages in bad harvest years, given the slow development of national and international communications and transportation systems. When Parisian administrators in 1819 reviewed the frequency of shortages since 1791, they concluded that the country could expect a poor harvest "with regularity," and that the government had to be prepared to intervene with little notice. The most recent shortage had been in 1817, and they believed it prudent to anticipate another by 1821.[36]

The possibility of hunger and unrest was ever-present, even in the mid-nineteenth century. Areas around cities clearly fared better that rural areas. They benefited from both rising crops yields and more productive milling technologies. In the Paris basin, demand drove gains in agricultural output, and there is substantial evidence of improvement in other densely populated northern areas, such as the Cambrésis.[37] Individual yearly bread con-

[34] Roger Price, *The Modernization of Rural France: Communications Networks and Agricultural Market Structures in Nineteenth-Century France* (New York, 1983) and *The Economic Modernisation of France* (New York, [1975]).

[35] David Pinkney gives a good overview of the pace of change in the 1840s. Water traffic remained roughly the same from 1847 until 1880. Railroads, however, made rapid gains at the end of the July Monarchy and throughout the Second Empire, helped in large part by government subsidies. The ton-kilometers of freight transported in France increased from 102 million in 1845 to 1.53 billion in 1855. The densest networks of railroads were in northern France. Thus, even given the significant expansion of railways, some cities and regions in the south did not have adequate access to transportation until after 1860, although the system was improving dramatically. David H. Pinkney, *Decisive Years in France, 1840–1847* (Princeton, 1986), pp. 34–60. See also Xavier de Planhol, *An Historical Geography of France*, trans. Janet Lloyd (Cambridge, 1994), pp. 271–3. Rick Szostak emphasizes the inadequacies in *The Role of Transportation in the Industrial Revolution: A Comparison of England and France* (Montreal, 1991), pp. 49–90. The problems involved in the development of a modest canal system are recounted in Reed G. Geiger, *Planning the French Canals: Bureaucracy, Politics and Enterprise under the Restoration* (Newark, London and Toronto, 1994), pp. 32–63, 115–36, 247–62.

[36] 13 September 1819, AN F11* 3063.

[37] Morineau, *Les faux-semblants d'un démarrage économique*; Philip Hoffman, "Land Rents and Agricultural Productivity: The Paris Basin, 1450–1789," *JEH* 51 (December 1991): 771–805; Jean-Marc Moriceau and Gilles Postel-Vinay, *Ferme, entreprise, famille. Grande exploitation et changements agricoles. Les Chartier. XVIIe–XIXe siècles, Les Hommes et*

sumption in France overall may have risen from 200–250 kilograms in 1789 to 280–290 kilograms in 1860. Daily per capita consumption of all foodstuffs may have increased from 1,700 to 2,000 calories between 1789 and 1840, 85 percent of which was vegetable, yet these figures are only the best estimates, and hardly uncontested.[38] In other words, the nationwide gains may have been from miserable to roughly adequate, but they probably were not from miserable to reassuring on a regular basis or throughout the country. The fragility and uncertainty, although not the utter insufficiency, of provisioning networks marked the French psyche into the nineteenth century.

Thus, there is a certain teleology to arguments that "modernization" eventually guaranteed adequate food supplies year in and year out, albeit late in the game. Was capitalism's victory so complete after the Revolution that hitherto insufficient fields and markets sprang to life, liberated by capitalist ideologies and the seductive call of new possibilities and profits? We now know that was not the case. Harvest shortfalls, dry riverbeds, slippery roads, and unwilling merchants left cities and towns perpetually exposed to the possibility of starvation.

Given the state's inherent interest in maintaining order, it is surprisingly absent from much of the literature on the early nineteenth-century economy. The state's most important role was that of sending troops to confront market crowds. Despite the sporadic presence of those guns in marketplaces, repression alone frequently failed to quell market conflicts. Soldiers often joined crowds demanding foodstuffs at low prices, juries were slow to convict, and pardons were sometimes easy to secure.[39] Thus, while I agree that problems eased by the late century, a critical question remains: What was the French state to do in the meantime, with no sense of what advances the future might hold? How was it to handle the shortages of 1801–2, 1811–2, 1816–7, 1828–9, and 1846–7? It could neither relinquish its population to the forces of the market, nor impose heavy-handed controls. The utter failure of both extremes during the Old Regime and the Revolution made those options nearly unthinkable, especially in cities. The ministers, prefects, and mayors of the early nineteenth century harbored few illusions about the necessary role they should play in shaping markets and in shoring up commercial relationships. They would have agreed with Seine-Inférieure

la Terre 21 (Paris, 1992); George Grantham, "Agricultural Supply"; Liana Vardi, *The Land and the Loom: Peasants and Profit in Northern France, 1680–1800* (Durham and London, 1993), pp. 87–109.

[38] [Jean Toutain], Institut de science économique appliquée, *La consommation alimentaire en France de 1789 à 1964, Cahiers de l'I.S.E.A 5* (Geneva, 1971), pp. 1923–2031.

[39] See Cynthia Bouton's forthcoming contribution to John Bohstedt, Cynthia Bouton, Manfred Gailus, and Martin Geyer, *The Politics of Provisioning in Britain, France and Germany, 1690s–1850.*

Prefect Girardin's assessment of the regulations imposed during the shortage of 1812. "It is important for the prosperity of the state and for public tranquility," he counseled ". . . that the government become the supreme regulator of prices."[40] Without such intervention, there would be no security.

Drawing in the state so explicitly might make some economic historians uneasy, especially those using the neoclassical models that dominate the field. Yet other areas of the discipline – among them new institutional economics – are more optimistic about the impact of institutions such as the state, firm, or household. Here, institutions are presented as "efficient ways of organizing human activity when markets alone will not suffice," although scholars differ on what constitutes a market failure and how it should be remedied.[41] The literature on economic development also holds promise. This body of work identifies essential preconditions for growth, and offers insights into how states or other institutions create or damage trade networks, thereby shaping a country's entry into a world economy. While there is intense debate over whether the government's efforts are beneficial or not in either the long or short run, and for whom – and these works focus generally on twentieth-century examples – it is worth asking how such studies can illuminate the transformation of French provisioning trades of the eighteenth and nineteenth centuries.[42]

The Setting

I have taken these questions about the role of the state in economic development into the administrative chambers and markets of the northern Paris basin and the V-shaped area that extends northwest along the Seine River

[40] Prefect Girardin of the Seine-Inférieure (S-I) to [Réal], 20 September [1812], AN F7 3639.

[41] Cohen, "Institutions and Economic Analysis," in *Economics and the Historian*, p. 60, along with the other essays in that volume. The classic outline of this approach can be found in Douglass C. North, *Structure and Change in Economic History* (New York, 1981); and *Institutions, Institutional Change, and Economic Performance* (New York, 1990). In addition, see *Economics as a Process: Essays in the New Institutional Economics*, Richard N. Langlois, ed. (Cambridge, 1986).

[42] The following illustrate the range of models and conclusions regarding the state's impact on economic development, although they in no way constitute an exhaustive list: Alexander Gerschenkron, *Economic Backwardness in Historical Perspective: A Book of Essays* (New York, 1965); Robert Wade, *Governing the Market: Economic Theory and the Role of Government in East Asian Industrialization* (Princeton, 1990); Robert Bates, *Markets and States in Tropical Africa: The Political Bases of Agricultural Policies* (Berkeley, 1981); Bjorn Hette, *Development Theory and the Three Worlds: Towards an International Political Economy of Development*, 2nd ed. (Essex and New York, 1995); *From Classical Economics to Development Economics*, Gerald M. Meier, ed. (New York, 1994).

toward the Channel (Figure I.1). The capital itself depended on the rich, yet still insufficient, plains that surrounded it, including the Seine-et-Marne, the Oise, and the Seine-et-Oise. It also brought grain from regions further afield, such as Picardy, Burgundy, the Orléanais, and occasionally Brittany. While a study of Paris alone would be of interest, it is crucial to move beyond the capital to the areas with which it competed for supplies. The struggle to assert Parisian claims to fields farther away had begun in the late seventeenth century, and continued to provoke tensions and much resistance well into the nineteenth century.[43] An important area of opposition was found in the departments of the Eure and the Seine-Inférieure.[44] Here, one can see the uneven processes of evolving trade networks and an increasingly effective state. Just because a ministry, whether in the Old Regime or the nineteenth century, willed something, did not mean that the rank and file in the provinces snapped into action, nor that their efforts bore fruit. That resistance prevented the wholesale victory of Parisian demands and left its imprint on national policies. Moreover, local solutions to provisioning crises spurred the consideration of such measures at the national level. Thus, the means by which supplies were generated and stabilized was by no means smooth or uninterrupted. By examining the interplay of Parisian needs, ministerial pressures, and regional responses, one can identify critical dimensions of centralization and policy formulation.

Particularly revealing on this point is the strained relationship between Paris and its rival to the northwest, Rouen. That city was the *chef-lieu* of its Old Regime *généralité* (an administrative district), roughly Upper Normandy, which became the departments of the Seine-Inférieure and the Eure during the Revolution. Rouen itself housed about 80,000 people involved in textile manufacturing and commerce and who were wholly dependent on its grain halls and bakers. The city's location on the Seine between Paris and the English Channel made it one of the great commercial centers of the Old Regime, an essential *entrepôt* for goods moving between Paris, northern European trade networks, and the French overseas colonies.[45] Many generations of merchants and bankers had contributed to the development of successful commercial societies and guilds, giving Rouen an undeniable legacy of mercantile attitudes and interests. In the late seventeenth century, its commercial community had resisted Colbert's policies, and it was even claimed that one Rouen merchant, Thomas Legendre, had

[43] Usher, *Grain Trade*, pp. 46–84.

[44] The Seine-Inférieure was renamed the Seine-Maritime.

[45] *Histoire de la Normandie*, Michel de Bouard, ed. (Toulouse, 1970), pp. 287–318; Dardel, *Commerce, industrie et navigation*; Marc Bouloiseau, ed. *Cahiers de doléances du Tiers Etat du bailliage de Rouen pour les Etats généraux de Rouen de 1789* (Paris, n.d.), p. xliii.

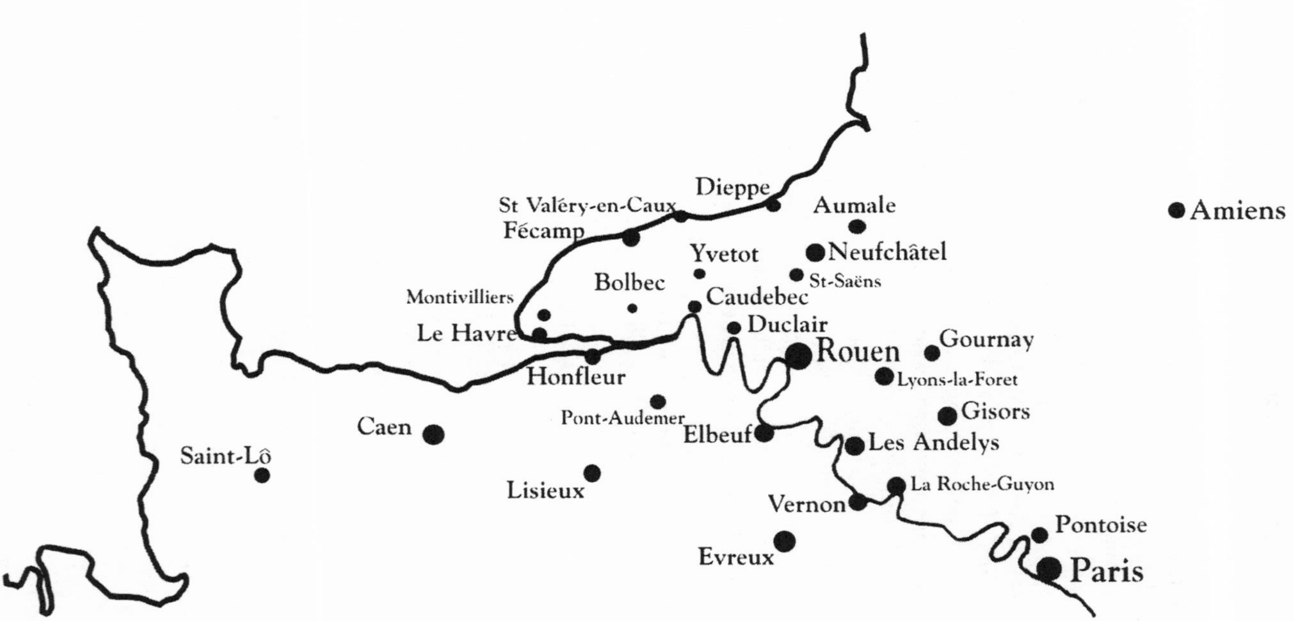

Figure I.1 The markets of northern France.

urged that minister, "*laissez-nous faire*," thus giving rise to Vincent de Gournay's later more famous formulation.[46] In the eighteenth and nineteenth centuries, textile production, especially cotton, increased, although not without suffering the effects of the agonizing business cycles that disturbed that sector of the economy throughout the period.

Upper Normandy had been part of the capital's provisioning hinterland for more than a century, but after 1800 the ties between the two regions became even stronger. Paris, growing from 500,000 to 700,000 during the Empire, steadily combatted Rouen's claims to the supplies of Upper Normandy, especially the wheatfields of the Vexin and the grain arriving in the Channel ports. The department of the Seine-Inférieure, among the most densely populated of France, did not yield easily and became an even more tenacious rival.[47] The department suffered from great variations in the quality of its farmlands, and often failed to produce even two-thirds of its needs. The Vexin normand, located in the department of the Eure, was the sole area of reliable surpluses, and thus was the prey of merchants from both Paris and Rouen. Further north, the countryside around Rouen was barren and sandy and could not support the city and its rapidly growing suburbs. More or less to the northeast and east of Rouen lay the Pays de Bray, where the rye and barley fields bore witness to the incessant struggle of the peasants to draw all they could from the land. The real riches of the Bray lay not in its inadequate cereal crops, but in the clover, oats, and beans that increasingly fed herds of dairy cattle. Wagons carted the Bray's milk, cheese, and butter to Paris, and the region increasingly looked to the capital to market its goods. The Pays de Caux, to the north and west of Rouen, rarely produced enough wheat, although its crops of oats were more abundant. Its people divided their time between farms and the looms and spinning wheels of the region's cottage industry. Here there were some advances in agricultural production, especially due to the decreasing amounts of land left fallow. According to prefectural surveys in the early nineteenth century, many of those lands had been replaced by alfala, clover, and colza. This change raised overall crop yields and provided some surpluses for Rouen and the other cities of the region.[48]

[46] Lionel Rothkrug, *Opposition to Louis XIV: The Political and Social Origins of the French Enlightenment* (Princeton, 1965), pp. 230–2, 395–9; Paul Butel, *L'économie française au XVIIIe siècle* (Paris, 1993), p. 21; J.F. Bosher, *The Single Duty Project: A Study of the Movement for a French Customs Union in the Eighteenth Century* (London, 1964), p. 34.

[47] Rouen itself increased from 73,000 inhabitants on the eve of the Revolution to more than 100,000 during the Restoration. Prefect Girardin to the MI, October 10, 1812, AN F7 3639; J.-P. Bardet, *Rouen aux XVIIe et XVIIIe siècles: les mutations d'un espace social* (Paris, 1983); *La Révolution en Haute-Normandie*, 2nd ed. (Rouen, 1989), p. 12.

[48] For information on agricultural and economic change in the region during the nineteenth

Despite those improvements, this section of northern France could not produce enough grain, and relied heavily on imports. These supplies came from neighboring Picardy and Lower Normandy, both unwilling sources, and intermittently from Brittany. Additional provisions arrived from outside France, primarily from the Baltic through Amsterdam, and occasionally from the Mediterranean, England, and North America.[49] The *généralité*'s location along the Seine, with its several Channel ports, made it an important waystation for such grain. Many of those shipments, however, were destined for the capital and not for the impoverished populations that lined roads and rivers to Paris. The sight of barges and wagons going to the capital had the power to enrage hungry communities and thus disrupt vital supply lines. The matter of meeting both Parisian and Upper Norman needs, then, was a delicate one and the subject of much administrative anxiety. In 1709, Controller General Desmarets cautioned Rouen Intendant de Courson not to allow the city's parlement to make greater efforts on behalf of Rouen's poor than were being made in the capital, for instance, as they might provoke greater unrest and perhaps heighten the sense of inequity.[50] In 1853, the prefects of the two departments, along with the Minister of Agriculture and Commerce, maintained constant communication, and sought to keep bread prices in balance so that one city would not drain the other of supplies.[51] Frustration, shortages, and uproar in either could spell starvation for both.

The coordination sought by Desmarets and subsequent ministers depended on the obedience and goodwill of the many authorities who supervised the grain trade on the local level. The officials belonged to many competing institutions – intendancies, parlements, the military, and municipal governments were but a few. Each of these rivals was present in Rouen, a city with a high "administrative density." That city was home to the intendant, to a prestigious and proud parlement, along with the region's Cour des Aides, several other courts, and a lieutenant generalcy of police. The

century, consult the prefectural reports in AN F^{1c} III Seine-Inférieure 8; AN F^{1c} V Seine-Inférieure 1 and 2; AN F11 514(B); AN F12 1269A; ADSM 6M 1070. See also Guy Lemarchand, *La fin du féodalisme dans le Pays de Caux: conjoncture économique et démographique et structure sociale dans une région de grande culture de la crise du XVIIe siècle à la stabilisation de la Révolution (1640–1795)*, Commission d'Histoire de la Révolution française, *Mémoires et Documents*, XLV (Paris, 1989); Jules Sion, *Les paysans de la Normandie Orientale: Pays de Caux, Bray, Vexin Normand, Vallée de la Seine. Etude géographique* (Paris, 1909); Gay L. Gullickson, *Spinners and Weavers of Auffay: Rural Industry and the Sexual Division of Labor in a French Village, 1750–1850* (Cambridge, 1986).

[49] Pierre Dardel, *Commerce, industrie et navigation à Rouen et au Havre au XVIIIe siècle, rivalité croissante entre les deux ports, la conjoncture* (Rouen, 1946).

[50] CG to de Courson, 14 May 1709, AN G7 15.

[51] See the correspondence in AN F11 2802.

jurisdictions of these institutions overlapped, and their officers and judges often fell into bitter conflicts. They were joined by a multitude of subdelegates, magistrates, lawyers, merchant corps, guild officers, and police, who challenged or supported these many authorities and who each claimed some right to intervene in the economy. In the nineteenth century, they were supplanted by prefects, subprefects, municipal administrations, numerous courts, and several kinds of police. The presence of so many competing authorities allows the historian to trace the difficult process by which rivalries and jurisdictional conflicts gave way to a workable level of centralization and coordination.

While this book focuses on one section of northern France, the problems it outlines prevailed elsewhere for several centuries – sporadic shortages, escalating rivalries between cities and farms, and tensions surrounding the state's centralization. When possible, I have drawn on the wealth of secondary works produced by several generations of scholars who have studied *subsistances* in other regions. In general, these works chronicle shortages and disturbances in the eighteenth century and pay only slight attention to the nineteenth century.[52] Nonetheless, they can illuminate the broad development of changing visions of the economy and the steady growth of the French state's capacity to influence provisioning. Thus, while not every department battled Paris for food, and only a few benefited from the enormous resources that state brought to bear on those problems, the policies adopted across France reveal the clear influence of northern French, and particularly Parisian, strategies.

Rouen and Paris – together they offer an opportunity to assess the tensions and achievements that attended the many attempts to reform the grain trade. Viewing the process over the long run, this book follows not the riots and failings that accompanied changing provisioning strategies, but rather charts the steady reformulation of the boundaries of state and economy during a critical period in France's history. Such an exploration reveals the state's increased competence and the resulting calm in times of shortages. Without food there could be no peace; even with food, peace was a tenuous and fragile thing. That, at the very least, any administrator knew. With every upward movement of bread prices, the possibility for violence

[52] For example, J. Letaconnoux, *Les subsistances et le commerce des grains en Bretagne au XVIIIe siècle: essai de monographie économique* (Rennes, 1909); Louis Stouff, *Ravitaillement et alimentation en Provence aux XIVe et XVe siècles, Civilizations et Sociétés* 20 (Paris and La Haye, 1970); Georges Afanassiev, *Le commerce des céréales en France au XVIIIe siècle* (Paris, 1894); Charles Desmarest, *Le commerce des grains dans la généralité de Rouen à la fin de l'ancien régime* (Paris, 1926); Jean Bouvet, *La Question des subsistances en Maconnais à la fin de l'ancien régime et au début de la Révolution (1788–1790)* (Macon, 1945).

increased. The carriages of Old Regime intendants, the halls of the National Convention, and the doorways of nineteenth-century mayors were the terminuses for many an altercation that began in markets and bakeries. The state dared not abandon the grain trade to the forces of the market. Instead, it had to master it.

PART ONE

The Market of the Enlightenment, 1720–1789

How was stability to be wrested from turmoil? If the market was to be tamed, who would do so, and how? For the 20 million inhabitants of the French kingdom in the early eighteenth century and for the thousands of public officials, no problem was so crucial as distributing the grain from those who grew it to those who purchased it. The demands of restive markets, the contending interests of grain producers, bakers, and consumers, the need to mobilize resources for war, and the centrality of economic concepts in the discourses of legitimacy ensured that the grain trade lay at the core of administrative and, hence, political calculations. The question remained, however, what to do? How could supplies be secured?

The tumultuous events of 1709 offered a comprehensive catalogue of the difficulties confronting public authorities. These had been numerous, and are familiar to any historian of Old Regime France – crippling shortages that drove prices up, unreliable transportation, provincial authorities intent on retaining local supplies, jurisdictional jealousies (some created by the process of centralization) between the many public officials involved in provisioning, and riots that caused even the best-intentioned owners to flee markets. The strategies on hand to combat them ranged from the many traditional regulations on the grain and baking trades to other exceptional measures. Without exaggeration, any public authority could have listed hundreds of ways – searches, price-setting, requisitions, the lifting of market fees, fines, imprisonment, among many others – through which he could try to influence the trades and secure foodstuffs for his townspeople.

During the eighteenth century, that lengthy list began to be winnowed down to a handful of more workable measures. Certain strategies were identified as potentially more beneficial to the dual aims of feeding the hungry and maintaining public order. Those measures had some common characteristics. First, they tended to rely increasingly on an understanding of market forces. The administrators' goal was to influence supply and demand rather than to force the participants to act against their self-interest. For instance, emergency grain supplies or subsidies were used to combat high prices, rather than price ceilings or requisitions, although officials found it hard to predict the effect of their calculated measures. These tactics dated from the seventeenth century at least, and gained momentum in the eighteenth. After the free-trade legislation of the 1760s, they accelerated more quickly. Second, the evolving provisioning policies depended less on dramatic public displays of administrative authority and leaned more toward behind-the-scenes coercion, directed especially at grain owners and bakers. Administrators sought to give the appearance of following routine and well-tested strategies that could be announced with little ado. They did not want consumers to be a party to their negotiations with producers and merchants. Open showdowns often led to riots. By the mid-century, the police *commissaire* who donned his robes and marched from bakery to bakery in search of infractions, trailing a crowd in his wake, as just one example, was called to order and prosecuted for his inflammatory acts.[1] Third, the evolving policies required the goodwill and coordination of the many contending officials involved in supervising the trades. Advance planning and discussion would have to replace the independent and contradictory actions of the men in the competing Old Regime institutions. Parlementary presidents, intendants, governors, lieutenants general of police, seigneurial agents, and their many subordinates would have to lay aside their rival claims to authority in order to best serve their communities.

None of these three lines of development proceeded in a clear and uninterrupted fashion in the eighteenth century, although they were discernable in the administrative initiatives as early as 1708–09. Controller General Demarets might have been able to offer some predictions and advice to his sucessors, although he would have hesitated to pronounce with certainty what might work best. His administrative heirs, weathering the riots of 1725–6, 1738–41, 1757, 1768–71, 1775, 1784 and 1788–9, would have admitted that their policies yielded only partial victories. Yet, they would have noted, they had managed to take the bewildering array of market regulations and nascent free-trade theories and had begun to hone them to meet the exigencies of their markets.

[1] See the events in Caen, 1772. Archives Départementales du Calvados (AD Calvados) 1B 1665A.

CHAPTER ONE

The Structure of Mill and Market

"The grain trade is a rather hazardous business," explained Jacques Savary, the author of the 1675 guide, *Le parfait négociant*.[1] This volume, which appeared in numerous editions during the course of the eighteenth century, spoke a truth every merchant knew. An anonymous Rouen merchant was equally blunt: "We will be quite happy to be out of the grain trade, because it is very a risky affair."[2] From across France, public authorities could provide a list of the reasons commercial houses avoided the trade – grain was heavy and difficult to transport; in winter, ice blocked ports; in summer, droughts dried river beds; during shortage, angry townspeople attacked their wagons and warehouses; grain spoiled in ship holds and damp granaries; storms forced captains to toss barrels overboard; lengthy paperwork waited in ports and at markets; and once the grain had arrived, consumers and bakers complained that imported wares smelled bad and refused to buy them. In short, profits were unpredictable and scarce, whether one brought grain from the next province or across the ocean. In an average year, when supplies were more or less sufficient, some commercial houses and individual merchants were moderately inclined to dispatch shipments from city to city or to roam the countryside in search of supplies. But in times of shortage, when they were most needed, they deserted the trade, fully aware that their warehouses, homes, and lives might become the targets of the hungry populace's fury.

How were administrators to overcome such widespread repugnance for the grain trade? Failing that, how were they to make up for such short-

[1] Quoted in Charles Carrière and Marcel Coudurier, "Marseille sauvant Paris de la famine . . ." in *Histoire, économies, sociétés. Journées d'études en l'honneur du Pierre Léon (6–7 mai 1977)* (Lyons, [1978]), p. 59; Johan Hoock, "Le phénomène Savary et l'innovation en matière commerciale en France aux XVIIe et XVIIIe siècles," *Innovations et renouveaux techniques de l'antiquité à nos jours*, Jean-Pierre Kintz, ed., Association Interuniversitaire de l'Est, v. 24 (Strasbourg, n.d.), pp. 113–23.

[2] Carrière and Coudurier, "Marseille sauvant Paris," p. 59.

comings? Their dilemma lay in deciding between the two competing visions of how to address the merchants' obvious and unrepentant distaste for the trade. One view relied on extensive, but frequently ineffective, powers to compel bakers and merchants to supply the people. Public authorities could storm into barns, granaries, and ships, ordering owners to send their grain to the market. They could set its price and fine or imprison any owner who resisted. Yet, such tactics, while initially successful, eventually drove merchants away, and caused farmers to abandon troublesome markets or even cereal production altogether. The other alternative looked toward market incentives. The state could offer bonuses for grain imports or subsidize purchases, and hope that the prevailing high prices at markets would draw grain from near and far. Yet a reliance on such incentives offered little recourse if merchants declined to respond, as was frequently the case, or when genuine shortages sent prices spiraling upward. Other alternatives were needed, ones that allowed the government to make up for the combined failings of merchants and harvests. And, most important, these had to be undertaken without antagonizing the grain owners and merchants whose activities formed the basis of any long-term security.

The Marketday in Upper Normandy

In the early modern period, that security was seen to rest on the many regulations of the trade. While the literature on *subsistances* has provided detailed descriptions of those regulations, a brief summary of practices in Upper Normandy can anchor this book's exploration of evolving policies.[3] In short, several centuries of royal, provincial, and municipal legislation set out a basic framework for the grain and baking trades. It restricted sales to public marketplaces, where the hall controllers or other officials monitored the grain's quality and recorded its price.[4] The public market seemed the best way to assure that the local community had access to cereals. Thus, producers could not sell at their farms or warehouses, before the harvest (sales *en verte*), or by sample. A 19 April 1723 declaration reinforced those traditional prohibitions.[5]

[3] Kaplan, *Bread*; *Provisioning Paris*; Letaconnoux, *Subsistances*; Stouff, *Ravitaillement*; Usher, *The Grain Trade in France*; Desmarest, *Commerce*.

[4] For example, the *Sentence du bureau* (Paris), 12 November 1671, AN F11 264. Since a 1667 royal ordonnance, every town had been required to maintain a register of grain prices. A.-J. Sylvestre, *Histoire des professions dans Paris et ses environs* (Paris, 1853), pp. xvi–xviii, 26–7; Labrousse, *Esquisse du mouvement des prix*, 1: 27.

[5] "Déclaration du Roy portant que les Bleds, Farines et autres Grains, ne pourrant estre vendus, achetez ny mesurez ailleurs que dans les Halles et marchez," 19 April 1723, *French Royal and Administrative Acts, 1256–1794* (New Haven, 1978), hereafter cited as *Royal Acts*. The acts are arranged by date on microfilm rolls.

As elsewhere across the kingdom, the Upper Norman marketday began at dawn. Owners pulled their wagons and horses into the town square or grain hall, often having traveled at night. Some were local farmers, others were *blatiers*, the small-time merchants who traveled from town to town with several sacks on a horse, mule, or cart. Several hundred sacks, weighing often 325 pounds each, might appear in middling-sized towns such as Yvetot or Caudebec. In the autumn, most of the sellers were from the small holdings scattered nearby; they had rushed to thresh, dry, sort, and market their grain in order to pay the rent and taxes due in November. Later in the year, they would exhaust their own supplies and return to buy more expensive grain for their households. The region's more prosperous farmers, called *laboureurs*, generally rented impressive tracts of land, and could wait until the high prices of spring or summer to sell.[6] The largest farms lay to the south of Rouen, in the Vexin normand, and frequently exceeded 100 hectares (Figure 1.1). There, *laboureurs* hired the many landless and land-poor day laborers (*journaliers*) of the area, and the tensions between the two groups were increasing. In the less prosperous Caux, holdings of at least 40 hectares predominated, and much of the population had only the most meager and insufficient plots to farm. They made ends meet by hiring themselves out as *journaliers* or by home textile production. Often, buyers and sellers in the outlying towns knew each other. They frequented the same markets, traveled the same roads, and sometimes worked the same fields, one as the employer or estate agent, the other as the hired hand.

The market itself, occasionally sheltered by a roof and even marked off into stalls, but more often simply an open square exposed to wind, rain, chickens, and sheep, filled with buyers and sellers on the appointed days of the week (Figure 1.2). The first bell announced the hours reserved for local retail sales; later bells opened the sales to bakers and then to outside merchants and millers.[7] The parties were to "*marchander*," or haggle, over prices. If they did not, they aroused immediate suspicions of illegal prior agreements.[8] Once a seller had named a price, he or she could not raise it.[9] Nor could sellers decide midway through a marketday that prices were too low and retire.[10] Every sack of grain displayed was available for immediate sale, and municipal authorities took control of any cereals not sold

[6] The term *laboureur* did not indicate whether land was owned or leased. In the eighteenth century, *cultivateur* more often indicated poor farmers, although by the nineteenth century, it replaced *laboureur*, and came to mean an agricultural producer of any type.

[7] In Rouen, for example, a ten o'clock bell announced the start of merchants' and bakers' hours. Desmarest, *Commerce*, p. 39.

[8] CG to De Courson, 16 April 1709, AN G7 1708; 8 July 1740, AN Y 9499.

[9] Ordonnance, 11 August, 1661; Réglement, 14 November 1719, AN F11 264.

[10] This was clearly an ongoing problem, given the frequency with which the restrictions had to be reiterated. See the Ordonnances from 1661, 1667, 1681, and 1693 in AN F11 264.

Figure 1.1 A view of Rouen from the southwest, surrounded by wheat fields. This work was part of an eighteenth-century study documenting the city's climate and its inhabitants' lives, especially their illnesses, over a fourteen-year period. The artist's perspective flattens the fields, which were actually rolling hills. Beyond the city are the cliffs that run along the Seine River. Reproduced with permission from the Bibliothèque Nationale de France, Paris. Cliché number C 73910.

Figure 1.2 The *halle* at Lillebonne. Many market halls, such as this one in Lillebonne, were open to the elements and crowds. The *granges*, or storage areas where owners, merchants, or bakers would place their sacks of grain, were located in the second stories or under the eaves of the structures. Reproduced with permission from the Bibliothèque Nationale de France, Paris. Cliché number B 570.

within three market sessions. Also, buyers could not resell grain for a profit at the market where it had been purchased. When the sale had been made, the sack was ceremoniously closed, sealed, and identified to the controller or the measurer. Those market officials then poured out the grain, verified its quality, and recorded its price. In return, they received a small portion of the grain as part of their salaries. Another portion of the hall fees often contributed to local emergency granaries.[11] Some of the purchases went directly to nearby mills, which levied additional fees in kind, including seigneurial or municipal dues, or simply the miller's salary, for grinding the grain. Other purchases were ground in primitive hand mills in homes or small bakeries.[12] The poorest buyers would take their dough to communal ovens or to bakers, or used the grain to make a simple gruel, while the wealthier households baked bread in their own ovens.

The authorities knew that most sellers regarded this retail grain trade as a nuisance, but agreed that it was important to assure the poorest buyers' access to local supplies through the public markets, even if only a few could afford an entire sack. These buyers preferred smaller quantities, such as a *boisseau*, the unstandardized measure similar to a small bushel basket, or even the much smaller *pot*. A frequent sight at the hall was that of several women pooling their coins in order to purchase a *boisseau*, which they then divided, pouring their portions into their aprons, pockets, or small sacks. If they did not have coins, they might ask for credit. Given the confusion and tensions of the average day, it was no wonder that *laboureurs* and merchants viewed the legislation tying them to the market with resentment. Avoiding the *bousculade* of the marketplace, they retreated to the dark corners of inns and cabarets surrounding the market, and agreed to illegal sales. Administrators inveighed against those activities, but realized that they were difficult to prevent.[13]

Some portion of the grain at outlying markets made its way to the cities, such as Rouen. It came by barge, wagon, and horseback. Upper Norman barges regularly made their way between Rouen and the loading points near the four markets – Elbeuf, Caudebec, Duclair, and Les Andelys – where the city's grain merchants' guild had wholesale purchasing privileges. Other vessels arrived from Le Havre, laden with cereals purchased by the region's

[11] In general, these were not onerous. In Aumale, for instance, the hall fee was 1 boisseau for every 48 sold. In Rouen, pressure from *blatiers* had forced the city to lower it to approximately 2 deniers per mine (4 boisseaux) sold. Anon., "Salaires et revenus dans la généralité de Rouen, au XVIIIe siècle," *Bulletin de la société libre d'émulation de Rouen* (1885–6), p. 240.

[12] Rouen's milling fees furnished additional emergency supplies. The city's bakers were required to use the city's five mills, but were allowed slightly greater profits on their best quality white loaves to cover the fees. Milling fees in Upper Normandy were between 6 and 10 percent of the grain ground (9 percent in Rouen). Research in other areas of France indicates similar ranges. Archives municipales (AM) Rouen, Chartrier n. 262; Desmarest, *Commerce*, pp. 206–7; Anon., "Salaires et revenus," p. 24; Stouff, *Ravitaillement*, pp. 29–30; Alain Guerreau, "Mesures du blé et du pain à Macon (XIVe–XVIIIe siècle)," *Histoire et Mesure* 1988 (3): 175; Kaplan, *Provisioning Paris*, pp. 545–50.

[13] Boislisle, *Correspondance*, 3: pièces 153 and 226.

Figure 1.3 The Rouen grain market was located along the north side of the textile hall, the Halle aux Toiles. On marketdays, the square below the tower was full of activity. The city's guild grain merchants would haul some of their wares from the storage places beneath the building, and sell them from 110 stalls for which they had guild privileges. Reproduced with permission from the Bibliothèque Nationale de France, Paris. Cliché number A 69519.

négociants, the affluent merchants involved in commercial ventures that stretched from the Baltic to the New World. The grain merchants' guild could lay first claim to such supplies, although the city's bakers contested their privileges. Porters, belonging to other guilds, carried the sacks up the muddy shore, and then carted them to the warehouses that lined the crowded streets between the river and the Basse Veille Tour market (Figure 1.3). Three days a week, the approximately eighty-five members of the merchants guild stationed themselves in their stalls, and on other days, they alone were allowed to sell from their warehouses.[14]

The Authorities of the Grain Trade

While the intendant could claim some authority to supervise them, the guilds lay more directly under the purview of the police and parlement. The lieutenant of police had the immediate local authority, although the intendant or a subdelegate might nullify his orders. This police official administered the guild oaths of loyalty and collected the fees associated with membership. The police *commissaires*, along with guild officers, saw that members followed legal practices. They patrolled the docks at daybreak, entered bakeries, and opened sacks and barrels, looking for illegal supplies and workers. The parlement sanctioned their activities – it validated guild statutes, heard appeals of police rulings, and in times of shortages, ordered searches and organized prosecutions. The controller general and the intendants could try to overrule the parlements, yet such actions could lead to stalemates and protracted conflicts. The perennial question after food riots, for example, concerned which institutions had the responsibility to apprehend and try the guilty.[15] These overlapping jurisdictions also offered

[14] The Rouen guild received royal confirmation with the Edict of December 1692, although it had existed for several centuries. The statutes permitted as many as 112 members to fill the same number of stalls at the city market, although the actual membership seems to have fluctuated between 80 and 90 (30 percent females, many of them widows). The office of grain merchant was abolished by the Crown in June 1775 and, unlike other guilds, was not reestablished in the Necker reforms of 1778. "Catalogue des marchands de grains, Année 1755," ADSM C 144. Rolls for the *vingtièmes de l'industrie* from 1779 to 1788 are in ADSM C 518–528. Fragmentary guild records are in ADSM E 195–6. The police records of all guild receptions are in ADSM 4 BP L/60-71. There is a 1787 summary of the guild's history, ADSM C 144, p. 200; and scattered materials in ADSM 4 BP 5870–5963; C 345, 350–6, 360, 362.

[15] See the 1709 dispute between the Rouen intendant and parlement over the meaning of royal legislation. "Déclaration portant qu'il sera procédé à la visite des magasins . . . ," 27 April 1709, Isambert, *Lois*, 20: 539; CG to M de Bernière, Procureur général du Parlement de Rouen, 28 June 1709; Boislisle, *Correspondance*, 3: pièces 392 and 509; Bardet, *Rouen*, 1: 103.

numerous routes of appeal for guild members or other disgruntled parties when unfavorable rulings came from one institution or another.

Controlling Bread Prices in the Old Regime

Much of the energy of the authorities in Rouen, as in other cities, turned toward the baking trades. Affordable bread prices underlay any hopes for urban tranquility. While many inhabitants bought grain at the city's market, a fair number of them purchased loaves from bakers, and did so increasingly as the century progressed.[16] Like the grain merchants, the city's approximately 135 bakers had been organized in a guild for many centuries. The Rouen statutes, simliar to those throughout the kingdom, outlined the many controls on the kinds of bread that could be produced and sold within urban jurisdictions. They instructed bakers to sell at only police-sanctioned prices and to have their shelves well stocked even during shortages. Guild bakers alone had the right to bake and sell inside the city walls, although one day a week, outside bakers – forains – could enter the city with their cheaper loaves.[17]

The issue of prices was the one most often contested by the bakers, and resulted in lengthy clashes between the bakers, the parlement, and the police.[18] Throughout France since the late medieval period, bread prices were set by municipal authorities, generally the police officials.[19] These men

[16] This was also the case in England. Christian Peterson, *Bread and the British Economy, c1770–1870*, Andrew Jenkins, ed. (Hants, UK, 1994), pp. 44–50.

[17] The Rouen bakers' guild dated from at least 1256 and that of Caen from 1469. Information on baking guilds is contained throughout BN Joly de Fleury (BN JF) 1742; AD I 23A; AN AD XI 14; AN AD+ 856, 858, 859; AN G7 1872; and for Normandy in ADSM 5E 52, 195–6, 423, 431, 512; and AD Calvados 6E 23. See also Françoise Desportes, *Le pain au Moyen Age* (Paris, 1987); Guerreau, "Mesures du blé et du pain"; Kaplan, *Provisioning Paris*, pp. 446–75. For similar policies across Europe, see Maurice Aymard, *Venise, Raguse et le commerce du blé pendant la seconde moitié du XVIe siècle, Ports-Routes-Trafics* 20 (Paris, 1966); and Christian Peterson, *Bread and the British Economy*.

[18] See the discussion in Chapter 4.

[19] This system had originated in the late medieval period and remained in effect almost continuously until the 1980s, even after the freeing of the baking trade in 1863. 12 November 1545, AN Y 17058; BN JF 1107, p. 120; BN JF 1111, p. 179; BN JF 1742, pp. 15 and 38; BN Manuscrits français (Mf) 21638, pp. 240, 345, 463; ADSM 6 M 1310; Pierre Dardel, *Statuts des boulangers et Barème du Prix du Pain à Lillebonne vers 1461; comparaison avec le Barème actuel* (Rouen, 1934); Desportes, *Le pain*, pp. 147, 152; Jean Martineau, *Les Halles de Paris des origines à 1789: Evolution matérielle, juridique et économique* (Paris, 1960), p. 232; Witold Kula, *Measures and Men*, trans R. Sretzer (Princeton, 1986), pp. 102–105; and Jean Meuvret, *Etudes d'histoire économique; recueil d'articles, Cahiers des annales* 32 (Paris, 1971), p. 45, and Antonin Lefort, *La boulangerie et le décret du 22 juin 1863 proclamant sa liberté* (Paris and Beauvais, 1899).

would ascertain grain price levels at the central market, and consult the columns on the pages of the city's *tarif.* Those pages listed all the possible prices for grain, and indicated the corresponding rates for which bread loaves could be sold. When a Caen police official skimmed through his city's *tarif* of 1776, for instance, he found that whenever a sack of grain cost 35 livres, the legal price for a first-quality white loaf was 2 sous 6 deniers[20] (Figure 1.4). The police's regular practice of decreeing the bread price was called setting *la taxe du pain.* The prices of the *tarif* reflected the city's sense of fairness to both bakers and consumers. Municipal authorities, generally with the parlement's encouragement, ran assays to determine the cost of baking a loaf of bread, and used that information to draft the *tarif.* They included the prices of grain and rent and the costs of milling, wood, and labor, and then accorded bakers a modest profit.[21] Thus armed, the lieutenant general of police decreed the appropriate price for bread, and changed it each time grain prices rose or fell at the city market.[22]

While bread prices were based in theory on a straightforward calculation of the costs of production, the *tarifs* actually were a critical tool for distributing profits and losses over the course of a dearth. There were numerous hidden costs and benefits within each *tarif*'s list of bread prices. It was a common French practice, for instance, for best quality wheaten loaves to weight less than second- and third-class loaves. A pound of first-quality white bread in Rouen had only 14 ounces, while second- and third-quality loaves had 16-ounce pounds.[23] This meant that there were higher profits overall to be made on white loaves in comparison with the "working-class" loaves. Thus, wealthier customers paid more per pound

[20] 11 August 1785, ADSM 1B 298.

[21] The records of numerous assays are scattered in departmental and communal collections, generally in the police, intendancy, and parlementary series. Paris assays between 1435 and 1759 are recorded in BN JF 1107, pp. 122, 129, 141; BN Mf 11347, p. 195.

[22] Examples of the *taxe* are found in BN JF 1107, pp. 82–3; BN Mf 21638, pp. 149, 153, 180–84, 197–203; Statuts . . . pour les maîtres boullangers de Paris, 14 May 1719, AN AD XI 14; AD Calvados 6E 52; ADSM 5E 195–196; ADSM C 144; Archives Municipales du Havre (AM Le Havre) HH and FM F^4 8; AD Eure 1B 258; AM Rouen, Chartrier no. 262; Dardel, *Statuts des boulangers*; and *Cahiers de doléances du Tiers Etat du bailliage de Rouen pour les Etats Généraux de 1789*, 2 vols., Marc Bouloiseau, ed. (Rouen and Paris, 1960), 1: 92.

[23] In Rouen, this practice compensated bakers for the costs of having to mill their grain at the city's banal mills. A lighter "pound" for first-quality loaves was common across northern France. "Règlement," Veille St Michel 1199 and 27 November 1576, AM Rouen, Chartrier no. 262; Archives Départementales des Yvelines (AD Yvelines) 2B 1142; 26 May 1761, AM Le Havre HH 25; BN JF 1742, pp. 14, 55 and 61–2; ADSM C 102, pp. 44 and 45; Bouvet, *Maconnais*, p. 73; and Anne-Robert-Jacques Turgot, *Oeuvres de Turgot et documents le concernant . . .*, Gustave Schelle, ed., 5 vols. (Paris, 1913–22), 4: 491–2. Hereafter cited as Turgot, *Oeuvres*.

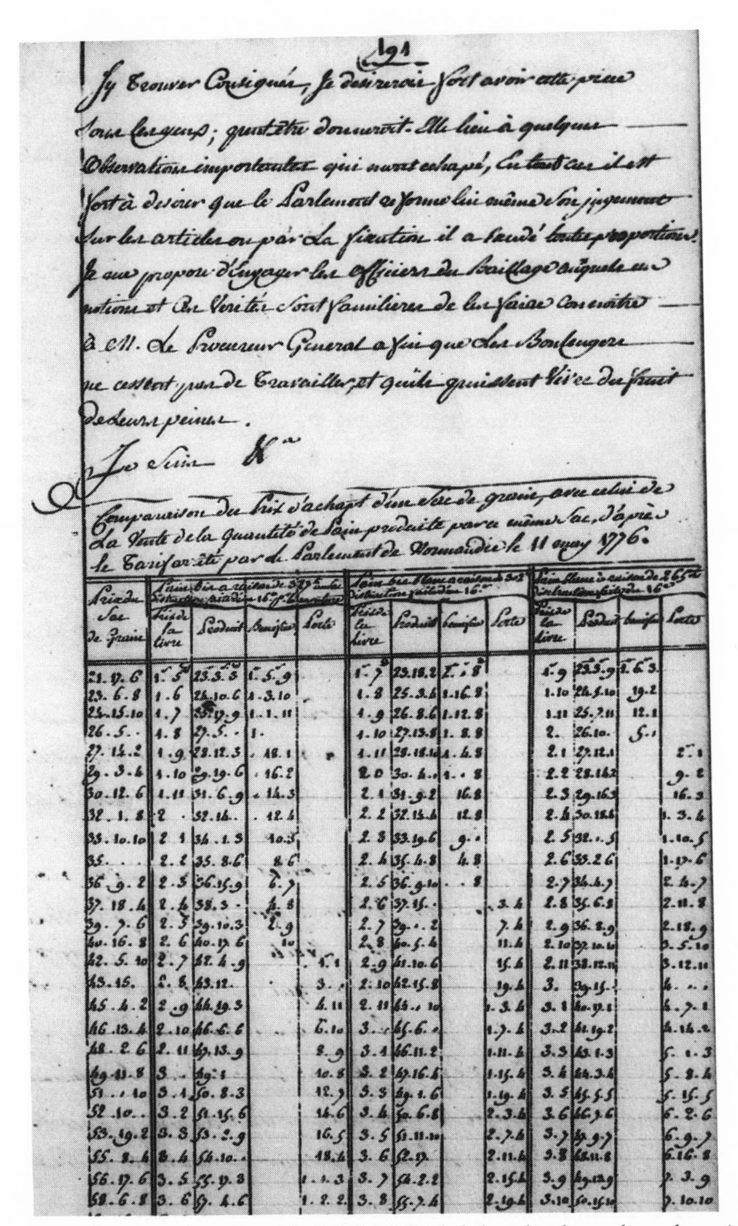

191

J'y trouver consignés, je désirerois fort avoir cette pièce sous les yeux; peut être donneroit elle lieu à quelques observations importantes qui m'ont echapé; en tout cas il est fort à desirer que le Parlement reforme lui même son jugement sur les articles ou par la fixation il a faussé toutes proportions. Je me propose d'engager les officiers du bailliage auxquels ces notions et ces vérités sont familieres de les faire connoître à M. le Procureur General afin que les boulangers ne cessent pas de travailler, et qu'ils puissent vivre du fruit de leurs peines.

Je suis &c.

Comparaison du prix d'achapt d'un sac de grain, avec celui de la vente de la quantité de pain produite par ce même sac, d'après le tarif arrêté par le Parlement de Normandie le 11 may 1776.

Prix du sac de grain	Pain bis [illegible] 16				Pain bis blanc [illegible] 16				Pain blanc [illegible] 16			
	Prix de la livre	Produit	Benefice	Perte	Prix de la livre	Produit	Benefice	Perte	Prix de la livre	Produit	Benefice	Perte
21.17.6	1.5	23.3.3	1.5.9		1.7	23.18.2	2..8		1.9	23.5.9	2.6.3	
23.6.8	1.6	24.10.6	1.3.10		1.8	25.3.4	1.16.8		1.10	24.5.10	19.2	
24.15.10	1.7	25.17.9	1.1.11		1.9	26.8.6	1.12.8		1.11	25.7.11	12.1	
26.5..	1.8	27.5..	1.		1.10	27.13.8	1.8.8		2.	26.10.	5.1	
27.14.2	1.9	28.12.3	.18.1		1.11	28.18.10	1.4.8		2.1	27.12.1		2..1
29.3.4	1.10	29.19.6	.16.2		2.0	30.4..	1..8		2.2	28.14.2		9.2
30.12.6	1.11	31.6.9	.14.3		2.1	31.9.2	16.8		2.3	29.16.3		16.3
32.1.8	2.	32.14..	.12.4		2.2	32.14.4	12.8		2.4	30.18.4		1.3.4
33.10.10	2.1	34.1.3	10.5		2.3	33.19.6	9..		2.5	32..5		1.10.5
35...	2.2	35.8.6	8.6		2.4	35.4.8	4.8		2.6	33.2.6		1.17.6
36.9.2	2.3	36.15.9	6.7		2.5	36.9.10	..8		2.7	34.4.7		2.4.7
37.18.4	2.4	38.3..	4.8		2.6	37.15..		.3.4	2.8	35.6.8		2.11.8
39.7.6	2.5	39.10.3	2.9		2.7	39..2		7.4	2.9	36.8.9		2.18.9
40.16.8	2.6	40.17.6	10		2.8	40.5.4		11.4	2.10	37.10.10		3.5.10
42.5.10	2.7	42.4.9		1.1	2.9	41.10.6		15.4	2.11	38.12.11		3.12.11
43.15.	2.8	43.12.		3..	2.10	42.15.8		19.4	3.	39.15.		4...
45.4.2	2.9	44.19.3		4.11	2.11	44..10		1.3.4	3.1	40.17.1		4.7.1
46.13.4	2.10	46.6.6		6.10	3..	45.6..		1.7.4	3.2	41.19.2		4.14.2
48.2.6	2.11	47.13.9		8.9	3.1	46.11.2		1.11.4	3.3	43.1.3		5.1.3
49.11.8	3..	49.1		10.8	3.2	47.16.4		1.15.4	3.4	44.3.4		5.8.4
51..10	3.1	50.8.3		12.7	3.3	49.1.6		1.19.4	3.5	45.5.5		5.15.5
52.10..	3.2	51.15.6		14.6	3.4	50.6.8		2.3.4	3.6	46.7.6		6.2.6
53.19.2	3.3	53.2.9		16.5	3.5	51.11.10		2.7.4	3.7	47.9.7		6.9.7
55.8.4	3.4	54.10..		18.4	3.6	52.17.		2.11.4	3.8	48.11.8		6.16.8
56.17.6	3.5	55.17.3		1.1.3	3.7	54.2.2		2.15.4	3.9	49.13.9		7.3.9
58.6.8	3.6	57.4.6		1.2.2	3.8	55.7.4		2.19.4	3.10	50.15.10		7.10.10

Figure 1.4 The Caen bread *tarif* of 1776. The left-hand column lists the prices of a sack of wheat. The other columns show the corresponding legal price for each type of bread along with the bakers' profits or losses. Archives Départmentales du Calvados, C 245. Reprinted with permission of the Archives.

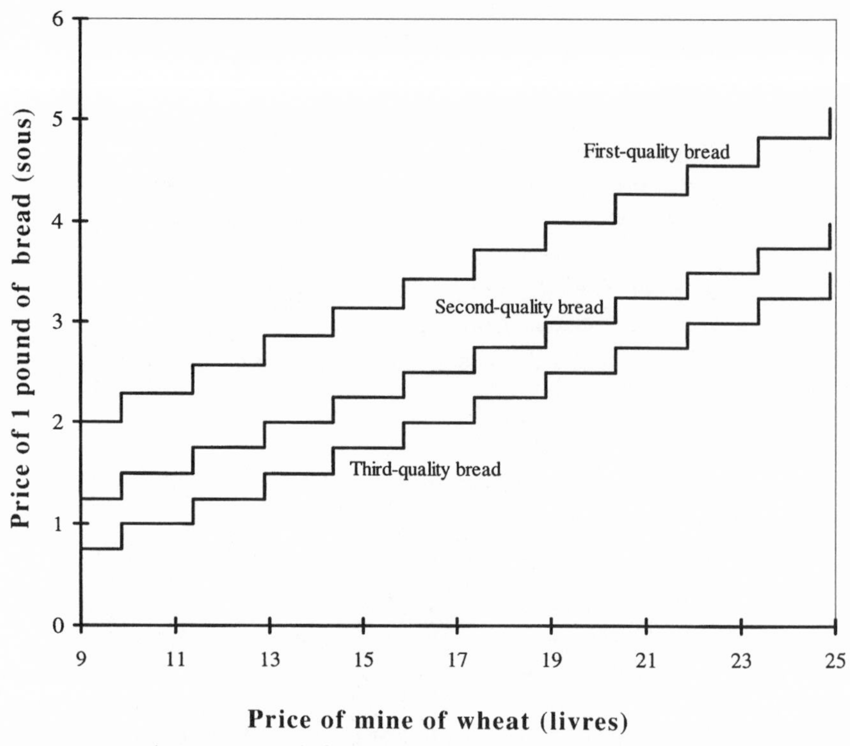

Figure 1.5 The Rouen *tarif* of 1725. A comparison of the legal price for bread as the price of grain rose. Source: ADSM 4BP LV 1/7, cahier 7.

for bread than the poor. Second, the *tarifs* often were structured to transfer the weight of rising grain prices progressively onto those more prosperous consumers. They did so by marking off comparatively slower rises in the price of second- and third-class loaves than for first-quality loaves[24] (Figure 1.5). All of this cushioned the poorest consumers from the worst effects of rising grain prices.

Beyond the way *tarifs* distributed the highest price rises onto the consumers who could best afford them, the application of the schedules provided other buffers to townspeople. Here, administrative judgment played an essential role. In a good year, when grain prices were moderate or did not fluctuate much, the *taxe* was reset only every few months and with little commotion. In periods of shortage, however, it became a much-contested cornerstone of local provisioning policies. The essence of these decisions

[24] Judith A. Miller, "Politics and Urban Provisioning Crises: Bakers, Police and Parlements in France, 1750–1793," *JMH* 64 (June 1992): 227–62. See also pp. 96–97 in Chapter 4.

concerned the timing and rate of bread price increases and decreases. When grain prices rose, the police generally rejected bakers' first petitions for a higher *taxe*. As a result, the bakers sustained losses at the outset, although because so many bought grain and flour off-market, legally or illegally, their losses were never what they alleged. Once the crisis had passed, the police were slow to lower the *taxe*, and thereby assured the bakers some compensation for their earlier losses.[25] Other strategic applications of bread policies included temporarily decreasing the size of loaves to mask price rises, making special distributions of bread to the poor, subsidizing bakers' purchases of flour or wood, lifting milling or measuring fees, and selectively prosecuting bakers' infractions.[26] In general, then, the practices governing the baking trade followed the police officials' sense of what would best protect the poor and promote public tranquility, without inflicting intolerable hardships on bakers.

When such efforts alone failed to offset grain price increases, the officials sometimes returned to the *tarif* itself, revising it downward to cut baker's profits overall. These revisions occasioned many battles between the bakers and the police. The recalculated *tarifs* sometimes came before the parlements, and the intendant might offer instructions that carried the controller general's endorsement.[27] Such efforts were astute and widely used moves following shortages. They not only effectively lowered future bread prices across the board, but communicated publicly that the authorities had sided with buyers. Parlementary and police records across the kingdom testify to the desire of French authorities to use *tarif* revisions to offset rising grain prices. In southwestern France, for instance, the neighboring towns of La Rochelle and Rochefort had several momentous disputes between bakers, administrators, and consumers during the eighteenth century.

[25] 20 April 1773, AD Calvados 1B 1665A; Armand-Thomas Hue de Miromesnil, *Correspondance politique er administrative de Miromesnil, premier président du Parlement de Normandie*, ed. P. LeVerdier, 5 vols. (Rouen and Paris, 1899–1903), 5: 4; Miller, "Politics and Urban Provisioning Crises," p. 238.

[26] BN JF 1161, pp. 10 and 12; BN JF 1742, p. 70; "Mémoire concernant l'affaire des boullengers (sic) . . . ," n.d., AD Somme C 458. The following cartons contain numerous additional eighteenth-century examples: AN G7 15 and 16; AN Y 9487; AM Rouen A 36, 38–40; AM Rouen B 17; ADSM 1B 290–291, 293, 297 and 3526; ADSM 5E 423; BN 4 Z Le Senne 30 and 2263.

[27] The assays were generally done by a committee of police officials and representatives from the bakers guild. They followed a fairly standard format. Three sacks of different quality grain would be bought at the local market hall and milled, the dough prepared and baked, and the resulting loaves counted and examined. The city's bakers would take issue with the each step, trying to show that the assay did not replicate the difficulties and expenses of their trade. BN JF 1107, pp. 122, 129, 141; BN Mf 11347, p. 195; *Correspondance des Contrôleurs généraux*, 3: pièce 392; AD Calvados, C 245, pp. 184–91; Petition from the Bakers of Evreux, 7 February 1785, AD Eure 1B 258.

There, authorities reduced the bakers' profits in two sets of *tarifs*, calculated in 1700–03 and again in 1708–09, each time in the aftermath of a shortage. The conflicts over these revisions persisted from many decades, and eventually led to three days of riots in 1784.[28] Eighteenth-century Norman intendants and parlementary *présidents* also were involved in numerous revisions, including those in Caen in 1743 and 1775, and in Rouen in 1775 and during the 1780s.[29] In the last decades of the Old Regime, authorities turned to such revisions with renewed hope. The free-trade reforms of that period prohibited other means of addressing grain price increases. Thus, *tarif* revisions, discussed more fully in Chapter 4, became one of the last acceptable means by which public officials could provide relief to their communities.

Administrative Leeway: Applying the Regulations of the Old Regime

Despite the many regulations on the grain and baking trades, actual practices in the eighteenth century encouraged their flexible application. In years of abundance, the host of participants in the grain trade followed their interests more or less as they wanted. It was widely acknowledged that the 1723 rules were to be used only *in extremis*. By the mid-century, the Paris Parlement's *procureur général* was well within accepted limits when he advised a *procureur fiscal* that the declaration "had as its principal object times of dearth and high prices . . . and thus one should not follow any rigorous execution except in such times." One should avoid prosecuting offenders "judicially except with great prudence and many precautions, when grain is at a reasonable price." Furthermore, such measures should be taken only after consultation with the *procureur général*.[30] Thus, off-market sales increasingly were allowed, if transacted discreetly and without disturbing local consumers. *Blatiers* roamed the muddy roads to the market stalls of distant towns, rarely registering their purchases with local officials. They fashioned themselves as *laboureurs et blatiers*, or as having one or another trade, depending on the needs of the moment – it mattered little.

Large landowners, *fermiers* (estate agents), and *laboureurs* also explored the ambiguities of their stations, adding purchased cereals to their stocks, storing and reselling at will. Wealthy urban residents – the *bourgeoisie* – exploited the permission they held to bring grain from their lands into town

[28] BN JF 1164, pp. 69–72; BN JF 1742, pp. 115–201; and Rochefort Dossier (1781–1782), AN U 1441.

[29] See the discussion later for a fuller account of the revisions.

[30] BN JF 1107, pp. 7–17.

for their households. They paid no entry fees as they passed by the *octroi* gates, nor did they pay fees if they resold their goods from their city homes. If they purchased other grain, carted it into the city, and then sold it, who was to know? Despite the 1723 edict's formal interdiction of off-market transactions, Parisian merchants, and eventually bakers, purchased grain with impunity at farms and ports near the capital, and were only occasionally called to order.[31] Bakers could produce shortweight loaves, providing they did not push beyond the accepted bounds, known as the *tolérance*.[32] As long as grain found its way to buyers, was transformed into bread, and suspected abuses provoked no great outrage, officials deemed it best to turn a blind eye.[33]

Overall, then, the many regulations offered administrators a number of points at which to monitor the grain trade, and gave them a certain leeway within which to exercise their judgment. They could waive or reduce milling fees and market dues, for instance, or even ask for exemptions from royal taxation for their jurisdictions. They could pressure grain owners to supply markets. If necessary, public officials could undertake searches for illegal stocks of grain and even levy fixed requisitions. They could limit sales at markets and, if supplies were especially short, bar outsiders. They could juggle bread prices and weights, or subsidize either the poor or the bakers. If unemployment were a problem, the government organized public works projects, such as textile *ateliers* or road crews. The Crown, the intendant, or a host of provincial and municipal institutions could fund or encourage the creation of emergency granaries. Even setting the price of grain – a more questionable activity – was accepted occasionally as a last resort. Between the field, the market, and the bakers' shelf, Old Regime authorities had many possibilities for supervising and influencing the trade.

The stakes were high. If authorities failed, violence resulted. Eighteenth-century harvest failures shook the kingdom no less than six times before the Revolution – in 1708–9, 1725–6, 1738–41, 1768–70, 1774–75 and 1788–9, and regionwide shortages placed additional burdens on administrators, such as in 1752, 1757, and 1784 in northern France. The general course of provisioning disturbances has been well-established by an extensive literature, and need only be summarized here.[34] Briefly, when grain

[31] Kaplan provides an excellent analysis of the ambiguities and stuggles over such restrictions, *Provisioning Paris*, pp. 171–83.

[32] Lefort, *La boulangerie*, p. 9.

[33] Meuvret, *Le problème des subsistances à l'époque Louis XIV*, *Texte*, 3: 97–143; and *Etudes d'histoire économique*, p. 205; Usher, *Grain Trade*, pp. 93–4, 320–2; Letaconnoux, *Subsistances*, pp. 86–91. See also Kaplan's analysis in *Provisioning Paris*, pp. 31–2, 291–2, 373–4.

[34] The most recent and thorough treatment is Bouton, *The Flour War*. See also *Mouvements populaires et conscience sociale, XVIe–XIXe siècles*, Jean Nicolas, ed. (Paris, 1985); George

owners brought too little grain to market, or demanded high prices, tensions rose. Would-be buyers, often female, assembled in markets to protest the owners' pretensions. If neither side would yield, someone generally ran to find an official, such as the lieutenant general of police, the mayor, or a subdelegate. Typically, the official stepped in to "reason" with the crowd, and then might be convinced to negotiate a price. In such cases, he often helped distribute the grain and collected payments. If possible, his assistants noted the names of each participant so that owners might be compensated for losses, and the guilty punished. This price-setting, called *taxation populaire*, reflected the buyers' judgment of what was fair, which they generally estimated at approximately 3 livres for a boisseau of wheat, and 2 sols per pound of bread.

As commercial networks expanded during the eighteenth century and owners sold grain outside of public markets – legally or illegally – distressed consumers congregated along roads, canals, and rivers to prevent local supplies from leaving their towns.[35] Many a wagoner feared the assailants hiding behind hedges or over the crest of a hill. Some of the hungry were bolder, and roamed in bands from farm to farm. If refused, they smashed locks and broke down doors and then loaded grain into sacks, aprons, sheets, and whatever else they had with them or could find. If the owners were lucky, they escaped unharmed. The few coins they received in return were small compensation for the silverware, linens, and cereals they lost in the attack. Often, they recognized their intruders – some had even been employed as day laborers. The owners' steady abandonment of public markets in favor of direct sales to millers or merchants caused these attacks to multiply in the late eighteenth century.

These market and roadside disturbances were difficult to prevent. The repressive apparatus of the Old Regime was insufficient and somewhat unreliable. The *milice bourgeoise* often failed to appear when called by the parlement or another authorized officials.[36] Public authorities could request military detachments, but soldiers could not be placed along every road or at every farm. If trouble were suspected at several markets and at nearby farms, such as in 1775, the troops had to move overnight. Arriving tired and hungry at their next site, they sometimes joined rioters. Moreover, even

Rudé, "La Taxation populaire de mai 1775 à Paris et dans la région parisienne," *AHRF* 28 (April–June 1956): 139–179 and "La Taxation populaire de mai 1775 en Picardie, en Normandie et dans le Beauvaisis," *AHRF* 35 (1961): 305–326; Tilly, "The Food Riot as a Form of Political Conflict in France."

[35] On this point, see especially Bouton, *The Flour War.*

[36] In March 1768, members of the *milice* ignored the Rouen Parlement's call to arms. They claimed disingenuously that the parlement's recent prohibitions against commoners carrying arms prevented them from performing such duties. De Crosne to CG, 23 March 1768; and Miromesnil to Bertin et al., 22 and 24 March 1768, ADSM C 107.

if willing troops could be mustered for several days or weeks, embittered buyers simply would wait for them to leave before taking out their anger on owners and authorities. Many swore, as did an incensed consumer at Etrepagny in 1775, "that there will not always be troops and that when they leave, [the buyers] would have their day."[37] Local authorities knew only too well that such threats could not be taken lightly, and tried to assure supplies at public halls and to protect bakers' access to grain. Preventing, not repressing, riots was their goal.

Attempted Reforms: The Failure of Physiocracy and Free Trade

If commerce and the marketplace alone did not bring grain in sufficient, steady amounts, how were local authorities to assure supplies and thus calm markets? They had many methods – outlined above – in their repertoire, but they were of uneven efficacy. The quest of these eighteenth-century administrators was to determine which would be the most successful. Already in 1709, authorities favored using inducements and persuasion to bring grain to market. They turned to threats and energetic intervention only as a last resort. In the next decades, these men concluded that they needed to rely even more on market forces, while sheltering their townspeople from the full impact of those mysterious mechanisms. Thus, they sought to influence prices and supplies, preferably by adopting low-profile means. When possible, they wished to avoid aggressive and highly visible measures. Drawing angry consumers into negotiations with merchants or bakers, for instance, only led to riots. The best methods, then, were clandestine ones. Nascent free-trade theories, along with the needs of growing cities, and concerns for French agricultural production overall, forced authorities to consider carefully the options before them. This mid-century combination of liberal theory and marketplace exigencies also offered further impetus to the choice already being defined – that is, finding a way to shape, rather than control, the provisioning trades.

The 1763 and 1774 free-trade decrees, which have received extensive treatment in the historical literature, were an important chapter in this century-long process.[38] These reforms formally lifted a number of controls on the grain trade. The Declaration of 23 May 1763, proposed by Controller General Bertin, who had circulated drafts to numerous intendants, opened the trade to anyone who wanted to participate. It allowed

[37] Courtois to intendant, 16 May 1775, ADSM C 109.

[38] Kaplan, *Bread*; Georges Weulersse, *Le mouvement physiocratique en France (de 1756 à 1770)*, 2 vols. (Paris, 1910).

grain to be stocked for more than one year, and abolished laws requiring merchants to register their transactions. It also permitted off-market transactions, which had been more or less tolerated. Moreover, it repealed the many tolls levied on grain in transit along rivers and roadways.[39] A year later, Louis XV allowed grain to be exported when its price was low.[40] Many of the provincial parlements had been clamoring for just such legislation, and welcomed the 1763–4 reforms.[41]

By 1768, however, exportation and then successive bad harvests produced famines and violence.[42] Parlement after parlement challenged royal authority by reimposing traditional regulations, and Louis XV forced numerous courts to rescind or modify their legislation.[43] By 1770, however, it was clear that free trade had failed both practically and politically. The new controller general, the Abbé Terray, oversaw the reversal of the free-trade policies. Louis XV suspended export, and finally revoked the 1763

[39] The Crown wished "to encourage *cultivateurs* in their labors . . ." through a "free and complete competition" in the grain trade and spoke of "interests," those of "*cultivateurs*" and "consumers." The declaration upheld market dues, however, and did not change any of the regulations pertaining to Parisian provisioning. "Déclaration du Roy, Portant permission de faire circuler les Grains, Farines . . . dans toute l'étendue du Royaume . . . ," 25 May 1763, *Royal Acts*. The preamble may also be found in Isambert, *Lois*, 22: 393. Kaplan has provided a full discussion of the political context in which such policies emerged. See *Bread*, 1: 97–163.

[40] Only twenty-seven ports were included in this permission and only French vessels could be used. If grain prices rose above 12.5 livres per quintal in the designated port of exit, exports were suspended. Import duties of 1 percent for wheat and 3 percent for lesser cereals were to be assessed at the points of entry. There was confusion over how ceilings should be calculated causing uneven implementation. "Edit du Roi concernant la liberté de la Sortie et de l'Entrée des Grains dans le Royaume," July 1764, *Royal Acts*; Miromesnil, *Correspondance*, 3: 294.

[41] Among the parlements registering the 1763 declaration were those of Provence, Grenoble, Bordeaux, Dijon, Toulouse, Lille, and Rennes. Many *présidents*, landlords themselves, perceived the possibility of personal gain from the legislation. In Rouen, the presidents hesitated for a year, possibly because they feared off-market trade would cost them revenues from the market dues many of them received. Elizabeth Fox-Genovese, *The Origins of Physiocracy: Economic Revolution and Social Order* (Ithaca, 1976), p. 26; Kaplan, *Bread*, 1: 181; Lemarchand, *La fin du féodalisme*, p. 233.

[42] Kaplan, *Bread*, 2: 497–519.

[43] Many of these measures prohibited off-market sales and exportation, and ordered requisitions. BN JF 1111, pp. 8, 23–5, 27; "Arrêt du conseil qui casse un arrêt du Parlement de Paris relatif au commerce des grains," 22 January 1769, AN F11 220; ADSM C 144; "Arrest du Conseil d'Etat du Roi, qui ordonne l'exécution de la Déclaration du 25 mai 1763, concernant la libre circulation des Grains dans le royaume . . . ," 31 October 1768, and "Extrait des Registres du Parlement [de Paris]," 20 January 1769, *Royal Acts*; Turgot, *Oeuvres*, 3: 257–8; Weulersse, *Le mouvement physiocratique en France*, 1: 182–5; Kaplan, *Bread*, 2: 513–4; Doyle, *The Parlement of Bordeaux and the End of the Old Regime 1771–1790* (London and Tonbridge, 1974), p. 204.

declaration with an *Arrêt du Conseil* on 23 December 1770.[44] His legislation reestablished requirements for merchants to register with local authorities, and once again restricted sales to markets. He left *libre circulation* between provinces intact, however, and furthered the cause of centralized authority by threatening any local official who meddled with grain shipments with fines that would equal the cost of the wares. For almost four years, the trade was subject to such regulation, and provincial authorities appear to have increased their supervision of markets.[45]

In 1774, with the death of Louis XV, the possibility for far-reaching reforms returned. Louis XVI placed Turgot, an ardent reformer, at the economic helm as controller general.[46] On 13 September 1774, Turgot reinstated the May 1763 free-trade declaration.[47] Additional articles permitted grain exports with the Crown's approval, although prices remained so high during Turgot's administration that export was never allowed. Later legislation decreased the number of formalities on merchants, abolished many market fees, eased coastal navigation, dissolved some of the municipal guilds involved in provisioning, and overturned many local ordinances.[48]

Turgot's attempts to free the trade were ill-timed, however. The 1774 edict was announced as the news of the bad harvest spread. By the next spring, the Flour War had erupted across northern France.[49] His many

[44] See Kaplan's analysis of Terray's position and of disagreements over his policy, *Bread*, 2: 510, 532–45.

[45] In the next years, some parlements and lower court officials ordered owners to supply markets, confiscated flour from ships in port, and printed sharp remonstrances against free trade, all extreme and wholly unaccepable measures even during Terray's ministry. Lesser measures were probably more the norm, however, and the prosecution of infractions throughout the kingdom may not have been very energetic. The Paris registers in AN Y 9648 identify eighty-six merchants or companies. The Bailliage of Bréteuil listed forty-nine merchants, most of whom described themselves as *blatiers*. "Enregistrement, Commerce des grains 1771," AD Eure 6B 207; BN, JF 1111, p. 31; 5 September 1771, ADSM C 106; Terray to Intendant Dupleix, 16 January 1770 [sic 1771], AD Somme C 83; and prosecutions in AD Somme C 87; Doyle, *Parlement of Bordeaux*, pp. 204–5.

[46] Turgot, *Oeuvres*, 4: 201–210; Gomel, *Les Causes financières de la Révolution française*, 2 vols. (Paris, 1892), 1: 127–31; Weulersse, *La physiocratie sous les ministères de Turgot et de Necker (1774–1781)* (Paris, 1950); Fox-Genovese, *Origins of Physiocracy*, pp. 66–7. For his attempts to restructure the royal treasury, see John Bosher's penetrating monograph, *French Finances, 1770–1795* (Cambridge, 1970), pp. 142–65.

[47] Turgot, *Oeuvres*, 4: 201–210.

[48] A flood of laws and instructions followed, suppressing many market personnel, municipal *octrois* (but not seigneurial *octrois*), and numerous dues and fees. The Six Edicts eventually required a *lit de justice* to register with the Paris Parlement. See the many pieces of legislation in AN E 2512. Turgot, *Oeuvres*, 4: 462–6; Edgar Faure, *La disgrâce de Turgot: 12 mai 1776* (Paris, 1961), pp. 409–36.

[49] The previous year's harvest had been good, making the prices of 1775 seem even higher by

reforms threatened entrenched elites, and he soon fell prey to court politics, ministerial rivalries, and widespread opposition to his programs. His sensational dismissal came in May 1776. Discredited but not repudiated, free trade limped on, however. Many of the articles allowing off-market sales remained untouched, although other free-trade policies were rescinded. One future financial minister, the Genevan banker Jacques Necker, a confidant of Terray, suspended exports on 27 September 1777, and restored many of the guilds.[50] One last time, Controller General Calonne reiterated the state's complete support for free trade in 1787. Within only a few months, however, the desperate shortages of 1788 forced his successor, the returned minister Necker, to curb export and trim free trade.[51] Thus, the grain trade existed in a kind of limbo after Turgot's fall in 1776.

The origins of this 1763 and 1774 legislation lay in mid-century concerns about the sluggish state of French agriculture and a general desire for a thorough reform of royal fiscal and economic policies. Plans for the restructuring of royal revenue collection, for road and canal building, and for improved industrial production animated provincial backwaters and Paris salons.[52] France's lackluster agricultural performance appeared particularly troubling to many of the educated elite. Sustained good harvests in the 1740s and 1750s had caused grain prices to plummet and production to stagnate. This awakened fears that landlords, frustrated by low profits, would abandon cereal production altogether. The *agronomes* – as an influential group of reformers was known – despaired when they com-

contrast. See Bouton, *The Flour War*; George Rudé, "La Taxation populaire de mai 1775 à Paris et dans la région parisienne;" and "La Taxation populaire de mai 1775 en Picardie, en Normandie et dans le Beauvaisis;" Guy Lemarchand, "Les troubles de subsistances;" Vladimir Sergeevitch Ljublinski, *La Guerre des farines: contribution à l'histoire de la lutte des classes en France, à la veille de la Révolution*, trans. François Adiba and Jacques Radiguet (Grenoble, 1979); and Robert Darnton, "Le Lieutenant de police J. P. Lenoir, la Guerre des farines et l'approvisionnement de Paris à la veille de la Révolution," *RHMC* 46 (1969): 611–24; Faure, *Turgot*, pp. 249–94; Lemarchand, *Fin du féodalisme*, p. 231.

[50] He also encouraged imports by offering bonuses to merchants. In addition, he reestablished and consolidated the guilds Turgot had abolished, producing a confusing and incoherent set of policies for much of the French economy. Antonio Carré, *Necker et la question des grains à la fin du XVIIIe siècle* (Reprint of 1903 edition, New York, 1979), pp. 205–7; Jean Egret, *Necker*, pp. 91–2; Edouard Chapuisat, *Necker (1732–1804)* (Paris, 1938), pp. 81–4.

[51] Chapuisat, *Necker*, pp. 146–67; Jean Egret, *Necker*, pp. 229–33; Jean Bouvet, *Maconnais*, p. 35.

[52] André J. Bourde, *Agronomie et agronomes en France au XVIIIe siècle*, 3 vols., *Les Hommes et la Terre* XIII (Paris, 1967); Jean Airiau, *L'opposition aux physiocrats à la fin de l'ancien régime: aspects économiques et politiques d'un libéralisme éclectique* (Paris, 1965), pp. 33–4; Hoock, "Le phénomène Savary;" Robert Forster, *Merchants, Landlords, Magistrates: The Depont Family in Eighteenth Century France* (Baltimore and London, 1980), p. 80.

pared the country with England's steady gains in productivity and that rival's taste for new methods and crops. The *agronomes* blamed the many regulations of the grain trade, especially the prohibitions against export, for low prices and the resulting slump. They predicted that soon worse would ensue.[53]

Increasing economic freedom had been a subject of discussion since at least the 1730s, when early liberal thinkers, including Claude Dupin, Jean-François Melon, and Richard Cantillon, had pressed for reforms. Unlike later theorists, many of these writers wanted to provide clear protections for the poor. They supported state-run emergency supplies, export prohibitions, and even occasional requisitions.[54] Their arguments for greater liberty, albeit mixed with sporadic emergency controls, laid the groundwork for the mid-century debates over economic policy. By the late 1750s, additional reasons for wide-ranging reform had become indisputable. The deficits caused by the Seven Years' War had not only disrupted the economy, but they had given the parlements an even weightier club with which to bludgeon the monarchy. By the late 1750s, the damage done by war, chaotic revenue collection, and suspicions of waste and corruption had driven parlements to demand a thorough investigation of the state's finances. The Rouen presidents were among the most outspoken in their calls for a full inquiry into the sources of royal deficits.[55] Thus, tax reform, grain production, parlementary politics, and economic liberalization were linked in elite and administrative minds.

This was the political and intellectual context that gave rise to Physiocracy, the prevailing form of economic liberalism in the late 1750s and 1760s. The extensive secondary literature on Physiocracy already gives a full analysis of the ideas of the school and of the reasons the Crown was drawn to its proposals.[56] But, given the impact of its thought on the poli-

[53] For a full description of eighteenth-century proposals for agricultural reform, see Bourde, *Agronomie*.

[54] Their writings circulated first in manuscript and eventually in printed form. Claude Dupin, *Oeconomiques*, 2 vols., Reprint ed. (Paris, 1913); Jean-François Melon, *Essai politique sur le commerce*, republished in *Collection des principaux économistes*, Eugène Daire, ed., (Reprint, Osnabrück, 1966); O'Brien, *The Classical Economists*, pp. 27–9; Antoine Murphy, *Richard Cantillon: Entrepreneur and Economist* (Oxford, 1986); Jean-Claude Perrot, *Une histoire intellectuelle de l'économie politique, XVIIe–XVIII siècle, Civilisations et Sociétés* 85 (Paris, 1992), pp. 45–6, 164. Simone Meyssonnier terms the mixing of free-trade theory and this concern for equity "*libéralisme égalitaire.*" *La balance et l'horloge: La génèse de la pensée libérale en France au XVIIIe siècle* (Paris, 1989).

[55] James C. Riley, *The Seven Years' War and the Old Regime in France: The Economic and Financial Toll* (Princeton, 1986), pp. 192–4; Georges Weulersse, *Le mouvement physiocratique en France (de 1756 à 1770)*, 1: 28–9.

[56] The standard collection of Quesnay's writing is *François Quesnay et la physiocratie*, 2 vols., (Paris, 1958), hereafter cited as *I.N.E.D.* Among the classic works on the Physiocrats are

cies of the grain trade, a brief summary is in order. The physiocrats, or the *économistes*, as they called themselves, argued unequivocally for lifting the controls on the grain trade. If owners could buy and sell where and how they wanted, these theorists reasoned, they would increase their profits and reinvest in their land. Eventually, France would boast surpluses rather than intermittent, incapacitating shortfalls. Following the lead of court physician François Quesnay, the Physiocrats desired not only increased production but increased *productivity*. This was a means to an important political end – increased royal tax revenues. The Physiocrats aimed to replace the complicated assortment of taxes that burdened the kingdom with a single imposition on landed property. A liberated grain trade would enable landowners to meet those heavy demands. Moreover, a constant flow of funds into the Crown's treasury would stem parlementary criticism and block the courts' mounting interference in royal policies.

What were the Physiocrat's specific proposals? At the very least, they wanted off-market sales legalized and exports permitted. Their sharpest critics feared that such freedom would allow prices to rise beyond the reach of the poor and working classes. While the Physiocrats admitted that prices would rise, they argued that the benefits would be shared by all. Stable, albeit higher, price levels, abundant harvests, and assured access to food would be welcomed by even the most destitute.[57] They opposed market fees and guild privileges, blaming them for needlessly interfering with competition and for raising prices, thus depriving producers of their rightful profits. Turgot, who had close ties to the Physiocrats, was particularly scornful of such privileges, and abolished many of them before being forced from office in 1776.[58]

the five studies by Georges Weulersse, *Les physiocrats* (Paris, 1931); *Le mouvement physiocratique*; *La physiocratie à la fin du règne de Louis XV (1770–1774)* (Paris, 1959); *La physiocratie sous les ministères de Turgot et de Necker*; and *La physiocratie à l'aube de la Révolution 1781–1792*, Corinne Beutler, ed. (Paris, 1984); Fox-Genovese, *The Origins of Physiocracy*; and D. P. O'Brien, *The Classical Economists* (Oxford, 1975).

[57] Their estimates of the "natural" or "base" price of grain ranged from 15 francs per setier (Mirabeau) to 18 (Quesnay) and even 24 livres in some cases. In general, they believed that 18 livres, approximately equivalent to English prices, was sufficient. Turgot departed somewhat from their views, believing that international free trade would lessen, but not eliminate, price volatility. Quesnay, "Fermiers," in *I.N.E.D.*, 2: 441–2; Weulersse, *Mouvement physiocratique*, 1: 480–86; [Anne Robert Jacques] Turgot, *Ecrits économiques*, Christian Schmidt, ed. (n.p., 1970), pp. 351–4; Karl Gunnar Persson, "On Corn, Turgot, and Elasticities: The Case for Deregulation of Grain Markets in Mid-Eighteenth Century France," *Scandinavian Economic History Review* 41 (1993): 37–50.

[58] Turgot, *Oeuvres*, 4: 201–210; W. Walker Stephens, *The Life and Writings of Turgot, Comptroller-General of France 1774–6* (London, 1895); Douglas Dakin, *Turgot and the Ancien Regime in France* (New York, 1939).

Despite the excited claims of the free-trade champions and minister-philosophes, the reforms did not bring about an immediate restructuring of the grain trade. The 1763–4 laissez-faire policies may have represented a breakthrough in theoretical rigor, and occasioned much public debate, but as a force for widespread, immediate change, they had a far more limited impact. First, many of their main tenets had already found a place in marketplace practices. As the next chapters will show, the public authorities who monitored the trade had been reaching an accommodation with liberal ideas since the early part of the century. They did so less on theoretical than on practical grounds, however. Those same practical concerns also dictated the limits they would place on free trade as it was envisioned in the 1763 and 1774 legislation.

Both before and after those free-trade reforms, local authorities were engaged in a search for genuinely workable policies that united their understanding of market mechanisms with the "subsistence imperative," as Steven Kaplan has termed it. Simply put – grain had to reach buyers. If commerce itself did not accomplish that elementary goal, then public authorities had no choice but to step in. The physiocratic-inspired policies simply added extra impetus – and a few additional difficulties – to the quest for practices that would assure supplies. The much-heralded physiocratic reforms were but part of a longer process that had begun in the early eighteenth century, and perhaps much earlier, and that continued well after their demise.

CHAPTER TWO

Simulated Sales: Shaping Supply and Demand in the Old Regime Marketplace

The search for strategies to feed the kingdom began in the marketplaces and administrative bureaus of eighteenth-century France. By mid-century, it resulted in methods that mimicked commercial practices and fooled merchants and buyers into believing, at least to some degree, that the government had retired a reasonable distance from the grain trade. These tactics created the illusion that supply and demand alone had drawn grain from areas of surplus to the markets where it was needed. The state was giving up its role of provisioner, or so these measures suggested. The illusion was not wholly sustainable, and the strategies' successes were incomplete. Nevertheless, well before the physiocrats' policies had issued from Versailles's chambers, local authorities were generating their own accommodation between free trade and hunger. They did not renounce their traditional role as provisioner of last resort, but masked their efforts in hopes of finding more effective means.

Free Trade Before the Physiocrats

From the outset, Old Regime hopes rested on *libre circulation*, the unhindered flow of grain between provinces, especially to the cities. A rudimentary notion of supply and demand was at work here – grain supplies were to move easily from places of abundance and low prices to those of shortage and high prices. Eventually, prices would level out to the greatest, if not complete, satisfaction of both buyers and sellers. The concept had been invoked during the 1708–9 shortages, when Controller General Desmarets had forced numerous parlements and lower authorities to rescind laws blocking grain leaving their jurisdictions. It remained so throughout the century. A 1751 *arrêt* established in unequivocal terms the Crown's support for the *libre circulation* of grain. "His Majesty has asked me to inform you," Controller General Machault's instructions to the intendants began, "that his intention henceforth is that the circulation and trade in grain from province to province cannot be interrupted except by higher authorities . . ."[1] The *arrêt du conseil* of 17 September 1754 then lifted the need for specific permissions or passports – although further legislation in 1757 reimposed some formalities.[2]

The fact was that even the Crown's insistence on the free flow of cereals failed to bring grain to those who needed it. Merchants did not rush to enter the trade. Thus, somehow, public officials had to apply enough force to ensure that markets were supplied, but not so much that owners – whether merchants or producers – would be alienated from grain trade and cereal production. In 1709 and later, the successive controllers general and numerous colleagues stressed the need to persuade owners to sell their grain, without antagonizing them, at least during the opening months of shortages.[3] Desmarets counseled against searches of warehouses and farms, explaining that they "always have a bad effect." Local authorities knew who had grain – the port and market records gave them that information – and thus armed, could meet with individual proprietors to "invite" them to bring grain to market. He instructed Intendant de Courson to pressure local grain owners verbally, and not to put anything in writing. In the Bas Dauphiné, officials were to use "remonstrances and verbal exhortations, accompanied by several threats, if necessary, in measured terms, carefully spoken," to convince known merchants to open their granaries. More forceful, but still avoiding marketplace theatrics, other instructions sent "the *maréchausée* to beat the countryside [near Soissons], to engage the grain merchants to sell grain to artisans and the poor for 52 écus per *muid* . . ." (a *muid* generally weighed 3,500–4,000 pounds, depending on the region). Eventually Desmarets did instruct local officials to scare owners with imprisonment, exile, and even death, although he did so only as a last resort in the spring of 1709.[4]

[1] The 1751 declaration maintained all the former conventions for merchants, such as the requisite paperwork to transport grain from town to town, and the 1723 prohibitions against off-market sales. Two items are worth nothing in his policies: first, that the king alone could order grain shipments interrupted; second, that Machault purported to present an immutable law. Most previous orders promoting the circulation of goods had been open to revision, or had concerned only a limited number of jurisdictions. CG to Intendant de la Bourdonnay, 12 December 1751, ADSM C 105; [Marcel] Marion, *Machault d'Arnouville: Etude sur l'histoire du contrôle général des finances de 1749 à 1754* (Paris, 1891), pp. 422–42; Desmarest, *Commerce*, p. 21; Schelle, *Gournay*, pp. 42–3, 72. Cf. *Arrêt du conseil*, 17 September 1743, AN AD+ 1260.

[2] 17 September 1754, AN AD+ 917; Draft, Circular from Intendant Feydeau de Brou, 3 February 1757, ADSM C 105, p. 62.

[3] See the numerous such admonitions in AN G7 15 and 16.

[4] 25 March 1709, and 28 April 1709, AN G7 15; Lachiver, *Les années de misère*, pp. 321–2; Boislisle, *Correspondance*, 3: pièces 153 and 226; Perrot, *Caen*, p. 245.

Desmarets' 1709 insistence that the harshest measures be reserved for last echoed the wisdom of many of his contemporaries who opposed the controls and privileges associated with "mercantilism."[5] Indeed, it now seems clear that even the regulated economy that many historians term "mercantilism" was far from rigid, and left room for the mechanisms of supply and demand to operate. In 1665, for instance, an anonymous official had counseled that fixing "the price of wheat is impossible; and everywhere one leaves prices free."[6] Not only did public authorities such as Desmarets concur with the advice of that anonymous 1665 author, but the liberal theorists of the early eighteenth century also proposed that greater freedom would benefit trade. The work of Pierre le Pesant de Boisguilbert, a Jansensist theorist and lieutenant general of police from a Rouen *robe* family, is particularly instructive.[7] His experience as a lieutenant of police of Rouen, among the other offices he held, convinced him that only increased liberty would end the shortages that had racked the kingdom.[8] He and his intellectual *confrères*, such as Vauban, did not reject government intervention completely, however. They believed that the export of grain should be restricted, and that royal and municipal officials should maintain emergency supplies bought when prices were low.[9] While it is difficult to trace a direct line from the desks of these theorists to the municipal halls and ministerial chambers of France (although in Boiguilbert's case, the connection is clear), both groups shared a commitment to allowing a limited form of free trade to prevail, at least at the outset of a shortage.

The officials' desire to encourage the workings of market mechanisms, and to avoid the use of force, was a recurrent theme in any eighteenth-century discussion of policies. In the late 1740s, Controller General Machault reminded Bordeaux *jurats* that they would be foolish to block

[5] Rothkrug, *Opposition to Louis XIV*; Heckscher, *Mercantilism*; Cole, *French Mercantilism, 1683–1700*; and *Colbert and a Century of French Mercantilism*. For a discussion of the problems of the concept of "mercantilism," see the articles in *Revisions in Mercantilism*, D.C. Coleman, ed. (London, 1969).

[6] The 1665 treatise nonetheless listed appropriate measures: prohibiting grain's storage and safeguarding its transport, outlawing commercial purchases near cities, and setting up warehouses where owners might be forced to place their stocks. "Police pour le blé et le pain," 1665, BN Mf 8127.

[7] Boisguilbert (1646–1714) held several bailliage and municipal offices. His works include *Traité de la Nature, Culture, Commerce et Intérêt des Grains* (2 vols.) (ca. 1707), *Détail de la France* (1695), and the *Dissertation de la nature des richesses, de l'argent et des tributs* (ca. 1707). G. Faccarello, *Aux origines de l'économie politique libérale: Pierre de Boisguilbert* (Paris, 1986).

[8] They hoped hoped that lower taxes and fewer controls would revive the economy. Harsin, *Doctrines*, p. 96.

[9] Faccarello, *Boisguilbert*; Bourde, *Agronomie*, 1: 130; Cole, *French Mercantilism*, p. 244; Meyssonnier, *La balance*.

supplies leaving their city: "You can count positively on abundance . . . [as long as the merchant's] trade is not hampered, but as soon as the complete *liberté* they need for the exercise of their trade is interrupted, their business will cease, and you shall find yourselves once again exposed to the evils and worries [of shortages that you created]."[10] Yet, even complete *liberté* – here meaning the merchants' right to "sell their grain, to buy some, and to move it from one province to another as they see fit," as he explained to the Rouen intendant at another juncture – often failed to draw grain to markets.[11]

Organizing Emergency Supplies

That failure forced authorities to embark on numerous operations to encourage and fund supplementary grain purchases. Parlementary magistrates, intendants, municipal councils, and all manner of officials organized such efforts, and sometimes paid for them with their own money. They dared not stand back and allow the vagaries of the market to starve their cities. In the absence of commercial grain, these state-sponsored operations eased the transition from local markets – with their insufficiency and regulation – to the free market that the liberal theorists had promised. These were nonetheless delicate endeavors. Supplies had to be located and merchants engaged to purchase and transport them on commission or convinced to buy grain for their own account. More difficult, the wares had to be protected along their treacherous path. These operations raised numerous questions about when and how to proceed. Was it best to buy in advance and then pay to have the grain stored, running the risk of losses? Establishing and maintaining granaries was costly and required much effort. Such purchases also stirred troubling beliefs that the monarchy and its ministers were hoarding grain, allowing it to spoil, or seeking to cause all manner of other ills to its subjects. By the 1750s, such fears had become so pervasive that the arrival of government-subsidized grain in Parisian warehouses sparked rumors that Louis XV desired only the worst for his subjects.[12] On the other hand, if the government waited until a shortage had been announced, it might be unable to find cereals or to arrange their timely arrival. And when numerous cities sent agents into the countryside or to

[10] Marion, *Machault d'Arnouville*, p. 427.

[11] 12 December 1751, ADSM C 105.

[12] For an analysis of these fears and of their role in the monarchy's desacralization, see Kaplan, *Famine Plot*; and Arlette Farge and Jacques Revel, *The Vanishing Children of Paris: Rumor and Politics before the French Revolution*, trans. Claudia Miéville (Cambridge, Mass., 1991).

Amsterdam and other ports, the competition only drove prices higher and set off increasingly speculative ventures. The resulting panic drove crowds to wharves and warehouses and government grain was an easy target for those intent on pillaging. Each possible strategy for government purchases, then, gambled with supplies and lives.

The efforts of Old Regime officials to fund grain purchases reflected their ambivalent sense of how to proceed. They alternated between amassing supplies in advance and waiting until shortfalls were apparent, and between organizing them entirely as government-run ventures and placing them wholly in private hands.

The question of the appropriate balance between private and public undertakings was not resolved until the nineteenth century. The overall, if inconsistent, trajectory of eighteenth-century efforts, however, reveals the desire to work in advance of shortages, rather than to cobble together last-minute purchases. The shortage of 1725 gave new impetus to such projects. Parisian and ministerial authorities had employed a range of traditional measures to maintain calm, such as keeping bread prices below the cost of production. Knowing these would be insufficent, the Crown had turned to financiers, while other officials, some from the parlement, commissioned additional ventures. The scattershot approach of 1725 resulted in chaotic and costly competition in the capital's hinterland. Moreover, once the grain arrived, it became clear that too much had been bought. To many administrative minds, the entire experience had been marked by hesitation, confusion, and corruption.[13]

Reviewing those attempts, Paris Lieutenant General of Police René Hérault agreed with the critics. The well-being of his city could not be left to last-minute purchases and good fortune. He established an emergency supply of grain in the Paris hinterland. He had to work economically. The royal treasury and merchants were still reeling from the collapse of John Law's scheme.[14] Herault turned to a likely source, the numerous religious

[13] 4 September 1725 AN AD I 23A; 22 September 1725, AN AD XI 14; [René-Louis de Voyer], Marquis d'Argenson, *Mémoires et journal inédit du Marquis d'Argenson . . .*, 5 vols. (Paris, 1857–80), 1: 199–200; [Edmond-Jean-François] Barbier, *Chronique de la régence et du règne de Louis XV (1718–1763)*, 8 vols. (Paris, 1857), 1: 398–400, 402–5; Kaplan, "The Paris Bread Riot of 1725"; *Provisioning Paris*, pp. 386–9; *Famine Plot*, pp. 10–23.

[14] On the devastating political impact of Law's system, see Thomas E. Kaiser, "Money, Despotism, and Public Opinion in Early Eighteenth-Century France: John Law and the Debate on Royal Credit," *JMH* 63 (March 1991): 1–28; "Public Credit: John Law's Scheme and the Question of *Confiance*," *Proceedings of the Annual Meeting of the Western Society for French History* 16 (1989): 72–81; "The Abbé de Saint-Pierre, Public Opinion, and the Reconstitution of the French Monarchy," *JMH* 55 (December 1983): 618–43. For an overall treatment of the politics of the Regency, see James D. Hardy, Jr., *Judicial Politics in the Old Regime: The Parlement of Paris during the Regency* (Baton Rouge, 1967).

communities within easy proximity of Paris, many with barns rumored to be overflowing. These communities were pressed into building up stocks beyond their needs. Many resisted or circumvented the orders. Some subcontracted the job to grain merchants and even bakers, and the totals never did reach the levels Hérault had wanted. Nonetheless, by 1729, more than seventy institutions had been induced to amass supplies. Whether in religious or commercial hands, Hérault had prepared emergency provisions at little cost to the government.[15] Additional declarations in the 1730s reiterated the instructions to all religious houses to keep a three-year supply of grain on hand for use in Paris. By the mid-1740s, though, it appears many no longer obeyed those orders.[16]

For the next fifty years, the ministers repeatedly changed their policies on when and how to undertake or encourage purchases, but they did not abandon them. In 1738, realizing that shortages loomed, Paris Prévot des marchands Turgot, the Assembly of Police, and the Paris Parlement outlined plans. The Minister of Finance, however, refused to believe reports of shortages or to sanction outlays. The strategies unfortunately stalled, only to be implemented too late.[17] Controller General Orry (1740–5), however, eventually did bring staggering amounts of grain to the capital from Italy, the Levant, England, the American colonies, and Holland.[18] In addition, he permitted cities across the kingdom to take out loans against future revenue and to fund their own undertakings. Rouen, for instance, drew together 301,005 livres from its *octrois*, and organized further expenditures of up to 700,000 livres.[19] Later, Controller General Machault (1745–54) resuscitated Hérault's idea of organizing granaries in advance of shortages. His 1749 *mémoire* described an ambitious plan to select one privileged company in each generality and to assign it to buy and store grain. If the grain was not needed, the companies could sell it later for a profit. Critics complained that this assumed too costly and imposing a role for the government, and Machault finally settled for a less grandiose system using a

[15] Kaplan, "Lean Years"; *Famine Plot*, p. 26. Similar stocks had been established to feed other French cities, such as Marseilles, Lyons, and Besançon.

[16] Marion, *Machault d'Arnouville*, pp. 429–30. See also the royal legislation, "Déclaration portant l'établissement . . . d'un grenier qui contiendra au moins dix mille muids de bled pour l'approvisionnement de Paris," 16 April 1737, *Lois*, 22: 1.

[17] The Minister of Finance was criticized harshly for being unwilling to organize supplies in advance of shortages, and suspicions of corruption clouded these operations. Richer d'Aubé, "Réflexions sur le gouvernement de France 1739 et 1740," AM Rouen, Ms 1172 U 75. (I am grateful to T.J.A. Le Goff for excerpts from this work.) D'Argenson, *Mémoires*, 2: 195.

[18] Although the totals are not known, much of the grain arrived in Marseilles and was then shipped to Paris, most through Le Havre and up the Seine. Carrière and Courdurier, "Marseille sauvant Paris," pp. 26–9; Kaplan, *Famine Plot*, pp. 37–42.

[19] Kaplan, *Famine Plot*, pp. 32, 36–9, 44. Henri Wallon, *La Chambre de commerce de la Province de Normandie (1703–1791)* (Rouen, 1903), pp. 193–4.

military supply company. This company failed to live up to the agreement, however, and in the 1751 shortages, Machault suspended the operations. He turned to more direct measures, attempting to purchase 200,000 quintals of grain in Holland, England, and the Lorraine through an intermediary, for resale at a loss. Here, too, his plans were thwarted, and public suspicions of Louis XV's intentions grew. Rumors ran that Machault himself, encouraged by Madame de Pompadour, was behind the shortage. Crown and controller general alike were tarnished by the successive operations' failures. Moreau de Séchelles (1754–6) abandoned these projects, believing them too costly. His successor, Moras, resurrected them in 1757, however, creating a stock of 40,000 setiers for the Paris region, some of which was directed to Upper Normandy.[20]

Despite the difficulties of these undertakings, the ministers agreed on their overall benefits. However, those benefits would accrue only if the numerous disadvantages could be overcome. Among the many problems, two stood out. First, the government's entry into the market invariably frustrated owners – both merchants and producers – who generally deserted the trade when government suppliers undersold them. Second, with those low-priced sales came heavy, even unsustainable, losses to local and royal coffers. Rouen had lost 33 percent of the money spent on 1752 purchases, for instance, and amassed almost 350,000 livres of debt. Some of these deficits were charged to the Chamber of Commerce, which wanted no further role in such ventures.[21] Thus, administrative attention was drawn to ways to distribute these supplies without exhausting either government funds or commercial goodwill.

Distributing Grain: Simulated Sales

In northern France, the administrators gradually refined pricing and distribution strategies and created tactics termed "simulated sales." Essentially, a simulated sale was one in which government grain was sold through a market or to bakers, often at a loss. The government's goal was to lower prices overall and to reassure townspeople who feared shortages. In some cases, the government provided the funds for the grain's initial purchase and transport and assumed all costs and losses. In others, it commissioned agents to search out supplies and bring them to market. The grain might be sold through an intermediary in order to diguise its origins, or it might

[20] Machault's 1749 plan instructed the company to purchase and store 139,000 sacks of grain, possibly a four-month supply for the capital. Marion, *Machault d'Arnouville*, pp. 430–3; Kaplan, *Bread*, 1: 347–8; and *Famine Plot*, pp. 50–51.

[21] Wallon, *Chambre du commerce*, p. 194.

be sold with great publicity to convince buyers that the government had taken their concerns into account. There were numerous means by which simulated sales could be used to influence grain prices, and local officials experimented with the possibilities the sales presented.

During the eighteenth century, local and royal authorities shaped these sales in ways they believed would offer the greatest long- and short-term benefit to their communities. There was a noticeable evolution in the skill and insight with which the sales strategies proceeded. The discussion of these sales moved from brief, inexact explanations to more detailed communications on specific ways to influence price levels and supplies at local markets. Most significant, the procedures that administrators chose progressed toward those that closely mimicked contemporary market practices. The simulated sales eventually were fashioned to present the illusion that some merchants had decided that increased supplies warranted asking a lower price. By the mid-eighteenth century, the intermediary, posing as an independent merchant, would stack his sacks in the market hall, pretend to survey the scene, and then ask a price somewhat below the going rate. He thus would feign a belief that the situation merited a lower asking price. All the while, though, that "independent" merchant actually was proceeding at the behest of government officials, who were fine-tuning techniques to conceal their hand. The development of these more covert methods contributed significantly to their success.

The correspondence from 1709 lays out in rather crude form how the early projects proceeded. Daguesseau, for instance, summarized the prevailing wisdom: "[t]he opposition of lower-priced grain to higher-priced grain has been, in all times, the most sure and efficient means to bring down the price of grain."[22] Here, he was arguing against placing a price ceiling on grain, and his discussion of exactly how to handle the grain the government would sell was somewhat cursory. He did not indicate whether the government would let it be known that it had ordered the carts arriving at market nor how low a price it would ask. Had Daguesseau been questioned more fully, he might have given detailed answers. Yet others involved in the 1709 ventures evinced the same limited concern with the complexities of those sales. When harvest shortfalls were first discerned, Desmarets requested simply that grain be imported so that local merchants would be scared and lower their prices.[23] In the spring, he commented in passing to D'Argenson that supplies could be "employed usefully to procure a decrease in the price of grain . . ."[24] Incoming grain could be "sold without a profit,"

[22] Boislisle, *Correspondance*, 3: pièce 380; Kaplan, *Famine Plot*, pp. 32, 36–9, 44; *Bakers*, pp. 546–7.

[23] Copy probably from CG to Hagnais, 11 September 1708, AN G7 15.

[24] CG to D'Argenson, 5 May 1709, AN G7 15.

although the bishop who was the intermediary would have to rely on the "goodwill" of the merchants involved.[25] Desmarets gave more instructive orders to the Tours intendant: He was to purchase grain through an *homme de confiance*, while taking care to act "very clandestinely and with many precautions" and resell it in the market "at a lower price."[26] The Marquis d'Argenson reported "secretly" engaging two merchants to sell grain from the king's warehouses in 1720, bringing down the price of grain "somewhat."[27] While these 1709 and 1720 projects contained the essential elements of eighteenth-century simulated sales – secrecy, increased supplies, and lower prices – the authorities wrote only vaguely of the procedures used or of the prices sought. The gradual refinement of simulated sales lay with their mid-century successors.

The Shortage of 1757

Shortages in Upper Normandy in 1757, 1768, and 1772 led to more thoughtful, precise discussion of how to dispose of government-sponsored provisions than had the earlier accounts. Interestingly, the mid-century administrators elected measures that bore a growing similarity to purely commercial activities.[28] They had several important goals. First, they sought to influence prices through increasingly hidden means. To do so, they worked through middlemen and avoided distributing the grain directly to consumers, although that strategy did not appear fully until 1768. Second, they tried to coordinate their efforts within the region, with a particular eye to Paris. Third, they avoided bringing disputes, such as those that bakers or disgruntled proprietors might make, into the open. When they adopted a visible profile, they did so in ways that created the illusion of routines, widespread accord, and even indifference. Their agreement on the value of such measures is evident in the 1757 activities, and became more pronounced in later efforts.

The shortage of 1757 itself was nothing exceptional. It could not rival the kingdom-wide crises of 1709 or 1738–41, nor the 1752 shortages that had led to violence in the streets of Rouen. The moderate level of the 1757 harvest deficits probably made it possible for authorities to take more mod-

[25] CG to Bishop St Mal, 22 May 1709, AN G7 16.

[26] Summary of letter, CG to Turgot de Tours, 12 February 1709, AN G7 15.

[27] D'Argenson, *Mémoires*, 1: 197.

[28] There may have been an instance of similar discussion in Paris in 1740. Controller General Orry and Paris Lieutenant of Police Marville, struggled with the issue of the pace and levels of price reductions they wished to effect. Orry wished to use small amounts of government grain to effect gradual price reductions at the hall; Marville wanted to work more quickly. Kaplan, *Bakers*, pp. 546–7.

erate measures to counter them. Genuine shortages were exacerbated by the Seven Years War. Upper Norman merchants were willing only to supervise the arrival of grain from Nantes and elsewhere, providing Rouen's intendant, the young, well-connected and talented Feydeau De Brou, could persuade commercial houses in those cities to send it.[29] The city commissioned purchases and offered bonuses to amenable merchants. A local miller, Roost, was engaged to bring 120 *muids* of grain each month to the Rouen hall from his establishments. He was to sell it at the "going rate" of the market each week between January and August 1757, with guarantees against losses. Similar agreements were made with estate agents and landlords near Magny and Chaumont. Other shipments – some commissioned by the city, others by the Rouen Parlement and the royal government – arrived from Bordeaux, Brittany, and the Loire Valley.[30]

These were not novel undertakings, of course, but the 1757 efforts were marked by a firm, if not uncontested, belief that the best results would come through shaping supply and demand rather than through coercion. This was in contrast to the policies of 1738–41, when the depth of the crisis had forced the government to resort to the known range of extreme measures – heavy fines, prison sentences, searches of mills and granaries, orders for owners to supply markets, and strict prohibitions against off-market sales. That was the last time before the Revolution that such aggressive means were used.[31] Those efforts did not solve the 1738–41 crisis, however, and there is some evidence that the Parisian prosecutions exacerbated the shortage.[32] In all likelihood, many authorities emerged from those shortages skeptical of the value of aggressive measures.

It is impossible to discern how the 1757 authorities viewed the harsh prosecutions of 1738–42. Their correspondence speaks of the troubled present and not of the lessons of the past. That present, however, was one in which the freeing of the grain trade was receiving serious attention. Intendant Feydeau de Brou had spoken forcefully in favor of liberal reforms

[29] Intendant to the Mayor and Echevins of Dieppe, and their reply, 27 June, 1 July and n.d. [ca. 5 July] 1757; Demoraz to Intendant, 25 June 1757, ADSM C 104, pp. 169–170, 172, 175.

[30] Roost's grain may have come from the reserves of the city's banal mills, which he leased. Draft [Royer] to de Belmesnil, 1 July 1757, ADSM C 104, p. 172/228; "Facture d'une cargaison de bled froment . . . ," Bordeaux, 2 July 1757, ADSM C 104, p. 48; Intendant of Rennes to Intendant Feydeau de Brou, 13 July 1757, ADSM C 104, p. 115; Intendant Feydeau de Brou to CG, 27 July 1757, ADSM C 104, p. 97; and "Procès-verbal of Incoming Ships," 3 December 1757, ADSM C 104, p. 13; Desmarest, *Commerce*, p. 92.

[31] The fines imposed ranged from 30 livres to 3,000 livres. See the many 1738–42 *sentences de police* in AN AD I 23A; AD+ 856, 857, 858, 859. See also Abel Poitrineau, *La vie rurale en Basse-Auvergne au XVIIIe siècle (1726–1789)* (Paris, 1965), pp. 92–3.

[32] These measures may have driven some *laboureurs* in the Paris zone from cereal production. Moriceau and Postel-Vinay, *Ferme, entreprise, famille*, pp. 232–4.

and had denounced the regulations that "obstruct [the landowner's] efforts and tie [his] arms."[33] His sympathies lay with *intendant du commerce* Jean-Claude Marie Vincent de Gournay and his followers, who were taking the first solid ministerial steps toward freeing commerce.[34] Gournay coined the adage "*laissez faire, laissez passer*," and, while not directly charged with supervising the grain trade, nevertheless emerged as a staunch advocate of reform.[35] Free-trade policies were deeply linked to the incendiary battles between the Crown and its sovereign courts. Gournay's proposals sought to free the embattled and impoverished monarchy from its parlementary enemies.[36] Encouraged by the progressive leanings of the new Controller General, Moreau de Séchelles, Gournay's group intensified its efforts, pushing in particular for a free trade in grain as a means of raising tax revenue and thereby warding off the parlements' demands to intrude in royal fiscal policy.[37]

Provincial authorities, especially those in the manufacturing cities and ports of Upper Normandy, would have been well aware of the activity of Gournay's offices. They may have read Claude-Jacques Herbert's searing 1753 critique of the grain trade regulations, *Essai sur la police générale des grains . . .*, which demanded that the trade be opened to all prospective buyers and sellers, safe from the meddling instincts of the misguided police.[38] The *arrêts* of 1751 and 1754, which offered more expansive and permanent definitions of *libre circulation*, and which lifted some of the paper work for shipments and allowed limited export, also had been implemented under the watch of these provincial authorities.[39] Thus, the general direction of economic policy would have been common knowledge in the

[33] Cited in Weulersse, *Mouvement physiocratique*, 1: 533.

[34] Gustave Schelle, *Vincent de Gournay*, Reprint ed. (Geneva and Paris: Slatkine Reprints, 1984); Meyssonnier, *La balance*, pp. 172–227; Antoine Murphy, "Le développement des idées économiques en France (1750–1756)," *RHMC* 33 (October–December 1986): 521–541; Jean-Claude Perrot, "Les dictionnaires de commerce au XVIIIe siècle," *RHMC* 28 (January–March 1981): 36–67; Butel, *L'économie française*, pp. 48–9.

[35] Writing to his superior, Daniel Trudaine, he cautioned that "the impediments placed on the grain trade, the searches carried out against the laboureur . . . , the obligations imposed on him to bring so many sacks to a market, tend to turn the king's subjects away from growing grain . . ." Cited in Schelle, *Gournay*, p. 72.

[36] Jean Egret, *Louis XV et l'opposition parlementaire 1715–1774* (Paris, 1970), pp. 50–132; Gomel, *Causes financières*, 1: 7–11; Baker, *Inventing the French Revolution*; James Riley, *The Seven Years' War*.

[37] Riley, *The Seven Years' War*, pp. 52–3, 58–9; Meyssonnier, *La balance*, pp. 228–32, 248–54; Kaplan, *Bread*, 1: 99–106.

[38] Despite his more expansive vision of free trade, however, he wanted numerous export restrictions. Claude-Jacques Herbert, *Essai sur la police générale des grains, sur leurs prix et sur les effets de l'agriculture* [1753]; Meyssonnier, *La balance*, p. 230.

[39] CG to Intendant de la Bourdonnay, 12 December 1751, ADSM C 105; 17 September 1754, AN AD+ 917; Marion, *Machault d'Arnouville*, pp. 422–42.

busy bureaus of Upper Normandy. Its political import would not have been lost for a moment on Intendant Feydeau de Brou, the son of the eminent *doyen* of the Council of State and future Keeper of the Seals, and related by marriage to an intendant of finances.[40]

The *généralité*'s 1757 provisioning projects originated in Feydeau de Brou's offices and focused on coordinating the pricing and sales of the government-sponsored imports. These activities bear the unmistakable traces of increasingly coherent free-trade theories and of many administrators' practical knowledge of market activities. The incoming shipments arrived at Le Havre and were to be sent to Rouen, Paris, and throughout Normandy as necessary. By early July, Le Havre's *procureur du roi de police sindic* Louis Plainpel determined there were unexpected surpluses. He forwarded them to Honfleur to avoid losses. There, subdelegate Le Chevallier and a relative, the Honfleur *négociant* Prémord, deployed the grain on the left bank of the Seine. Feydeau de Brou's secretary, Royer, issued instructions: The grain was to be distributed to "bring about a steady reduction in the price at markets." Small amounts of grain were to be sold in order to achieve an initial slight decrease in prices. If Le Chevallier and Prémord moved too quickly, Royer explained, prices would fall "all of a sudden" and "there [would be] every reason to fear the *laboureurs* and grain merchants . . . would, once discouraged, cease bringing wares . . ." A disaster would follow, Royer cautioned, because Le Chevallier and Prémord "would be unable to sustain the low prices . . ." their sales had produced. Balance and vigilance would be the keys to their success.[41]

Within days, Le Chevallier and Royer fell to arguing about the subtleties of pricing and timing the sales. At first, Royer had approved of Le Chevallier's management – the subdelegate had waited several days before beginning sales. "[D]uring that interval," the secretary had reasoned, "the news of the arrival of the ships spread and confidence was restored." Then Royer's praise waned. Apparently, Prémord had delayed the first sales until the end of the appointed marketday. The merchant had hoped that the appearance alone of the government sacks would unsettle other sellers. He expected that they then would reduce the asking price of their own accord, and that his assistants could step in as the market closed to maintain the more affordable levels. Thus, he would sell only a small portion of the grain on the first day and be able to continue more sales at reduced prices later in the week. Royer, however, had wanted him to begin selling at the first market bell "and then to defend the territory" by cutting the price by one- and two-livre increments per *somme*. It was important, Royer wrote, "to sell [quickly] and even abundantly in the first days in order to lower prices

[40] Barbier, *Chronique*, 8: 57.
[41] Royer to Le Chevallier, 2 July 1757, ADSM C 105, p. 88.

in surrounding markets." While earlier he had warned against sharp initial decreases, he now rebuked Prémord for his slowness.[42] Despite the criticism, Prémord continued more patiently than Royer desired. Perhaps his experience in the trade made him wisely wary of Royer's plans. Even then, his methods brought anxiety-stirring results. In late July, noting that area prices ranged from 24 to 37 livres per *somme*, Prémord asked 27 livres for his grain. The intendant had wanted prices somewhat lower, but Prémord resisted. As it was, news of 27-livre levels brought hundreds of buyers unexpectedly into Honfleur.[43] The grain sold in two days and Prémord's request for more sacks earned additional reprimands. The intendant himself intervened and suddenly proclaimed Prémord's prices too low.[44]

One might argue that these efforts were abysmal failures. After all, none of the parties had been able to calculate workable prices, and their attempts may have aggravated a moderate shortage by drawing additional consumers to already burdened markets. Plainpel, too, found it difficult to predict how price levels would affect demand. His Le Havre sales lured too many buyers from far-off towns. Yet, the authorities' combined actions carried the region through the crisis until more ships, and finally the harvest, arrived. By August, Le Havre reported numerous ships in port; the grain's price was high, but the supplies were at least available.[45] Thus, even with all these mishaps and frustrations, the undertakings sent much-needed food to areas that otherwise would have suffered. Significantly, while unrest frequently threatened, there were few riots – in all, maybe a confrontation in Rouen in early June, then disturbances in Honfleur and Les Andelys in mid-July, and "*un peu d'émotion*" in Rouen two weeks later. The second time violence threatened Rouen, the intendant had several *muids* of grain hauled immediately to market to be sold "slightly below the going rate" under the city's aegis.[46]

Surprisingly, the efforts to direct grain to needy markets do not appear to have chased many merchants away. Instead, the bonuses and commissioned purchases may have tied them more tightly to the fragile trade networks. When they did sustain losses, Feydeau de Brou did what he could to lessen them, explaining that "it is only by giving distinct signs of protection to merchants" that they would continue the trade. He even petitioned the controller general for funds to reimburse the owners of the

[42] Royer to Le Chevallier, 6, July 1957, ADSM C 105, p. 91.

[43] Originally, Prémord had anticipated having some difficulty selling the 1,100 boisseaux of wheat on hand, and had advertised some at 26 livres.

[44] See the many letters between Prémord, Royer, Feydeau de Brou, and Le Chevallier, ADSM C 104.

[45] See the lengthy correspondence on these matters during July and August 1757. ADSM C 104, pp. 76, 78, 79, 83, 129/170, 130.

[46] Feydeau de Brou to CG, 27 July 1757, ADSM C 104, p. 97.

incoming shipments who complained that the prevailing prices in Upper Norman markets were too low as a result of Royer and Plainpel's projects.[47] The combination of government-purchased and government-encouraged provisions fed the villages and cities of Upper Normandy without destabilizing commercial networks. While the intendant and his subordinates could claim only moderate success, fortunately that achievement was almost sufficient, and relative calm had prevailed.

The more skillful use of simulated sales did not progress smoothly, however. There were dissenters who challenged both the intendant's overall vision and his authority. Their actions were disruptive not only to the marketplace but also to the ongoing process of centralization. Among them were the expected suspects. The province's lieutenant general, the Duc d'Harcourt, for instance, an unyielding rival of both the parlement and the intendant, subjected ships to unwelcome searches for illegal cargo in January 1757. Of course, ships fled Norman ports, and this had intensified the shortages. Feydeau de Brou declared Harcourt's actions wholly "contrary to the views of the court and government, whose intention is that the grain trade . . . be absolutely free in the kingdom's interior. . . ." After several clashes, Harcourt backed down.[48] Later that spring, a *brigadier* of the Ecouis *maréchaussée* tried to force farmers to sell grain below the going rate at Pont Saint-Pierre, while the lieutenant general of police in Les Andelys sparked "*une émotion considérable*" by ordering grain in local warehouses carted to market. The parlement itself added to the tensions by issuing *arrêts* that ordered requisitions and forbade hoarding in June 1757, thereby encouraging the police's harsh measures.[49] The most spectacular display of insubordination occurred in Honfleur, where a magistrate contested Le Chevallier's plans to use the government grain to lower prices gradually. Rather than waiting for the simulated sales to have an effect – which would then clear the way for lower bread prices – the magistrate strode into the marketplace in full robes and lowered the price of loaves 5 sols before the first sack of grain was sold. The bakers refused to accept the new *taxe*. The subdelegate was furious. Summoning the guild officers to his chambers, he agreed that the magistrate's decrees were unconscionable, but warned that raising bread prices would bring riots. Instead, the bakers were to continue to sell bread for the low price of 22 sols per loaf or face prison. The subdelegate promised to compensate them for their

[47] Feydeau de Brou to CG, ADSM C 104, p. 76.

[48] Draft, Circular from Intendant Feydeau de Brou, 3 February 1757, ADSM C 105, p. 62; Desmarest, *Commerce*, pp. 93–4. On Harcourt's conflicts with the Parlement in 1763, see Barbier, *Chronique*, 8: 95–8.

[49] Feydeau de Brou to CG and M de St Florentin, 16 June 1757, ADSM C 104, p. 249/321; Desmarest, *Commerce*, pp. 37–8, 66–71.

losses later.[50] Despite the centuries-long process of centralization, there was always one more renegade local authority who could wreak havoc with Feydeau de Brou's fragile designs for the simulated sales. The actions of the Honfleur magistrate, and of others like him, implicitly contested both royal authority and moderate provisioning policies, but Feydeau de Brou and his subdelegates generally prevailed.

The need to curb wrathful magistrates, intemperant parlementary presidents, and jealous port authorities was intricately related to the need to coordinate the region's overall provisioning strategies. This was all the more apparent when the intendant reviewed the confusion that had accompanied purchases encouraged by two parlement *conseillers*. The men, influential local seigneurs, convinced a minor merchant house to bring grain to Rouen by promising substantial commissions and funds to cover any losses. There were no such funds, however, and Feydeau de Brou had to scramble to address the merchants' petitions requesting their rightful reimbursements. To avoid alienating them, he covered such losses as best he could, and then drew up plans to avoid further embarassments.[51] He fervently wished to establish order in the provisioning projects and to draw all involved – subdelegates, police, and the parlement, among others – into accord. Centralization and coordination formed an important basis for the successful navigation of crises.

Recovering from the tense moments of 1757, the intendant proposed arrangements that followed the three main lines that had been evident in that year's plans. First, he wanted the government to encourage merchants to bring grain supplies from outside Normandy in advance of shortages and then to exploit them to influence prices through simulated sales. Second, the integration of efforts appeared ever more necessary. Independent ventures by authorities had to be checked. Third, the emerging consensus in the intendant's offices held that public displays of force were inappropriate and dangerous. In order to maintain control, officials had to operate outside the turbulence of the marketplace. The cautious use of grain supplies

[50] The subdelegate promised to keep the bread price artificially high for a few days once the price of grain had fallen. The magistrate's actions probably sparked the July 13 uprising in Honfleur. Le Chevallier to Royer, 10 July 1757, ADSM C 104, p. 137/181; Le Chevallier to Intendant, 13 July 1757, ADSM C 104, p. 124/164.

[51] The *conseillers* had pressured two *commis* into buying grain from Orléans. The *commis* had no experience in the grain trade, and believed that the city's *échevins* would cover their costs and commissions. They claimed losses of 1952 livres (11 percent), not including the commissions they expected. Petition of Robert and Louis Durand to the intendant and "Account of Purchase, Costs and Sales," n.d. [ca. late 1757], ADSM C 104, pp. 60 and 277. There is no indication that they received the funds requested, although the intendant did honor other such requests. See ADSM C 104, pp. 6, 22, 23, 25 and 131/172.

through the simulated sales allowed administrators to satisfy the three conditions, without discouraging merchants and producers. These operations formed a bridge between an imperfect present of shortages and confusion and a far-away future in which commerce and agricultural production alone might suffice.

It took several months to decide on the appropriate strategies. Feydeau de Brou enlisted the parlement and the chamber of commerce in his discussions. He particularly wanted the chamber to offer its funds from *octrois* as bonus payments to any merchant bringing grain to the city from outside the province. The chamber disagreed, arguing that any news of subsidies would only incite rumors of further shortages; the chamber's funds had been depleted in 1752. It appears that the intendant won out, however, and established a three-livre bonus for every *muid* of grain brought from elsewhere. He did not want the government to take ownership of the grain – an important point – but instead to provide strong encouragement to merchants inclined to help the city. In other words, he wished to help commercial houses bring grain to Upper Normandy, rather than compete with them. Once the grain had arrived, he could push merchants to carry it to market when needed. While it is unclear to what extent the Rouen authorities were able to persuade merchants to generate and maintain those supplies, their intentions were clear. The intendant's offices would coordinate the creation and distribution of emergency grain – but would avoid undertaking the ventures itself – and then use the grain to shape prices.[52]

Simulated Sales in the Age of Physiocracy

In 1763, the free-trade legislation appeared with great fanfare. One might expect Feydeau de Brou's projects and others like them to have been vanquished by the physiocrats' triumph. Instead, simulated sales continued. The ministers and local authorities saw such strategies as necessary, and even logical, complements to the reforms. The Parisian operations have received the fullest historical treatment, in part because of the critical role they played in the monarchy's collapse. Briefly, Bertin placed the Parisian warehouse projects in private hands to spare the impossibly over-burdened treasury. He called on master baker Pierre Simon Malisset to assume the

[52] 23 June 1757, ADSM 1B 276; Desmarest, *Commerce*, p. 82; Wallon, *Chambre de Commerce*, pp. 203–4. In 1759 and 1760, the city officials sold 50 to 70 *muids* of grain to help distressed workers, although the details are not given. The grain may have come from the purchases Feydeau de Brou hoped to encourage; more likely, it came from other municipal supplies. 18 March 1759 and 15 October 1760, AM Rouen A 38.

operations and to maintain the 40,000 setiers believed essential for Parisian security (a Paris setier contained 156.1 liters).[53] The government was to direct the grain's sales and Malisset was to see that the warehouses were replenished. Malisset feverishly built and restored warehouses and mills, particularly around Corbeil. He ran afoul of accusations of fraud in late 1767, which caused Controller General Laverdy to revoke the contract. Laverdy's successor, Maynon d'Invau (1768–9), a skeptic who favored a wholly free trade, nonetheless later honored the contracts. In 1769, Terray restructured the Malisset system as a *régie*, placing it in the experienced hands of Sorin and Doumerc, the one a grain merchant, the other a financier. He ordered increased supplies for an extensive network of warehouses surrounding the capital.[54]

The royal provisioning projects are best known for their role in heightening the public's belief that the monarchy was knowingly and contemptuously starving the French people by permitting its ministers and agents to hoard grain, drive up prices, and allow wares to spoil. As royal authority became further discredited in the 1770s, emerging public opinion pointed repeatedly to these operations as evidence of the king's willingness to place his subjects' well-being in the dubious hands of merchant companies who would as soon starve the city as feed it. The accusations aimed at the projects contributed to the pre-revolutionary malaise and have enabled historians to identify in them one of the many sources of the monarchy's collapse.[55] Yet, for all the wrongs blamed on Malisset, Sorin, Doumerc, and all the others involved in these systems, the uses to which the grain from these and other undertakings was put can illuminate the continuing refinement of eighteenth-century simulated sales, and give some indication of the value of these projects.

Rouen officials contended with both the shortages that stretched from 1768 to 1772 and with the worsening political climate. By 1768, numerous parlements had begun to demand a return to regulation and a full investigation of the monarchy's finances. Rouen's magistrates were among the most insolent, and the year was particularly trying in that city.[56] Rising grain

[53] This was probably a three-weeks' supply. See Lieutenant of Police Marville's 1740 estimates, D'Argenson, *Mémoires*, 5: 174.

[54] The records are held in AN F11* 5. See also Miromesnil, *Correspondance*, 5: 162–9. For information on imports coming through Le Havre and Rouen, consult Pierre Dardel, *Navires et marchandises dans les ports de Rouen et du Havre au XVIIIe siècle* (Paris, 1963), p. 164. The full story of these complicated contracts and their political impact is in Kaplan, *Bread*, 1: 349–407 and 2: 614–60.

[55] Kaplan, *Famine Plot*; Léon Cahen, "Le prétendu pacte de famine: Quelques précisions nouvelles," *Revue historique* 176 (1935): 173–216; "Le pacte de famine et les spéculations sur les blés," *Revue historique* 152 (May–June 1926): 32–43.

[56] This account is drawn from Laverdy to intendant, n.d. [1768] and intendant to CG, 3 July

prices incited an uprising – Laverdy called it a "sedition" – in late March. The Rouen authorities had only a handful of *cavaliers* and officers from the *milice bourgeoise*, and were helpless against infuriated crowds. Additional assistance was mustered, including help from the Duc d'Harcourt, but only after numerous warehouses holding grain for Rouen and Paris had been pillaged.[57] The Rouen parlement hanged several of the allegedly guilty without trial, then halted further prosecutions while reimposing regulations. These actions were in sustained opposition to the wishes of *Premier Président* Miromesnil, Intendant Thiroux de Crosne, and the controller general.[58] Throughout that year, Miromesnil and De Crosne met late into the night, deliberating respectfully over drafts of letters and planning for a better means to feed the city. Their shared sympathies, even genuine friendship, can be attributed to their deep commitment to free trade and to their growing conviction that behind-the-scenes, well-coordinated operations were most effective. Economic theory and pragmatic concerns were united in the 1768 vision of Rouen's two most powerful authorities.

The *premier président* and the intendant, along with the ever-present Royer, fine-tuned the methods that had been used in 1757. They adhered to commercial methods closely, and attempted to remain wholly out of view. In the fall, they quietly engaged several merchants to bring a total of five to six *muids* of grain to the market hall, and instructed the Rouen subdelegate to do likewise. Later agreements brought more sacks to market through middlemen. "Of all the measures I could take," De Crosne explained, "the most certain is not to appear to be mixed up in the [grain trade]. . . ."[59] Miromesnil would never reveal how merchants' grain had come into his disposition. He received reports from "discreet" men sent in disguise to observe incoming ships. He barred the primary merchant involved from coming directly to his bureau, and his emissaries privately contacted the bakers. The two men assured the controller general that all would proceed covertly and that no one would guess their role.

A fine distinction in the strategies for pricing the grain separated the 1757 plans from the 1768 plans. While Feydeau de Brou, Royer, and Le Chevallier had indicated specific prices for the 1757 sales, Miromesnil instead planned "to bring about a decrease [in the price] by abundance alone, without hampering the merchants' *liberté*" in 1768. By this, he meant

1768, ADSM C 105, pp. 171–172; De Crosne to CG, 23 March 1768; CG to De Crosne, 27 and 28 March 1768; Bertin to De Crosne, 5 April 1768; and Trudaine to M. de la Michodière, 22 June 1768, ADSM C 107; Miromesnil, *Correspondance*, 5: 4, 45–207.

[57] De Crosne to CG, 23 March 1768; and Miromesnil to Bertin et al., 22 and 24 March 1768, ADSM C 107; Miromesnil, *Correspondance*, 5: 153.

[58] Thiroux de Crosne was intendant from 1768 to 1785.

[59] Miromesnil, *Correspondance*, 5: 100–1.

that he would not place the grain below the going rate of the hall and he would not suggest a price himself. Miromesnil instructed Feray, the most active merchant involved, to "sell neither too high nor too low" and to stay within the bounds of the prices at each day's market.[60] Feray, however, did not follow the directions and claimed losses that the *premier président* refused to reimburse.[61] Miromesnil was adamant: "Far from bothering [him] about the price," he wrote Laverdy, "I have always told M. Feray he should sell at the going price and at such a price as his agents indicated . . . and I never demanded anything against their wishes." He and De Crosne "never took any measures . . . to bring down the cost of wheat," he protested. "All our care was to see that the hall had enough grain and that the merchants were completely free to sell at whatever price they judged appropriate," he reiterated.[62] Along with the projects underway with Feray, Miromesnil advanced grain and flour through intermediaries to several bakers so that they could sell bread below the city's *taxe*. This multifaceted plan would force merchants "to sell grain at the public hall for a lower price, and also would decrease the price of grain of the country *blatiers*." However – and this is the point – Miromesnil did not want to meddle directly with prices, nor to have anyone guess his role in the sales. An increased supply alone was to work its magic.[63]

Further evidence of these Upper Norman preferences to rely on market-mimicking sales is found in Miromesnil's rejection of more sensational schemes. These designs included an offer of grain from *intendant des finances* Trudaine de Montigny, who appeared not to grasp the methods – neither the extreme discretion nor the pricing strategies – that Miromesnil and De Crosne wanted to pursue. Trudaine, who had been an avowed advocate of free trade, planned to send 1,200 to 1,300 sacks of flour to the Orléans market in order to lower prices in that city. He hoped to stun merchants, causing them to dispatch their wares quickly to Paris at lower prices.

[60] By and large, the 1768 grain belonged to the merchants or had been commissioned by Malisset, whereas the ownership of the 1757 wares was less clear. Some of the 1768 grain originated in Laverdy's projects and was being stored in Upper Normandy. After the riots, Laverdy may have placed Feray's grain entirely at Rouen's disposal. Feray, a Protestant, was offered a *brévet* conferring all the privileges of nobility for his role in the projects. Laverdy to intendant, n.d. [1768] and intendant to Laverdy, 3 July 1768, ADSM C 105, pp. 171–2.

[61] Merchants' complaints of such losses were frequent. See Montaran to the Intendant of Caen, 6 December 1777, An F11* 1; and M de Lessart to M Planter le jeune, 26 February 1791, AN F11* 3.

[62] Turgot, then the Intendant of Limoges, was supervising similar sales at market prices. Dakin, *Turgot*, p. 110.

[63] In fact, Miromesnil, De Crosne, and Laverdy believed that attempts by local police officials to intervene aggressively in markets had contributed to that spring's riots. CG to De Crosne, 10 April 1768, ADSM C 107.

In addition, he offered to send grain to Miromesnil so that prices could be forced downward in Rouen. Miromesnil declined, and cautioned against these plots. The *premier président* rebuffed other propositions to have grain sold directly to the needy. He reorganized them so that the grain was sold quietly through an intermediary at the hall. In that way, Miromesnil explained to Trudaine, "one will still believe that it is the continuation of commercial operations."[64] In every situation regarding the grain trade, Miromesnil and De Crosne preferred prudence and discretion, and tried to create the illusion that commerce alone was supplying the city.

That discretion underlay the appearance, if not the fact, of the authorities' neutrality. It was crucial to the success of Miromesnil's market. Merchants could not be scared away. They had to be persuaded that other merchants were supplying the market and that prices were falling slowly – "naturally," Miromesnil wrote – as a consequence. De Crosne had brought off precisely such sales immediately after the March riots. He had sent merchants quietly to the wharves and had had them supervise the transport of grain to the hall, as if it belonged to them. The market's "abundance alone prevented price increases, without harming the merchant's liberty."[65] Such methods had less *éclat*, but more promise than the steep, risky, and all-too-obvious reductions that Trudaine and others intermittently proposed. Some of that promise was fulfilled. By mid-summer, ships lined the Seine, many brought by Feray and Malisset-connected merchants, but many others by purely commercial transactions. Miromesnil and De Crosne had not been able to prevent the violence of that March, but they had at least managed to find grain to carry the city through the *soudure* of the late summer, and had done so without brutalizing merchants.[66]

The extant correspondence of 1772 is less rich than that of 1757 and 1768, but confirms the Upper Norman authorities' intentions to continue their simulated sales. Some of the actors, such as Miromesnil and De Crosne, could claim over a decade of experience with these activities. Royer, the intendant's secretary, could attest to longer service. Throughout 1772, Terray arranged for the merchants involved in the Parisian warehouses, among them the Feray family and one local pariah, Sieur Planter, to direct supplies to Upper Normandy.[67] Paris Lieutenant General of Police Sartine

[64] Miromesnil, *Correspondance*, 5: 189. See also his advice to Laverdy on carefully handling the Rouen Parlement's desire to restore regulation. Miromesnil, *Correspondance*, 5: 100–1, 159.

[65] Miromesnil to the Chancellor et al., 24 March 1768, ADSM C 107.

[66] Miromesnil, *Correspondance*, 5: 48, 59, 64–5, 110–1, 207–8.

[67] Planter had a long-standing feud with the people of Vernon and Petit Veronnet, where his warehouses were located. His warehouses were attacked on several occasions, including in 1768, in 1775, and finally in 1789, when he was forced into hiding. Letter from Desfeine to an unknown recipient, Rouen, 26 October 1765, ADSM C 105; intendant to CG, 21

sent his assistant, Joannin, to supervise this mission. Joannin organized the clandestine sale of grain via intermediaries to Rouen's guild bakers or merchants, while De Crosne vouched for the operation's secrecy. The sellers were to ask the prices the current bread *taxe* would allow – in other words, within the range of market prices, but neither above nor below them. Joannin and De Crosne hoped that the increased supply of grain would bring down the average market price, so that eventually the bread *taxe* could be reduced. Here, Terray disagreed, and probably revealed his ignorance of the workings of urban grain and bread markets. He suggested that Joannin sell small amounts to household buyers through the Rouen hall, in order "to prolong the period of this assistance," so that it would work "entirely to the advantage of the people." There is no fuller discussion of Joannin's reasoning, but he and the intendant nonetheless chose the greater discretion of the covert sales in an anxious city where most consumers bought bread.[68] Their decision to work invisibly to bring about lower bread prices was very much in keeping with the strategies of 1768. Terray's solution, on the other hand, was not consonant with the buying habits of the city, and its obvious nature could draw apprehensive buyers to the hall or ports.[69] Thus, in 1772, the Rouen intendant and the *premier président* once again oversaw the somewhat concealed sale of grain, at prices close to those of the market, to quietly ease averages downward. The government had entered the market through direct or indirect means and had used simulated sales to maintain calm.[70]

The simulated sales of the mid-eighteenth century offered an effective means to address the inherent contradictions of prevailing policies and commercial inadequacies. Unfortunately, neither *libre circulation* nor its later counterpart, free trade, had induced grain owners to furnish markets. The sales were a critical administrative way station – halfway between the direct intervention of many previous centuries and the market economy envisioned in some fortunate future. Over the course of the eighteenth century, the loose methods of the simulated sales of 1709 were rendered more precise

January 1772, and CG to intendant, 28 January 1772, ADSM C 106; Draft, De Crosne to Bertin and Choiseul, 8 April 1768, ADSM C 107; AN DXXIX bis, doss. 2, n. 17.

[68] An additional concern was the bitter civil proceeding between the bakers and the merchants over access to grain in port that threatened to increase dock-side disturbances. Draft, intendant to CG, n.d. [ca November 1771], ADSM C 106, p. 55.

[69] Joly de Fleury to De Crosne, 19 January 1772; CG to De Crosne, 28 January 1772; De Crosne to CG, 5 December 1772, and Joannin to the Intendant of Rouen, 26 January 1772, ADSM C 106.

[70] The Dijon *élus* wanted the government to found simulated sales in 1770–1. See their similar analysis of the risks and benefits. Léon Blin, "Notes sur une disette de grains en Bourgogne (1770–1), in *Actes du 93e Congrès national des sociétés savantes. Tours. 1968*, 2 vols., (Paris, 1971), 1: 245–66.

and more consistent with actual commercial practices, even mimicking them. They were less obviously disruptive to commercial ventures, better coordinated, and stirred fewer dramatic confrontations in marketplaces than the wider range of tactics that had been used throughout the early modern period. They allowed prices to be eased downward without arousing merchants' fears or validating the buyers' insistent demands for low prices. As liberal theories gained credence – and the physiocratic assurances remained unfulfilled – simulated sales sustained fragile trade networks. Authorities underwrote or subsidized operations. When local merchants despaired, the intendants' constant correspondence with the Controller General brought news of grain and prices, and indicated the merchants disposed to load barrels onto ships destined for distant ports. Simulated sales not only helped ease shortages in critical moments, but offered merchants information, connections, and support that could be used in the future. Thus, in the markets of Upper Normandy, the authorities maintained a constant, if less visible, presence, plotting to draw grain to their markets and to ensure its sale at affordable prices. These sales made it possible for authorities to claim publicly that free trade might make good on its promises, while attempting privately to ensure that markets would be supplied. The free trade of the mid-century – both before and after 1763 – relied on the silent working of simulated sales.

CHAPTER THREE

Scripting "Free" Trade

The arrival of Turgot brought an abrupt repudiation of the simulated sales and forced further clarification of the meaning of free trade. The carefully refined strategies of De Crosne, Miromesnil, and their subordinates across Upper Normandy were banished in the space of only a few months. Turgot was unwavering – market forces alone were to feed the country, even if shortages across Europe were readily apparent. "Not only are rising prices inevitable," his 2 November 1774 *arrêt* explained, "but they are the only possible remedy for scarcity." He attacked the government's former role as provisioner, charging that such undertakings had nurtured the "illusion" that the king "could keep grain cheap." The king could do no such thing.[1] In part, Turgot had hoped to set himself and Louis XVI apart from the scandalous accusations that had troubled Louis XV and Terray. The new controller general's other motives were less political: He fervently believed that the government's sporadic entry into the grain trade had driven merchants away in utter disgust. The only solution, he argued, was to free merchants to ply their trade. Abundance would follow.[2]

Turgot's predictions of abundance were not fulfilled, and the shortages that occurred during his ministry brought rapid denunciations of his policies. That Necker's best-selling work condemning full-scale free trade appeared only months after Turgot's *arrêt* added greater fuel to the debates surrounding royal policies.[3] The attacks were not confined to an enlightened readership, but included the resistance of the newly restored par-

[1] "Lettres patentes sur arrêt du conseil," 2 Novembre 1774, ADSM C 103; Turgot, *Oeuvres*, 3: 102–10.

[2] "Memoires remis par M de Montaran à M de Calonne," 25 November 1783, AN F11 265; Faure, *Turgot*, p. 413; Charles Musart, *La Réglementation du Commerce des grains en France au XVIIIe siècle* (Paris, 1922), pp. 135–47.

[3] *Sur la législation et le commerce des grains*, 2 vols. (Paris, 1775). The work, which went through at least four more editions before the Revolution, vehemently attacked free trade.

lements. The Rouen *présidents*, for instance, refused to register the *arrêt* for over a year, and then inserted a phrase that undermined Turgot's intent: "The Court and the magistrates within its jurisdiction shall continue, as in the past, to see to it that the halls are sufficiently supplied with grain." They posed a question that was both rhetorical and accusatory and that went immediately to the heart of the problem. If merchants did not respond to the lure of profits, the Parlement begged "His Majesty to indicate . . . what direction it should follow if halls and markets are not supplied."[4] Turgot, furious with all such challenges, indicated only that there was no further call for parlements to interfere with the trade.

Turgot's Hard Line

Most important, the Controller General declared that the government was no longer a grain merchant. In other words, there would be no more government purchases and certainly no more simulated sales. Merchants could ply their trade safe from government competition. To press home that point, Turgot formally barred municipal corps in April 1775 from undertaking purchases.[5] Even in June, when markets were still restless, Turgot did not soften his stance. Writing the intendant of Caen, he reported that two merchants, the familiar Feray and Daugirard had shipments arriving in Le Havre. The intendant was to take no steps save tax exemptions to ensure that some of those cereals reached his *généralité*. The intendant would have to hope that high prices would bring the grain to the region, Turgot wrote.[6] Adding greater difficulties for administrators and their townspeople, the controller general deprived cities of their traditional emergency supplies by abolishing the many municipal market fees and *octrois* collected in kind.[7] Thus, during the rocky months of the Flour War and its aftermath, authorities were left with neither emergency supplies nor suppliers. Because of Turgot's unequivocal policies, they had only their wits and some soldiers to see them through until the harvest.

This did not mean that Turgot would not seek to shape the market, however. The controller general turned his energy toward sustaining grain

[4] After complicated negotiations, Louis XVI forced their compliance. 27 January 1776, AN E 2523, p. 31; Miromesnil, *Correspondance*, 5: 149–55; Floquet, *Histoire du Parlement de Normandie*, 7 vols. (Rouen, 1840–1842), 6: 423–5. Turgot, *Oeuvres*, 4: 216–9. See also the dossier prepared for Controller General Calonne in December 1786 concerning past problems with the Rouen magistrates in AN F11 208.

[5] He did honor some of the contracts for Paris, however. 24 April 1775, AN E 2515, fols. 496–8.

[6] Turgot, *Oeuvres*, 4: 460–1.

[7] See the legislation in AN E 2512.

owners' profits so that French supplies would be assured in the long run. For instance, he attempted to rouse commercial spirits by offering moderate bonuses for any imported wheat that arrived between 15 May and 1 August.[8] The trick was to enable the poor to pay high prices, or somehow to make up the difference in the short run. To that end, he lifted many market dues, granting mild relief to buyers.[9] Even then, the price of grain might remain out of reach, and so he created modest (some would say insufficient) public work projects in numerous cities.[10] Such programs were not new. Turgot had made use of them in Limoges, and they also had been organized in Rouen during 1768, although to no good effect there.[11] He refused to offer any direct relief, though, such as the traditional distributions of bread that often accompanied shortages. "If [your] province decided to create some form of distribution," he wrote a bishop, "I believe it would be better [to distribute] money than grain."[12] The poor could tramp from workshop to bakery with a few extra sous in hand, purchase a costly loaf, and the baker in turn could afford to pay the miller or merchant the full price of the year's expensive grain. Stable, albeit high, prices – and not artificially low ones of Malisset's warehouses and simulated sales – were the basis for success in Turgot's eyes.

In a sense, the new controller general's subsidies to the poor represented a further elaboration of the ideas that underlay simulated sales. His measures did not visibly change the price at market, for instance, even if they did help sustain the high prices merchants desired. Moreover, they helped bridge the gap between the free market Turgot desired and the realities that confronted him. Thus, even though he wished to break with the past, with its intervention and simulated sales, his methods reveal a continuing commitment to shaping the market through measures that mimicked its routines.

Throughout his years as Controller General, Turgot was mindful of the role of public opinion and of the need to check criticism. During the 1750s,

[8] He offered 18 sous per quintal for wheat. "Arrêt du conseil . . . qui accorde des gratifications à ceux qui feront venir des grains de l'étranger," 24 Avril 1775, *Royal Acts*.

[9] See the legislation in AN E 2512 and 2515.

[10] Rouen received 20,000 livres for its workshops in April 1775, a small, but still helpful, sum. 24 April 1775, AN E 2515, fols. 496–8; Turgot, *Oeuvres*, 4: 407–10, 213–4, 226, 231, 460–2; Faure, *Turgot*, pp. 229–30, 247.

[11] The 1768 workshops only increased the supply of fabric, putting still more people out of work. The intendant then assigned workers to build roads, but found they were too weak to work. Controller General to intendants, 5 November 1770; and "Résultat des travaux publics établis en 1768 pour le soulagement et subsistance des pauvres," n.d., AN F11 1191–1192. Projects in 1739 failed because they were organized too late and were underfunded. Richer d'Aubé, "Réflexions sur le gouvernement de France 1739 et 1740," pp. 3–5, AM Rouen, Ms 1172 U 75.

[12] Turgot, *Oeuvres*, 4: 231.

the royal economic bureaus had begun presenting policies as reasoned responses to the demands of the French subjects, and one can discern in the preambles and press runs of these documents evidence of the expanding enlightened public that historians have found elsewhere.[13] Turgot's efforts went well beyond those of his mid-century predecessors. He had determined that a crucial factor in the collapse of the 1763–4 policies had been the ministries' failure to persuade local authorities and some larger public of the incontrovertible logic of free trade.[14] He appended lengthy circulars to his 1774 *arrêt*, addressing any lingering questions. "I desire to make finally this truth so commonplace that none of my successors could contradict it," he explained.[15] When objections surfaced, especially from the reestablished parlements, he replied that the king and his ministers could not be blamed for high prices: "That would be to credit men with God's secret. I do not order the rain or frost." The limits of the Crown's responsibility lay in establishing free trade and providing public works – in short, everything "that is permitted a wise administration." While Turgot was "very upset that [wheat] is expensive," he nonetheless concluded that "no human power would know how to prevent grain from being expensive when it is scarce." In fact, "that dearness is a remedy, bitter no doubt, but necessary against scarcity."[16]

Fearing that his arguments would not prevail in the midst of a shortage, he redoubled his efforts. He not only wanted to persuade the French people of the truth of his cause, but also to reassure grain owners that the government's sympathies lay with them. That last point was critical. As long as owners believed the government would not protect their shipments, they would have every reason to flee the trade. Throughout the Flour War and its aftermath, Turgot issued a torrent of letters, pamphlets, and affiches that reached even outlying towns. When Louis XVI offered amnesty to those willing to repay the owners for their losses, for instance, Turgot drafted an accompanying "Instruction aux curés." In it, he contended that the distur-

[13] The 17 September 1754 *arrêt du conseil* confirming *libre circulation* had explained that His Majesty desired to yield "finally to the wish of those provinces which have asked for a long time" to be able to export grain – although he then listed reasons for restrictions. In contrast, the legislation reiterating *libre circulation* in the 1740s had announced simply that shortages made it essential to transport cereal between provinces. *Arrêt du Conseil*, 17 September 1743, AN AD+ 1260; *Arrêt du conseil*, 17 September 1754, AN AD+ 917.

[14] See, for example, the lengthy explanation of one subdelegate, Lemarié at Magny. He believed it more expeditious to convince people that the low prices accompanying government imports in June 1774 were a sign of royal benevolence, rather than the natural forces of the market. "After all," he protested, "it would be impossible to convince the people never to blame the king for a rise in the price of grain . . ." Lemarié to intendant, 9 June 1774, ADSM C 106, p. 3.

[15] Turgot, *Oeuvres*, 4: 233; Dakin, *Turgot*, p. 178.

[16] "Letter to Abbé . . . on the freedom of the grain trade," Turgot, *Oeuvres*, 4: 224–6.

bances were not caused by high prices – "the people" therefore had no reason to complain – and that low prices constituted "theft." The *curés* were to add their endorsements to Turgot's assertion that "no pretext can exonerate" anyone from the obligation "to make restitution for the entire amount [stolen] to the true master of the [goods] usurped."[17] Subdelegates were to post news of the amnesty in their markets. The controller general sent troops into marketplaces and organized chilling roundups. He then printed up lists of restitutions to broadcast their "salutary effects," and encouraged intendants and subdelegates to distribute leaflets that promised protection and reimbursements to wary grainowners.[18] He even circulated a gruesome affiche from Paris's Châtelet criminal court announcing the hanging of two men guilty of pillaging.[19] The subdelegates responded. One arranged a "glorious" ceremony to thank local *laboureurs* who had supplied the town of Gisors faithfully.[20] Several others made a public display of posting Turgot's offers of amnesty and restitutions, although De Crosne discouraged these announcements in the immediate aftermath of the Flour War. The intendant instructed his subdelegates instead to inform owners privately of the government's willingness to reimburse them.[21] Publicity – whether to explain economic theory, subdue rioters, or reassure owners – was central to Turgot's attempts to bring *liberté* and order to French markets.

His energetic campaigns did not fully allay the confusion of the intendants, subdelegates, and the other officials who oversaw the trade, however. Several issues perplexed them, despite their long experience and at least mild enthusiasm for free trade. First, given the growing acceptance of market forces even before 1763, the distinction between the physiocratic market and contemporary practices was not necessarily clear. Even the two main differences – that anyone could enter the trade and that off-market

[17] Turgot, *Oeuvres*, 4: 437–42. The "Ordonnace du Roi, Amnistie," 11 May 1775, is reprinted in Turgot, *Oeuvres*, 4: 443.

[18] In order to be acquitted, rioters had to produce an *attestation* signed by their curé. Des Ervolus, an experienced police official from Evreux, led the efforts in Upper Normandy, where Planter had lost a Paris-bound barge to an attack from islands in the Seine at La Roche-Guyon. Faure estimates that the damages reached 610,000 livres and that Planter received 50,000 immediately from the government. Des Ervolus to De Crosne, 24 June 1775, ADSM C 108, p. 310; "*Etat* of grain stolen, lost and of restitutions," folio in ADSM C 107; "*Etat* of requests that the maréchal d'Harcourt proposes to make," [n.d. 1775], ADSM C 107; Turgot, *Oeuvres*, 4: 443; Faure, *Turgot*, p. 278. For attempts to restore order in neighboring Picardy, see AD Somme C 88. Numerous requests for amnesty are in AD Eure 1B 398.

[19] AD Somme C 88; Turgot, *Oeuvres*, 4: 443.

[20] Courtois to intendant, 12 May 1775, ADSM C 109, p. 31.

[21] Paterelle to intendant, 14 June 1775, ADSM C 106; Draft, Intendant to Paterelle, 9 June 1775, ADSM C 109, p. 247.

sales were allowed – did not usher in irregular practices. The permission for off-market sales had only confirmed activities that were becoming widespread in years of plenty, for instance. That they would accompany a period of shortage, however, was another matter entirely, and did occasion great concern in both 1763 and 1774. Moreover, the legislation of neither decade addressed the intricate workings of the market halls. From the administrators' vantage point in 1774, many essential elements had been left unchanged: Many guilds remained intact, although Turgot eventually dismantled them; the boundaries between merchants, millers, and bakers were still heeded, if only in theory; the police and guild syndics continued their inspections of grain in port; bells still sounded to mark off the hours for retail and wholesale transactions; and there was no pause in the struggles over legal bread prices.[22] Moreover, the people were still hungry, and in 1775 there was once again not enough grain.

That last problem was the stumbling block for the myriad officials hunched over their desks with Turgot's *arrêt* spread before them. (He had accounted for the unusual length of the text: "One will find it wordy and dull, but I wanted to render it so clear that every village magistrate can make it understood to the peasants."[23]) Despite the controller general's confidence in free trade, local authorities gazed uneasily at the market squares that held no supplies. If merchants did not rush to take advantage of the *liberté*, and later the bonuses that Turgot promised them, what was to become of their town? The grain, of course, did *not* arrive, and the administrators' hands were tied through the worst days of 1775. Forbidden to undertake purchases, deprived of emergency supplies from market dues, and having been told to remove themselves from regulating the trade, there was little they could do. Thus, the insistent tones of Turgot's policies and of the physiocrats' earlier tirades added new difficulties to the authorities' search for surer ways to bring wagons and barges laden with grain to their towns.

The Neutral Administrator

The search yielded a vocabulary and a set of practices that became the working definition of free trade in Upper Normandy. A central element in

[22] For further information on some of these problems, including the quandaries posed by the parlements' reimposition of regulations in the late 1760s, followed by Terray's policies, see De Crosne to CG, 7 May 1772, ADSM C 106, p. 71; Blin to intendant, 21 May 1775, ADSM C 110; Lantier, "Saint-Lô," p. 15; Desmarest, *Commerce*, p. 205; Kaplan, *Bread*, 1: 240.

[23] Turgot, *Oeuvres*, 4: 223; Dakin, *Turgot*, p. 178.

this terminology was the administrator's purported neutrality in the market. Before and after 1763, even the most cursory familiarity with liberal theories and the strategies in use made it clear that authorities could not interfere too openly with market forces. That consensus had been evident in the 1757 attempts to shape prices, particularly in Le Chevallier's frustration with the magistrate who had reduced bread prices dramatically. In 1768 and 1772, Miromesnil and De Crosne had created the illusion of neutrality in their simulated sales. That illusion became the requisite reality of the markets of 1775.

What did neutrality imply? Without doubt, it meant that public officials were to refuse to set prices or break agreements, even when pressed by crowds. Instead, they were to urge the parties to buy and sell "*de gré à gré*," by mutual agreement. The 1775 officials argued and reasoned with buyers. Unlike their predecessors, however, they were limited to hollow declarations that they would take no role in the market, save to maintain order. That they had no grain to distribute through hidden channels left them with only the power of their own strained voices, occasionally backed by soldiers. Dorian, a military official posted at Pavilly in June, for instance, refused to place a price on grain. He informed crowds that his sole "mission was to prevent trouble in the marketplace."[24] Leporquier, the subdelegate at Chaumont, would only urge buyers to refrain from pillaging.[25] These phrases were not entirely new. In 1768, Controller General Laverdy had directed the police at Bourges in the far-off Berry "to limit themselves uniquely to preventing disorder in the market."[26] Yet in 1775, that expression became not only the administrator's repeated incantation, but the principal standard by which to judge his actions.

The "Master of the Price"

The authorities' neutrality – meaning their refusal to set a price on grain, especially in the midst of a disturbance – was a central feature of the emerging definition of free trade. An elaboration of this neutrality frequently included a related, familiar phrase, "the master of the price." The phrase had a long history as a pejorative term, and meant that someone had disrupted the normal practice of bargaining over the price of grain. It was leveled at anyone who refused to negotiate peaceably in the marketplace, who defrauded the other parties, or who wanted to provoke price rises or reductions. Thus, food rioters, hoarders, and even kings and ministers who

[24] Dorian to intendant, 14 June 1775, ADSM C 110.
[25] Leporquier to intendant, 6 May 1775, ADSM C 108.
[26] Kaplan, *Bread*, 1: 226.

organized warehouses were charged with seeking to become the "master of the price."[27] After the mid-century, its pejorative meaning was joined by a more favorable one, a significant step in the growing solidity of marketplace definitions of free trade. Here, the term indicated that owners had been able to sell their grain for the prices they wished – generally, "*de gré à gré*" – and implied the accord of the buyer, as was consonant with liberal theory.[28]

The buyer's acquiescence, however, frequently disappeared from the authorities' post-1763 formulation. During the worst days of 1768, the subdelegate at Elbeuf, for instance, explained to an irate mob that the owner should be "the master of the price of his [merchandise], just as every other merchant is of the price of his wares."[29] By 1775, similar expressions appeared in report after report. In the first riot of the Flour War, the acting lieutenant general of police of Beaumont-sur-Oise told a crowd "the sale of grain was free at all times" and he would not "force down the price of grain."[30] In Les Andelys, the *laboureurs* were "absolutely masters of the price."[31] Subdelegate Leporquier expressed his satisfaction that the *laboureur* enjoyed "the *liberté* to sell his grain without being bothered by the people" in Chaumont.[32]

Intendant de Crosne reinforced the definition. When one subdelegate recounted that he had encouraged buyers who could not afford an entire boisseau to purchase smaller amounts, De Crosne praised his actions: "Because the sales [took place] according to mutual agreements and you made them pay the going rate, your actions are entirely within the views of the Council. . . ."[33] Deviations received unequivocal rebukes. An official at Les Andelys merited censure for trying to force an important grain merchant, the much-hated Planter, to sell grain for a low price. The owner "is the master of his goods," De Crosne declared, "and he should have the

[27] D'Argenson, *Mémoires*, 4: 99.

[28] For other examples, see Miromesnil, *Correspondance*, 5: 66; "Questions sur les causes de la cherté actuelle . . . ," Subdelegate of Abbeville, 20 December 1773, AD Somme C 86; Kaplan, *Bread*, 2: 501, 518–9.

[29] Blin to Michodière, 23 June 1768, ADSM C 107.

[30] Cited in Bouton, *Flour War*, p. 83.

[31] Other examples can be found in Hamelin, for Ouldart, 16 May 1775, ADSM C 108, p. 166; Leporquier to intendant, 6 May 1775, ADSM C 108; and Bouton, *Flour War*, p. 96.

[32] The town had been very hard hit by the riots of the first week of May. Over 300 people had gathered in the marketplace and had taken all the grain for 16 livres per setier. They then proceeded to force landholders, including the *fermier* for the Prince de Conty, to sell them grain for the same price, well below the going rate of 29 livres. See also the comments of local officials in Evreux in 1768 and in Louviers in 1784. Procès verbal of 16 April 1768, AD Eure 1B 392 (5); Bosquier, fils to intendant, 21 May 1784, ADSM C 106, p. 204.

[33] Clericy to intendant, 25 June 1775, ADSM C 110; intendant to Clericy, 27 June 1775, ADSM C 110.

liberty to have it taken or not taken to market to sell it at the *prix défendu* [the price the owner chooses]. . . ."[34] His subordinates reported that they too struggled to uphold the emerging rules of free trade in the face of the occasional immoderate representative of the police and courts. At Neufchâtel, for instance, the subdelegate described the unacceptable actions of a *procureur* ("*un homme faible*," in the subdelegate's estimation) who had pressured a *laboureur* to sell grain at a reduced price to a knife-wielding woman. The subdelegate claimed to be furious with the *procureur*. "I swear to you," he informed the intendant, "that I would not have behaved as the *procureur* in the least, that I would have begun by putting the woman in prison." Lest there be any lingering doubt about his understanding of his responsibilities, he added, "I would have left the *laboureurs* absolutely the masters to sell at the price they wanted."[35] The phrasing in these episodes was familiar: The "masters of the price" were to be protected, authorities were to remain neutral, and eventually free trade would prevail.

So heavy an emphasis was placed on this one element of free trade – the protections against price-setting – that other possibilities seemed to disappear. Condorcet criticized precisely this restricted meaning of free trade: "With the best intentions in the world, [the police and troops} commit one act of stupidity after another and it is solely because they do not know what liberty is. They believe that because they have not set a price on wheat, they have fulfilled the law completely."[36] Condorcet's complaints to the contrary, this narrow definition of free trade persisted in post-1775 formulations. In 1784, the intendant cautioned Dieppe authorities to see that "the *laboureurs* and landowners remained masters of their merchandise," as one example.[37] A 1782 query from Argenton in the far south of the kingdom indicates perhaps some increasing latitude for owners' demands and would have merited Condorcet's approval. The question concerned the owner's right to raise the asking price during the course of the market – an action generally deemed illegal before the liberal reforms. In this instance, however, the Paris Parlement *procureur général* instructed the *avocat fiscal* not to interfere, contending that "it is in [the owner's] interest to sell for the highest price possible. . . ."[38] While some authorities may have been willing to sanc-

[34] Doré to intendant, 11 May 1775, ADSM C 110; intendant to Doré, 13 May 1775, ADSM C 110.

[35] For his many efforts, this subdelegate was recommended for a 1,000-livre reward after the Flour War. Bequel to intendant, 6 May 1775, ADSM C 100; "*Etat* of requests that the maréchal d'Harcourt proposes to make" [n.d. 1775], ADSM C 107.

[36] *Correspondance inédite do Condorcet et de Turgot (1770–1779)*, Charles Henry, ed. (Paris, n.d.), p. 212.

[37] Intendant to Mayor and *Echevins* of Dieppe, 27 May 1784 ADSM C 106, p. 223.

[38] 9 June 1782, BN JF 1162.

tion such actions in 1775, the extant records from Upper Normandy give no indication that they did. Nonetheless, this 1782 understanding may reflect a broadening sense of what it meant to leave the owner the "master of the price."

Empty Markets

Still, free trade, however defined, did not bring the much-needed abundance to French markets. Neither Turgot's aggressive campaigns in support of owners, nor the authorities' deft efforts to guarantee the *liberté* of their halls ensured abundance. Instead, the owners' reactions to liberal policies only confirmed Savary's seventeenth-century conclusion that few merchants would ever want to enter the trade. Perhaps these unwilling traders feared riots, perhaps profits were still too erratic and shortages too genuine. Perhaps owners *did* wish to hoard supplies, drive prices up, and reap the rewards. Whatever the reasons, the free-trade decrees seem to have left many public halls sporadically and troublingly deserted. Few new faces arrived at markets, and many that did vanished as quickly as they had appeared. *Laboureurs* took advantage of their newly found right to sell outside the marketplace and millers and merchants roamed backroads and farmyards in search of grain. The markets were sporadically empty and townspeople and their officials were persistently uneasy.

Reports from Upper Normandy lamented the *laboureurs*' desertion of halls during shortages and the authorities' inability to counter it. In the autumn of 1767, for instance, owners near Rouen held back their grain. A credible rumor alleged that they would boycott the hall the coming spring in response to the parlement's decision to reimpose market regulations. Equally troubling, there were reports that Elbeuf's *blatiers*, who had once frequented the hall reserved for city's retail trade, would sell only to wholesale merchants.[39] The Lyons-la-Forêt subdelegate complained in 1775 that the laws left the "*cultivateur* the master [of his merchandise] at all times to sell the wheat from his property without the obligation to bring it to market."[40] The Aumale subdelegate outlined one familiar ploy: First, *laboureurs* would see that only a sack and a half arrived at the weekday market; then, they would bring thirty or forty sacks to the Saturday market.

[39] The parlement's arrêt of 15 April 1768 restored regulations prohibiting any off-market transactions and the storage of grain. Miromesnil, *Correspondance*, 5: 48, 65–6, 181, 183.

[40] Havet to intendant, 17 May 1775, ADSM C 109, p. 128. See also, Havet to intendant, 1 June 1775, ADSM C 109, p. 140; Havet to intendant, 20 and 21 May 1775, ADSM C 109, pp. 133–4.

Anxious, the townspeople would buy them all for 8 to 10 sols more per boisseau than at the preceding market so that "prices rise everyday."[41]

Clearly, some of the grain was moving directly from farmer to merchant to miller or baker, bypassing the public market, and finally reaching consumers in the form of loaves of bread. It is hard to discern the paths it took or what amounts were hidden away or secreted out of the kingdom.[42] Yet the specter of empty markets was disconcerting, not only to restive buyers, but even to Turgot. "Why don't the *fermiers* and the landowners understand their own interest better," he groaned, scanning reports of the barren halls.[43] Explanations for the grain owner's absence abounded. A Gournay official blamed the presence of "troops everywhere" for encouraging *fermiers* to demand high prices and refuse to come to market. Turgot's much bruited promises of protection and restitution had encouraged them to hold out longer still.[44] Some *laboureurs* claimed the reverse, however: They feared more violence and preferred to stay away – certainly understandable given the events of that spring. The Seine, for example, had been blocked completely for almost a month at the critical bridge at Pont-de-l'Arche between Rouen and Vernon (Figure 3.1). No merchant was willing to brave the townspeople who had stormed one of Planter's barges and refused passage to others.[45] A host of authorities, many in the military, promised escorts to market, but found few proprietors willing to risk confrontations along roads and in marketplaces.[46]

Thus, administrators pressed for more effective solutions. In 1768, Miromesnil had counseled them only to have "patience." He advised local authorities to "let [merchants] get used to this type of commerce with which they are unfamiliar and that they undertake fearfully." He wished to wait for better years before further defining legitimate practices.[47] Yet, while he had advised patience, he and De Crosne nevertheless had orchestrated the simulated sales. By 1775, Miromesnil had departed to become Louis XVI's Keeper of the Seals, and De Crosne, with Turgot's instructions unavoidably clear, could venture only extremely circumscribed advice. Most significantly, he had no grain to offer the subdelegates who begged for help. "What will

[41] Tricornot to intendant, 7 June 1775, ADSM C 108. See also Havet to intendant, 4 June 1775, ADSM C 109, p. 141; Hamelin for Ouldart to intendant, 16 May 1775, p. 166; and Hyffe to intendant, 9 June 1775, ADSM C 108.

[42] Blin to De Crosne, 21 May 1775, ADSM C 110; Lantier, "Saint-Lô," p. 21.

[43] Cited in Faure, *Turgot*, p. 232.

[44] Guillaume de Deuxpont to Intendant, 5 August 1775, ADSM C 109, p. 279.

[45] Saint Hillier, commander of the detachment at Pont de l'Arche to intendant, 6 June 1775, ADSM C 110.

[46] Clericy to intendant, 25 June 1775, ADSM C 110; intendant to Leporquier, 17 May 1775, ADSM C 108, p. 326.

[47] Miromesnil, *Correspondance*, 5: 116.

Figure 3.1 The Pont de l'Arche bridge was a particularly vulnerable place for grain shipments moving up the Seine to Paris. Reproduced with permission from the Bibliothèque Nationale de France, Paris. Cliché number C 1518.

we do?" inquired the Gournay subdelegate when *laboureurs* near Gournay refused to supply the market. "The prohibitions of the *arrêt* . . . of 13 September [1774] tie our hands completely."[48] The intendant replied cryptically that he saw "with great pain the difficulty in which you find yourself. It is not impossible to get yourself out of this difficulty by the precautions that are in your power."[49] To Doré, the Vernon subdelegate, De Crosne stated flatly that he was "quite persuaded you are not neglecting anything in order to supply your market and you have . . . reason to hope your actions will have the same success in the future as they have up to the present."[50] To another: "I cannot recommend enough that you not neglect any workable means . . ."[51] What were those "precautions," "actions," and "workable means?"

De Crosne described but one possibility, the familiar path of "invitations." That strategy had been an important element in Demarests's 1709 plans and had been echoed by myriad eighteenth-century authorities. In 1775, it was all that was left of former policies. To Leporquier, De Crosne suggested that he "exhort the *laboureurs* to bring their grain to market."[52] In greater detail, he explained to Paterelle how this was to work. Evidently, that subdelegate had gotten off to a bad start. He had written letters to the sixty most substantial candidates and only three responded, bringing a distressingly low total of 30 boisseaux when 300 were needed.[53] The intendant, exasperated, spelled out his desires more fully. Paterelle was to meet with the proprietors. "A letter does not work like conversation." De Crosne explained. "You shall reason with them," he advised. "[T]hey will open means to you, they will . . . bring to your knowledge the *blatiers* whom you [shall contact] and whom you shall engage to provision your markets."[54] Certainly De Crosne was too experienced an administrator to expect such plans to work. Given his earlier role in the simulated sales, he may have been sympathetic to his subordinates' pleas, yet he dared not contradict official policy. Turgot himself, practiced in such matters, also proposed invitations as "the most suitable means" to bring proprietors to market and assured the Caen intendant that it should be "easy to persuade them. . . ."[55]

[48] This was one of the few times that the free trade policies were invoked by name. Paterelle to intendant, 3 May 1775, ADSM C 109, p. 237.

[49] Intendant to Paterelle, 20 May 1775, ADSM C 109, p. 240.

[50] Intendant to Doré, 2 June 1775, ADSM C 110.

[51] Draft, intendant to Havet, 25 May 1775, ADSM C 109, p. 129.

[52] Draft, intendant to Leporquier, 17 May 1775, ADSM C 108, p. 326.

[53] Paterelle to intendant, 17 May 1775, ADSM C 109, p. 238.

[54] Draft, intendant to Paterelle, 20 May 1775, ADSM C 109 p. 240.

[55] Turgot, *Oeuvres*, 4: 229. See also the instructions of the Paris Parlement's *procureur général* in 1788, explaining that even in times of desperation and revived regulation, it was best not to order *laboureurs* to bring grain to market. BN JF 1163, p. 11.

The controller general even sanctioned introducing implied threats of the buyers' potential for violence into these invitations. "[O]wners are, in fact, interested in lending themselves to whatever will maintain the freedom from which they are the principal beneficiaries," he explained. Riots would bring losses and perhaps a return to regulation, something no owner wanted. It was the authorities' task to outline the potential consequences of the grain owners' ploys, the controller general suggested.[56] Yet, these "invitations," with their accompanying threats, had little effect.[57]

Stretching Neutrality: Local Officials in the Flour War

The constraints of official policy – including the prohibitions against government grain purchases – and the intendant's clear intention to follow Turgot's orders, placed Upper Norman authorities in a desperate situation. The murmurs from the marketplace could be heard from their windows. They dared not feign indifference as their carriages rode through the jeering crowds. Moveover, they still had the responsibility to stride into the hall, draped in whatever robes, sashes, ribbons, and medals confirmed their powers, to address the hungry villagers seeking food. "The people reason with their stomachs," went a contemporary saying. Unfortunately, Turgot's long-winded circulars and harangues would be of no use when women drew knives from their aprons and pointed them at sacks; his troops, if they arrived, might be. In the interim, it was up to the authorities to navigate between the demands of the poor and wretched gathered in their markets and outside their gates, and the strict limitations of the royal policies. How could local authorities transform their supposed "neutrality" and the corresponding injunctions against price setting and forced sales into strategies that would calm their towns?

The skill of local authorities showed in their ability to take the emerging definition of free trade and use it to guide their actions in their marketplaces while intervening on behalf of buyers. Where possible, they constructed transactions according to the limits they understood: no forced sales, no set prices. Afterward, despite the extremes, even the illegalities, that some of their measures represented, they managed to describe what they had done as if it actually conformed to accepted practice. They negotiated sales, for instance, bringing buyers and sellers into uneasy agreement.

[56] Turgot, *Oeuvres*, 4: 229–31. See also Royer's 1757 permission to use ruses to counter owner's demands for high prices: "You could even allow the hope of other shipments to arise, without making many guarantees." ADSM C 104.

[57] Hamelin to intendant, 30 May 1775, ADSM C 108. See also Clericy to intendant, 25 June 1775, ADSM C 110; and ADSM C 108, pp. 331, 333.

They forced sacks open, gathered buyers, and decreed the amounts each could have. These were strategies of last resort, ones of accommodation, arm-twisting, and anxiety. While they would argue that they had no choice, their reports also reveal their ongoing struggle to construe their actions as falling within the prescribed boundaries. Even in the breach, they honored the emerging definition of marketplace free trade. This was the uneasy but necessary confrontation of the "subsistence imperative" and free trade in times of shortage.

The concept of the sellers' right to be the "masters of the price" held firm in these episodes, no matter how far the authority strayed from lawful practices. Listen to a subdelegate's account of the wisdom and generosity of the Duc de Montmorency, Seigneur of Gournay. Sellers that day, "terrified" of one woman who "was rash enough to gather together a number of women and to recount, as if it were a grand event, the story of the revolt at Gisors," had begged Montmorency "to fix the price at which he wanted them to sell their grain." He had refused, but advised them to sell cheaply to the poor and for "the price it was worth" to the wealthier. He then "had the goodness to offer the sellers 120 livres" [the price of three or four sacks of grain] as restitution for their losses on sales to the poor. No riot followed, although the market was tumultuous, which compelled the sellers to lower their prices sharply. The subdelegate credited Montmorency's "prudence and generosity" for averting violence.[58] The intendant offered no criticism for Montmorency's intervention, nor for the low prices, even though the seigneur's actions had not demonstrated absolute neutrality. Instead, the intendant praised the duke's willingness to maintain the sellers' asking prices for the wealthier buyers, and for the subsidies that supported the levels they demanded. Montmorency had fulfilled all that the government desired.[59] De Crosne expressed similar approval to Dorian, a militia commander at Pavilly, who twice took coins from his own purse to help buyers meet the prices owners wanted, even offering to break a sale to an unwelcome *blatier*. In this case, the intendant cautioned Dorian to avoid dispensing such largesse on a regular basis, since it would encourage the people to expect authorities to side with them each time prices rose. Yet, because Dorian had not set the price of grain at that market, he was within the permissible practices for late eighteenth-century free trade, and merited the intendant's guarded approbation.[60]

There were other such episodes that elicited De Crosne's approval in 1775. Clericy, the subdelegate at Yvetot, for instance, refused calls to set a

[58] Paterelle to intendant, 3 and 17 May 1775, ADSM C 109, pp. 237–8; Bouton, *Flour War*, pp. 128–9.

[59] Paterelle to intendant, 3 May 1775, ADSM C 109, p. 237.

[60] Dorian to intendant, 14 June 1775; intendant's reply, 18 June 1775, ADSM C 110.

price or to bar *revendeurs* from the marketplace. With some ingenious arguing, however, he persuaded the merchants to sell a few measures of grain to the poor for lower rates. In return, he insisted the other buyers meet the sellers' prices – precisely the Duc de Montmorency's strategy.[61] The intendant supported Clericy's tactics: "Because [the sales] took place according to mutual agreement and you made them pay the going rate, your actions are entirely within the views of the Council, and this means for maintaining tranquility and preventing riots can only be approved."[62] The concept of mutual agreement – *de gré à gré* – figured in another more extreme case in Bacqueville, where owners were pressured to reduce their rates. Two lone mounted policemen, distressed by the crowd supporting a woman who was arguing with a *laboureuse*, "advised the *laboureuse* to sell it for the price offered, to avoid a great *rumeur*." The owner complied, and the others, "scared, sold all the grain they had at much below the price of the preceding halls." One of the policemen explained pointedly that he had *not* set a price, but had kept the peace on a day when things could have turned much worse.[63] In another case, sellers in Les Andelys had been so frightened by the irate villagers that they sold their wheat for prices that careened between 18 and 27 livres. The subdelegate, ever mindful of how free trade was understood, was able to construe the situation to meet the official definition of free trade. He stressed that the "*liberté*" of the price had been left to the proprietors.[64] The requisite "mutual agreement" in each of these instances had clearly come about under duress. But what mattered more to the intendant and these many subordinates was that the transactions could be interpreted as the result of genuine accord.

A more strained, but still sanctioned, act of intervention on prices occurred in Gaillefontaine. There, the subdelegate, De Franconville, had presided over numerous uneasy market days. He had rejected all buyers' cries for price ceilings. Finally, when a woman armed with a knife and her accomplices were dragged to his office, he heeded their complaints that the rye they wanted to buy was of very poor quality and vastly overpriced. Taking the town market controller with him, De Franconville examined the grain, and declared the owner's high prices unjustified. He asked the controller to name more appropriate levels, "according to the price of the hall," and then crafted a deal that set the rye slightly below rates for better grain. Calm returned, and De Franconville and the controller continued to settle disputes in a like fashion throughout the day. When rain sent buyers and

[61] His letter hints that he also told sellers they could refuse to extend credit. Clericy to intendant, 25 June 1775, ADSM C 110.

[62] Intendant to Clericy, 27 June 1775, ADSM C 110.

[63] Courtois to intendant, 16 May 1775, ADSM C 109.

[64] ADSM C 106.

sellers scattering, the subdelegate took the operation to a nearby inn. The *laboureurs*, according to De Franconville, "went away . . . very happy and thanking me, as if I had bought them their lives." By using the controller's expertise and authority, he had imposed levels that reflected arguably fair market prices. Far from undermining free trade, he and the controller had restored it, and had allowed sales to be concluded. With complete honesty, they could argue that they had not broken their pledge of neutrality, nor had they truly set prices.[65]

The authorities' careful use of the narrow definition of free trade not only allowed them to influence (and even set) prices, it also gave them leeway to force owners to sell their grain. For example, at the peak of the Flour War, news that Louviers buyers had begun imposing prices reached owners en route to the market. Apprehensive, the owners "turned on their heels and hid," and the mounted police rode out to find them. The brigade pledged to allow owners to sell at whatever price they wished, but instructed them to sell all that they had. On the next market day, the police reported that the owner had been "more or less master of the price, although with difficulty and while having to yield somewhat on the price he wanted." The intendant appears to have offered no protest to this reasoning.[66] Captain Tricornot, stationed in Aumale, upheld similar guidelines later that summer. Angered when sellers brought several sacks to market and then opened only one – "as if the others were not for sale" – he commanded the sellers to open all their sacks, signaling they were for sale, and invited prospective buyers to step forward. As long as the proprietors were willing to sell all they had, he allowed them to name their price. Explaining his actions, he focused on how he had refrained from setting prices rather than on how he had coerced owners to sell every sack: "I believe it is the intention of the government that the *fermiers* be required to sell the wheat they bring to market, *as long as one pays the price they ask. . . .*" Presumably, the intendant agreed.[67]

There were examples later in the century. In Saint-Lô in Lower Normandy, the subdelegate and a magistrate intended to force owners to sell their grain in 1784, but were uncertain about whether they should sanction any price the owners demanded, especially if they were not "reasonable." The officials finally decided that any attempt to persuade an owner to reduce prices placed "obstacles in the way of this trade, which requires a great deal of freedom." Apparently for the subdelegate, this "freedom" could not prevent owners from being coerced to sell their grain, but it did

[65] De Franconville to intendant, 6 May 1775, ADSM C 110.
[66] Dagoumer to intendant, 6 and 9 May 1775, ADSM C 109, pp. 34–5.
[67] Tricornot to intendant, 22 June 1775, ADSM C 108. Italics mine.

protect their right to high prices.[68] In like fashion, a Rouen *blatier* was sentenced to a three-livre fine when he refused to keep his sacks open during market hours in 1788.[69] And, in the far-away Maconnais in the spring of 1789, the intendant pressed local merchants to make their grain available to townspeople. "It may be necessary to stop grain shipments," he explained, "but not to set their price, unless by mutual agreement."[70] Each episode confirmed the clear consensus that the owner's control over the grain's price, and less so the conditions of its sale, were at issue.

Overstepping Boundaries: Planter's Warehouses in 1775

There was one distinctly troubling incident of administrative confusion and violated boundaries, however. It occurred in Vernon, where Planter had his elaborate clusters of warehouses, and involved Doré, a capable, determined, and faithful subdelegate. At the outset, it resembled De Franconville's adventure in which he and the hall controller crafted bargains in Gaillefontaine – in fact, the two episodes fell only a day apart. First, there had been a violent attack on Planter's "tower" warehouse, including gunfire that injured four to six people. Then there had been a raid on Planter's barges as they passed by La Roche-Guyon on their way to Paris. Local inhabitants, including several substantial *laboureurs*, hid on islands in the Seine, then jumped into boats and paddled out to seize more than 50,000 livres of grain that Turgot had ordered to be protected.[71] The maréchal d'Harcourt, now the province's governor, eventually recommended Doré for a 1,000-livre bonus for his valiant efforts to protect Planter and his grain.[72]

The intendant was more stinting in his praise, for Doré, however well-intentioned, had created a disconcerting mixture of acceptable and unacceptable actions. From the first moments, Doré had realized he was facing an impossible situation: The market was overrun with buyers and there were only seventy sacks. He requisitioned an additional fifty sacks from

[68] Cited in Maurice Lantier, "La crise des subsistances en 1784, à Saint-Lô," *Annales de Normandie*, 25 (March 1975), p. 15.

[69] Desmarest, *Commerce*, p. 222.

[70] In a second incident that spring, grain was confiscated briefly from merchants in Macon and then sold for less than the going rate. The confiscation came at the intendant's orders, but he rebuked the *échevins* for selling the grain below the merchants' price. Bouvet, *Maconnais*, pp. 51, 54–5.

[71] Planter to intendant, 7 May 1775, ADSM C 110; Bouton, *Flour War*, p. 89.

[72] "*Etat* of requests that the maréchal d'Harcourt proposes to make," [n.d. 1775], ADSM C 107.

local merchants and then reorganized the marketplace, pushing everyone out and marking an entry and exit. To prevent any arguments over prices, he called in fifty buyers and sellers and told them to agree on a range of prices. Explaining the process to the intendant, Doré emphasized that the prices had been reached "after a due hearing of both parties [*contradictoirement*], between the seller and the buyer." Like De Franconville, Doré had recreated the bargaining process, or at least that was how he described his actions. In this case, he not only solicited information on prices and then pronounced acceptable levels, but he also established the physical parameters of the market itself.[73]

The intendant reacted quickly to news from Vernon, including protests from Planter. De Crosne could applaud Doré's actions within the market hall. No price had been set, and Doré had brought buyers and sellers into agreement. The intendant nonethless sharply criticized the decision to take grain and flour from Planter's warehouses (Figure 3.2). Doré protested the reprimand. He insisted instead that he had been following De Crosne's express commands. His arguments had some merit, especially within the context of the late eighteenth-century definition of fair practices. As part of a failed campaign of "invitations," encouraged by De Crosne, Doré had approached Planter about selling some of his flour in the hall; Planter had declined. Doré, however, had a written order from the intendant giving permission for bakers to buy directly from Planter's tower warehouse – a measure that recalled the simulated sales of the previous decade. The subdelegate broadened that permission to include bringing 50 sacks to market against Planters' opposition. Here was Doré's error, according to the intendant. In all, a total of 120 sacks was sold, one boisseau at a time, in order to satisfy as many buyers as possible. Planter's blistering objections brought quick action from the intendant. De Crosne reprimanded Doré and prohibited the use of any more of Planter's supplies.[74] Here again, important elements of the emerging definition of free trade were being confirmed by De Crosne – Doré could bring buyers and sellers into accord, enforce reasonable prices as long as they reflected going rates, and even delineate the boundaries of the marketplace, but he could not direct owners to bring grain to market. That Planter's wares were desperately needed in Paris added to the magnitude of the mistake.

[73] Doré to intendant, 7 May 1775, ADSM C 108, p. 89.

[74] Planter's objection, certainly plausible, was that bakers would abuse his goodwill by buying up his flour at low prices. In order to satisfy Planter, Doré expressly forbade the bakers to buy any, so that Planter would not feel they had taken advantage of the situation. Doré to intendant, 7 May 1775, ADSM C 108; Doré to intendant, 10 and 11 May 1775; intendant to Doré, 13 May 1775; "Excerpt from the letter from Planter le jeune of Vernon of 12 May 1775," ADSM C 110.

Figure 3.2 The Vernon bridge, where some of Planter's warehouses were located. Reproduced with permission from the Bibliothèque nationale de France, Paris. Cliché number H 119883.

Doré's errors aside, there were numerous variations on the marketplace definition of free trade, but they more or less amounted to the same thing. Prices could not be set, nor could owners be directed to cart grain from their farms or warehouses to market. Once in the marketplace, a nebulous concept of mutual agreement was to prevail. The episodes discussed here conformed to such guidelines. Other, less dramatic choices made by local authorities followed the same sense of the limits of free trade. It was frequent, both before and after 1763, for example, for an official to distribute grain in small amounts to retail buyers. In this way, he could ensure that buyers could afford the grain, without coercing an owner into lowering a price. This had been one of Prémord's 1757 tactics; it occurred through Upper Normandy during the Flour War and afterward.[75] By limiting sales in this way, each of these officials was able to maintain high prices, and could argue that he had remained within the bounds of free trade.

[75] Dagoumer to intendant, 6 and 9 May 1775, ADSM C 109, pp. 34 and 35; Theroulde to intendant, 15 May 1784, ADSM C 106, p. 299. Theroulde also forced the sellers to agree to selling the grain *raz*, or levelled off, rather than with the traditional *comble*, or heap, thus further lowering the price of a boisseau.

The Accepted Script

No requisitions, no set prices, administrative "neutrality," and sellers who were the ostensible "masters of the price" – those were the standards that defined late eighteenth-century free trade. Even when pushed by circumstances to violate those boundaries, the Upper Norman officials observed the requisite limits by "scripting" their activities to fall within the lines of the accepted definition they had helped to create. They built on the understandings that had informed simulated sales, and in particular on the notion that direct interference with prices would be disruptive and ineffective. The price – and its supposed sanctity – thus became the focal point for the 1775 authorities. Turgot, by halting simulated sales, not only had stripped authorities of an important means for ensuring some tranquility, but had conveyed unequivocally that interference with prices was an unacceptable error. Rather than abandon their cause and their markets altogether, however, the authorities kept those limitations in view as they crafted their own market-based standard for free trade. If owners were to remain "masters of the price," then it was up to authorities to redefine that concept in ways that did not leave them utterly powerless. They were tested by the deep shortages of 1775, yet rallied, and found tactics that both upheld Turgot's demands and offered the prospect of peace and as much abundance as could be mustered in that troubled year.

CHAPTER FOUR

Narrowing the Focus: Bakers and Bread, 1760–1789

> La taxe du pain peut être comparée à un veritable labyrinthe dont on n'a pu essayer de sortir sans risquer de rentrer dans un encore plus difficile.[1]

The much-celebrated free trade reforms of the 1760s and 1770s had stripped local officials of a number of their tactics for countering shortages. Turgot's policies, especially those prohibiting simulated sales, had placed exceptionally tight limits on the authorities' activities. To a certain extent, the officials in more rural areas had managed to overcome the constraints through their deft implementation of the legislation. Dorian, Tricornot, Doré, and countless other more anonymous authorities had sought to curb the impact of the liberal policies in their marketplaces. Their redefinition of free trade offered them some leeway to continue to intervene in transactions in their villages and small cities. Such efforts, while helpful in cities, could not fully address the problems of shortages in urban areas, however. Skillful negotiations between restless townspeople, a handful of buyers, and several officials at an outlying market could restore order, especially if backed by some soldiers. Cities presented greater difficulties. Most urban consumers bought bread, not grain. Moreover, the sheer number of the poor and workers gathered in cities would have rendered marketplace negotiations, speeches, and subsidies only moderately effective. Most important, urban officials needed to control bread prices, not grain prices, although certainly the price levels of the two commodities were related.

Like their colleagues in more rural regions, urban officials took the possibilities that the prevailing policies offered, and then reworked them to fit the needs of their cities. Given that their populations bought bread, public authorities turned to the regulations of the baking trade to offset the combined impact of intermittent harvest shortfalls and the liberal reforms. The

[1] BN Mf 11347, p. 198.

distinction between bread and grain is critical to understanding the nature of the free-trade reforms and the actions of urban officials. Even though many of the controls on the grain trade had been lifted in the 1760s and 1770s, those on the baking trades were very much intact. Moreover, only a few authorities seriously considered lifting them.[2] The sporadic but unrelenting price rises of the late eighteenth century, with sharp spikes during the shortages of 1757, 1768–71, 1774–5, 1784, and 1789–90, convinced authorities they could not abandon their cities to the full force of free trade. Thus, the regulations of the baking trade held the last chances for appeasing hungry urban buyers. These efforts centered on the use of the *taxe du pain*, or bread price setting, described in Chapter One. While grain prices might fluctuate at the local market, bread prices were set by the local authorities, and nothing in the free-trade reforms suggested that the *taxe* had been abolished. Thus, officials focused their energies on exploring the possibility that the *taxe* could still be used to counter high grain prices.

Throughout the eighteenth century, the police had used the regulations on the baking trade, and especially the right to set bread prices, as a means of preventing unrest. Primarily, the officials refused to raise bread prices in the midst of shortages. Generally, when reports from the local grain market indicated that prices had risen, the police simply delayed increasing the official bread price for a few days. These delays did not cease after 1763, and there were both routine and extreme examples of such actions.[3] The Carcassonne *consuls* in the far south of France, for instance, set prices as much as 19 percent below the legal level on their price schedule, or *tarif*, from 1749 to 1783. They even stubbornly disregarded unequivocal

[2] There were occasional proposals to allow bakers to sell by weight. These proposals came from Le Havre, Paris, La Rochelle, and Châlons-sur-Marne, and were quickly rejected because the police anticipated bakers would use inaccurate scales. Turgot was the lone Controller General opposing the structure of the baking trade. In 1770, when the Limoges bakers challenged bread price levels, Turgot suspended them, opening the trade to all comers in that city. He generally urged intendants to reduce the bakers' profit margins, believing the guild monopolies yielded unnecessarily high prices. Turgot, *Oeuvres*, 4: 491–5; "Procès verbal des séances de l'Assemblée provinciale de la généralité de Rouen," November–December 1787, ADSM C 2111; BN JF 1742, pp. 38–45 and 138–9; Tillet *Expériences et observations sur le poids du pain . . .* (Paris, 1781), BN Collection Le Senne, 8 Z 4275; "Mémoire pour la communauté des maîtres boulangers," 19 February 1789, in *Les Elections et les Cahiers de Paris en 1789*, 4 vols., ed. Ch.-L. Chassin (Paris, 1888–1889), 2: 553–62.

[3] For information on such practices throughout France, see ADSM 46 BP 45; AD Eure 1B 258; AN E 2515, folios 544–545; BN JF 1161, pp. 10 and 28; BN JF 1163, p. 37; BN JF 1164, p. 111; AM Le Havre FF 66, pp. 53–5; and AM Le Havre PR F^4 153; Lemarchand to the Intendant of Rouen, 14 November 1788, ADSM C 102; Miromesnil, *Correspondance*, 5: 80.

instructions from the Parlement of Toulouse to desist.[4] There were other extraordinarily flagrant abuses, such as the 1772 attempts by Caen's lieutenant of police to extort additional revenue, flour, and loaves for his household from bakers. He had been unable to afford the price of his office, and expected the members of the city's many guilds to contribute extra funds. The bakers refused, and he vengefully lowered the bread *taxe*. Eventually, he was prosecuted by the Parlement of Rouen, but the whole affair only added to the deep strains between bakers and the police in that city.[5] Political calculations and private grievances, not grain prices alone, then, were reflected in the prices French consumers paid for bread. While there were limits to how far a local official could push the bakers, the fact that the police controlled bread prices gave those authorities an important means of protecting their cities against the high prices that shortages brought.

Bread Prices in Rouen, 1775–1785

The significance of late eighteenth-century *taxes* and *tarifs* lay not only in the way they enabled authorities to lower bread prices legally, but also in the means they provided urban authorities to lessen the impact of free-trade decrees of 1763 and 1774. The use of such measures in Rouen is a revealing case study of the ways the baking trade regulations cushioned the impact of grain price increases, especially after Turgot had forbidden simulated sales. The Rouen events stretched from the spring of 1775, when the Flour War rocked cities and towns across northern France, until 1786. The police's schemes provoked much opposition from bakers. The resulting disputes were the material for numerous parlementary rulings. In the end, however, the courts upheld the police's decision to hold bread prices down. There were two sets of measures. The first concerned the application of the *taxe* itself in 1775; the second, the subsequent revisions of a 1725 *tarif* that the police believed offered bakers unreasonable profits.

In March 1775, knowing of the disturbances in surrounding markets, the Rouen police examined their city's practices. They desired measures that would be effective, yet that could be taken quietly. Disagreements, which they knew would arise, were to be handled between the Lieutenant General of police and the bakers in his chambers. The city officials did not want to cause lengthy battles outside shops and in markets. Those altercations could

[4] BN JF 1742, pp. 13–20; "Mémoire pour le corps des boulangers de Carcassonne . . . ," AN AD XI 14.

[5] See the materials from the 1775–8 inquiry into Dutouchet's actions since 1772. He was finally ordered to resign in 1778. ADSM 1B 290, 291, 293, 3786.

become riots. Among the more promising and least conspicuous measures was an increase in the weight of a loaf of bread, without increasing its price. The city's first-quality bread had a fourteen-ounce pound, and thus gave bakers better profits than second- and third-quality loaves. Thus, the police ordered bakers to increase the size of the best one-pound loaves to sixteen ounces. In return, the city lifted the collection of fees at the municipal mills. To offer the poorest inhabitants more aid, the city reduced the lower bread qualities by three sols per pound. Those measures lowered the price of an ounce of white bread 19 percent, second class bread 10 percent, and third class by 9 percent, despite a 29 percent increase in grain prices.[6] These reductions were insufficient, though, and finally the police suspended the *tarif* altogether. They pushed the price of first quality bread down 12 percent and held the price of lesser qualities even[7] (Figure 4.1). These policies, frustrating to bakers, reassured consumers and helped to maintain order. As riots exploded in outlying markets and in Paris during April and May 1775, Rouen remained restive, but calm, by comparison.

As the weeks passed and shortages continued, the municipal officials felt increasingly uncomfortable with the lengthy suspension of the *tarif*. The price schedule in use dated from earlier in the century, probably from 1725. Both the police and the Parlement hoped they could rewrite it to provide lower bread prices. Thus, they undertook a new assay.[8] The resulting *tarif* of July 1775 reduced the price of a one-pound white loaf sharply, by as much as 9 percent at some grain prices. Changes in the ways average grain prices were to be calculated allowed further decreases.[9] The Parlement also decreased the rate at which the price increases occurred, so that overall, the bakers lost significantly.[10] No doubt, the news of the Parlement's willingness to cut bread prices helped to defuse tensions in Rouen. There was a certain amount of public theater in the May announcement of the assay

[6] 3 April 1775, AM Rouen A 39; 30 May 1775, ADSM 1B 3526; *Sentences*, 8 April and 6 July 1775, ADSM 4 BP LV 1/7, cahier 7.

[7] 30 May 1775, 1B 290; ADSM 4 BP LV 1/7, cahier 7.

[8] The benefits of *tarif* revisions were discussed throughout the spring and an assay was ordered by the parlement on 23 May 1775. 6 April, 22 and 23 May 1775, 1B 290. "Arrest de la Cour du Parlement de Rouen," 13 July 1775, ADSM 4 BP LV 1/7, cahier 7.

[9] The data are thin for the next decade, but the extant prices follow the 1775 *tarif* closely. ADSM 1B 290; ADSM 4 BP LV 1/7, cahier 7; and ADSM C non-côtés; Miromesnil, *Correspondance*, 5: 127; Trugard de Maromme to Intendant, 15 May 1784, ADSM C 106. I am grateful to Albert Hamscher for suggesting prison records for indications of bread prices. These prisons' notations on the price of bread purchases confirm that the *tarif* was followed. ADSM C 929–932.

[10] Using a linear regression model to examine the *tarifs*, it is clear that the 1775 revisions sharply decreased the rate of change for bread prices as prices of grain rose, and then maintained the lower rates of change in 1785. The regression coefficients for the prices on the pre-1775 *tarif* [facing table] are significantly higher than those of the July 1775 and 1785 *tarifs*:

and the subsequent drafting and publication of the *tarif* in July. The Parlement's refusal to countenance the baker's protests probably furthered the cause of order within the city. Working together, the magistrates, police officials, and the intendant had crafted a long-term solution to urban woes. They understood thoroughly the importance of granting what relief they could, and also of giving the impression of goodwill toward the people of Rouen during a period of unrest and instability. During the same weeks that Dorian subsidized buyers at Pavilly, and Doré and other subdelegates negotiated prices and supervised sales in outlying markets, the visible officials of Rouen kept some semblance of order by controlling bread prices carefully.

Further Rouen revisions came in the next decade. The new *tarifs* redistributed some of the gains and losses of the 1775 *tarif*. After 1775, the bakers lobbied incessantly for a higher *tarif*.[11] Each appeal was rebuffed until 1784, when the cost of wood soared and rivers and mills froze. Finally, a freak summer hailstorm destroyed most of the crops in northern France. Four bakers in the Rouen guild of about one hundred declared bankruptcy, and then a questionable police ruling in November provoked others to close their ovens and dare the police to imprison them. The Parlement yielded to their petitions and ordered an assay.

The 1785 revisions held both good and bad news for the bakers, although in general the new *tarif* differed very little from that of July 1775. The bakers' profits on first-quality bread were diminished slightly, most notably at the lowest prices for grain. The baker's profits on second- and third-quality loaves, however, were increased an almost imperceptible amount, especially when flour was dearest. Overall, a baker would have made a bit more on second- and third-quality loaves in 1785 than in the preceding years. If buyers chose the lesser quality loaves during a shortages, then bakers would have fared moderately better during shortages. While the Parlement offered no reasons for its willingness to give the bakers these modest improvements at some grain prices, it is likely that it wished to compensate them for the cost of buying grain on credit during shortages. This was a perennial problem, and one that bakers complained of bitterly. Further 1788 revisions also produced meager bread price increases at some grain prices. The bakers' gains most likely were offset by changes in the

[11] 24 January 1777, ADSM 1B 292; 25 November 1777, ADSM 1B 293.

REGRESSION COEFFICIENTS FOR THE PRICE OF BREAD
AS A FUNCTION OF THE PRICE OF WHEAT

	Before 1775	July 1775	August 1785
1st Quality	0.190	0.154	0.159
2nd Quality	0.167	0.135	0.139
3rd Quality	0.167	0.135	0.139

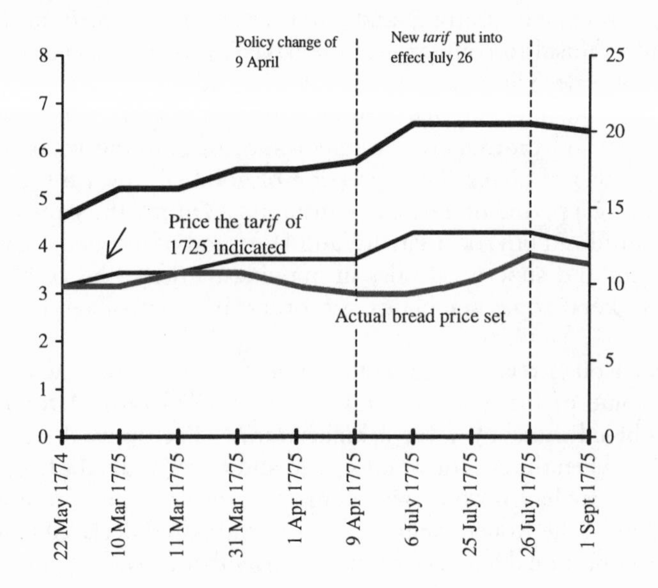
First-quality bread
Price of bread (sous)
Price of wheat (livres)
Policy change of
9 April
New tarif put into
effect July 26
Price the tarif of
1725 indicated
Actual bread price set
22 May 1774
10 Mar 1775
11 Mar 1775
31 Mar 1775
1 Apr 1775
9 Apr 1775
6 July 1775
25 July 1775
26 July 1775
1 Sept 1775

(1)

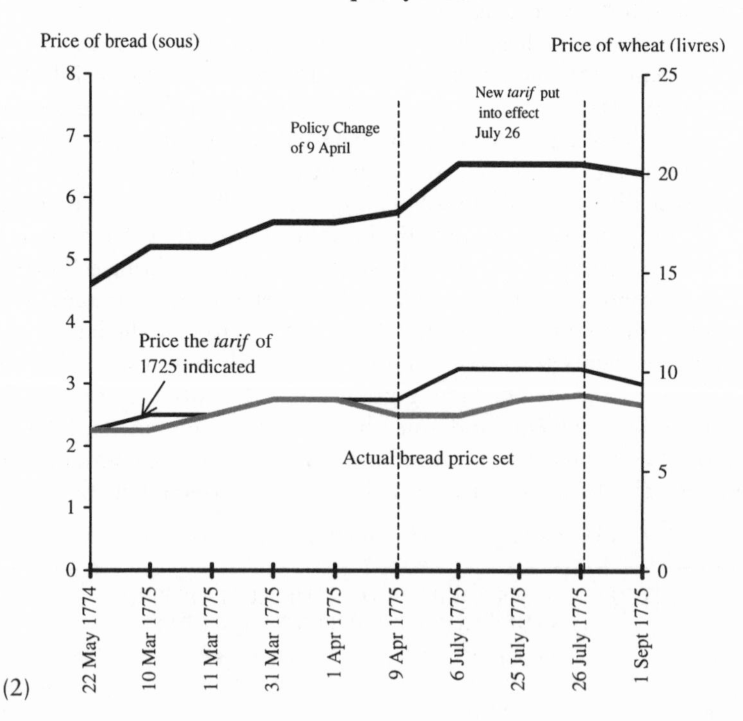
Second-quality bread
Price of bread (sous)
Price of wheat (livres)
Policy Change
of 9 April
New tarif put
into effect
July 26
Price the tarif of
1725 indicated
Actual bread price set
22 May 1774
10 Mar 1775
11 Mar 1775
31 Mar 1775
1 Apr 1775
9 Apr 1775
6 July 1775
25 July 1775
26 July 1775
1 Sept 1775

(2)

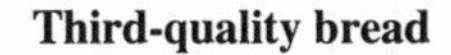

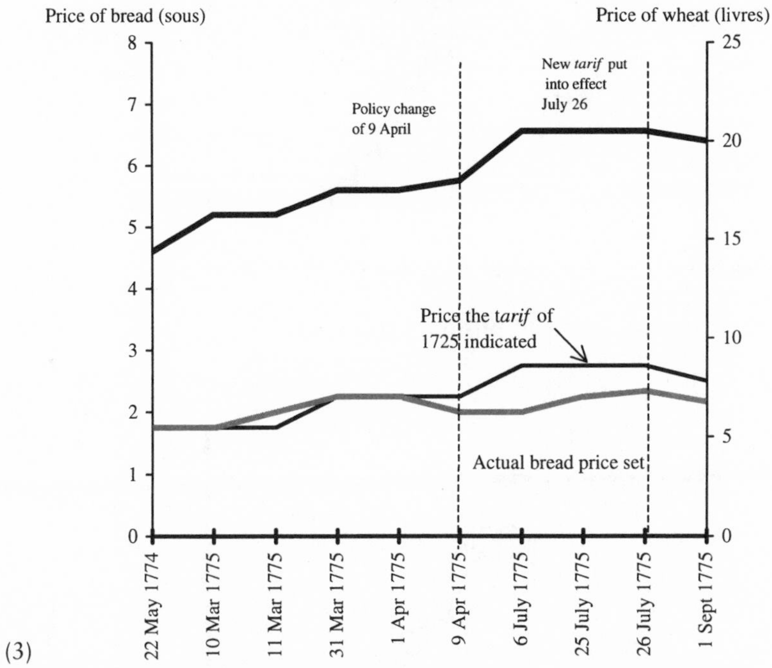

(3)

Figure 4.1 Rouen bread and grain prices in 1775. A comparison between the legal bread prices that the *tarif* of Rouen indicated and the prices the police actually set. The police held the prices of bread especially low between 9 April and 26 July 1775. Sources: ADSM 4BP LV 1/7, cahier 7; ADSM 1B 290; AM Rouen A 39.

ways that the police calculated the price of grain at the hall, however. In any event, the revisions of the 1780s did not bring bread prices back up to the 1725 rates. In the long run, the bakers had lost[12] (Figure 4.2).

Tarif Revisions Across the Kingdom

Rouen was not the only city in Upper Normandy that used revisions to counter rising grain prices. A nearly forty-year battle developed in Le Havre when *procureur du police sindic* (and future subdelegate) Plainpel decided to bring bread prices down in 1755 by changing the way average grain prices were calculated. He knew that bakers bought most of their grain at

[12] ADSM 4 BP LV 1/7, cahier 7; ADSM 1B 290, 292, 293, 296, 298, 299; ADSM C 929–932; and ADSM C non-côtés.

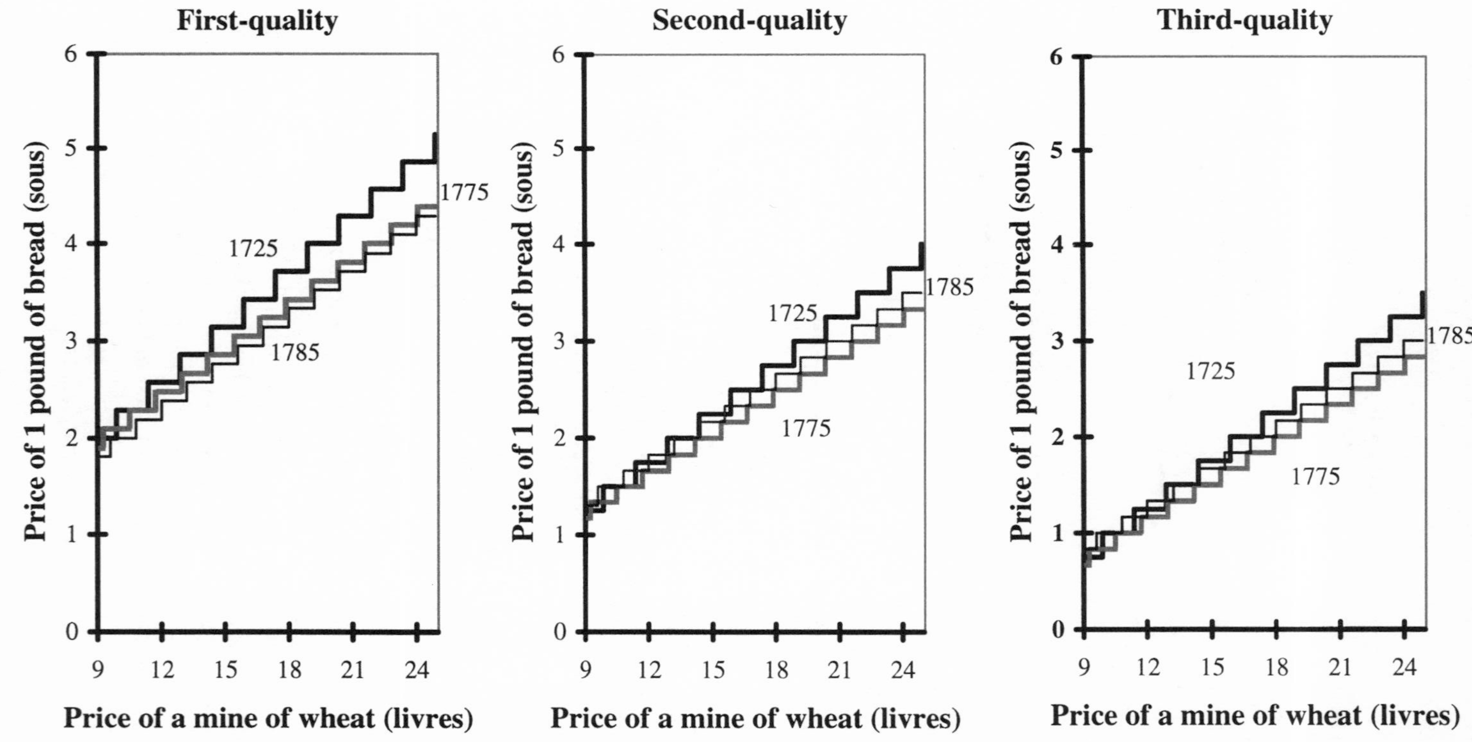

Figure 4.2 A comparison of the Rouen *tarifs* between 1725 and 1785. The two *tarif* revisions of 1775 and 1785 pushed the price of most loaves of bread down sharply from the 1725 rates. Sources: ADSM 4BP LV 1/7, cahier 7; ADSM 1B 290; AM Rouen A 39.

outlying farms and markets, and wanted Le Havre's bread prices to reflect those lower prices. Plainpel, not given to wholly legal tactics, rode roughshod over jurisdictional boundaries numerous times during the succeeding years. In 1761, the Rouen Parlement determined that his attacks on the bakers failed to present even the facade of administrative neutrality and that he had far exceeded his authority. That in no way meant, however, that the parlementary présidents supported the bakers' demands for a more favorable *tarif*. The Parlement sanctioned four further assays – a second one in 1761, and others in 1768, 1769, and 1775 – and adopted many of Plainpel's formulas. The bakers did not succeed in their mission to return to the pre-1755 practices until June 1791, when the National Assembly took their side. Their victory was short-lived, however. When grain prices rose a week later, the city officials again refused to countenance the bakers' demands for higher bread rates.[13]

Despite the rancorous court battles that revisions provoked, the process was in wide use throughout the kingdom in the late eighteenth century. Parlements and ministers encouraged the police to use revisions to limit the profits of bakers in times of shortage. There were notable struggles within the Paris Parlement's jurisdiction, which extended to much of central France. The revisions in far-off Clermont-Ferrand from 1771 to 1779, as one instance, are among the most fully documented and reveal many similarities to the Rouennais efforts to curb bread price rises. During those eight years, the Clermont-Ferrand bakers and authorities wrangled over a series of revisions to the price schedule. A municipal commission first had imposed an exceptionally low set of prices. Responding, the bakers demanded a return to the previous *tarif* and its much higher prices. Various authorities weighed in, agreeing that the commission had been over-zealous in its price cuts. The final *tarif*, adopting the counsel of a local magistrate, reflected a compromise, but nonetheless lowered prices noticeably. It pushed the price of second- and third-quality loaves down more than 20 percent in some instances, while the decreases for first-quality loaves ranged from 4 percent to 13 percent (Figure 4.3).[14]

The most incendiary revisions in the parlement's jurisdiction took place in the 1780s in Rochefort and La Rochelle, to the southeast along the Atlantic coast. (The two cities' *tarifs* had been linked since at least 1703.) After debilitating confrontations with municipal officials, the Rochefort bakers petitioned the Parlement to intervene in 1784.[15] The bakers could

[13] The many papers pertaining to these disputes are in AM Le Havre HH 25.

[14] Clermont-Ferrand Dossier (1771–1779), AN U 1441; BN JF 1162, pp. 14–5; Miller, "Politics and Urban Provisioning," p. 248.

[15] BN JF 1164, pp. 69–72; BN JF 1742, pp. 115–201; and Rochefort Dossier (1781–1782), AN U 1441.

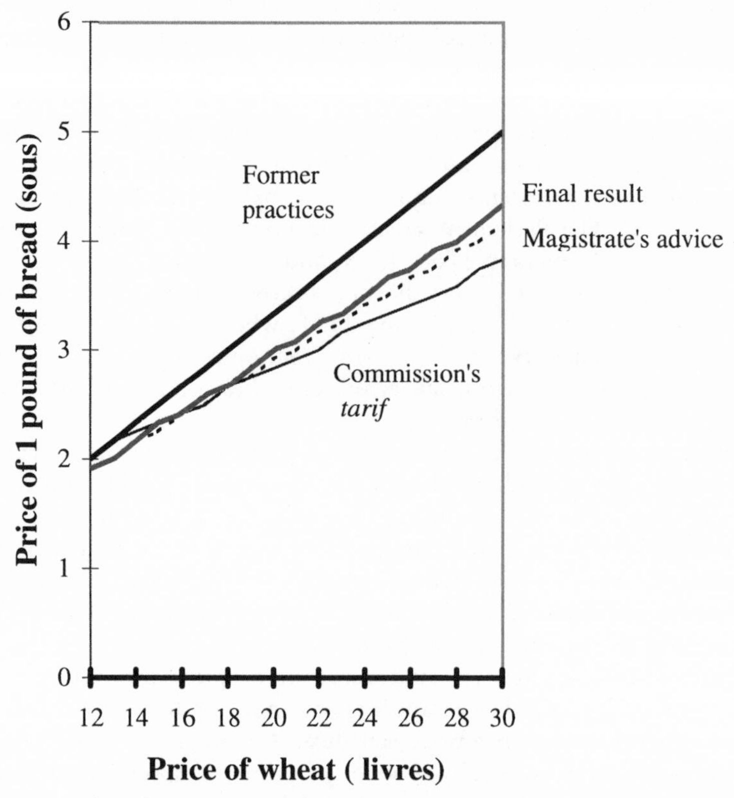

Figure 4.3 The Clermont-Ferrand *tarif* revisions. Sources: AN U 1441; BN JF 1162.

claim with a certain legitimacy that historical precedent favored an earlier set of *tarifs* with higher rates. Seeking to avoid such a conclusion, the Paris Parlement recruited the services of the Royal Academy of the Sciences for accurate assays. To the great dismay of the Parlement, the Academy determined that the bakers had been abused grievously. The Parlement reluctantly instructed Rochefort and La Rochelle to return to the earlier, higher price schedules. Several months later, when grain prices spiked in the summer of 1786, though, La Rochelle officials once again forced bakers to sell below cost. The baker syndics marched to Paris, returning triumphant not only with a parlementary ruling in their favor, but with an entirely new *tarif* that granted even higher prices. The city, filled with 10,000 migrant workers who had come to work on the harvest, exploded. For several days, the crowds rampaged, targeting the shops of the syndics and beating up the wife of a baker who had gone to Paris. The omninous warnings of the city's *procureur* had come perilously close to true. If prices rise, he had cautioned, "we will see massacres, streets filled with cadavers, my house burned and my life threatened."[16]

The controllers general and other finance ministers clearly understood the use of *tarifs* in keeping urban peace. Laverdy, the controller general who had permitted exports in his 1764 decree, had proposed to the Rouen Parlement that it hold the line against bakers, for instance.[17] Turgot, too, urged cities to review and lower the prices of their *tarifs*. When grain price rises had become noticeable in Caen in 1774, he recommended the intendant to have the *tarif* in use – it dated from 1743 – recalculated: "If you can find a way to have *le peuple* pay eight deniers per pound less than in 1743, they will not notice the increase in the cost of grain." The bailliage officials reviewed the 1743 *tarif* and declared it too high. Stunned, the Caen bakers petitioned the recently restored Parlement of Rouen, alleging not only that the revised schedule was unfair, but that the lieutenant of police was exacting revenge for the bakers' refusal to pay him additional fees to cover the price of his office. (Those allegations proved to be true, and the Rouen Parlement eventually punished the lieutenant of police for abusing his authority, and especially for attempting to extort additional fees from many Caen guilds.) As for the bread prices, however, the Parlement sided against the bakers. It oversaw an assay, and the resulting price schedule slashed prices between 4 and 22 percent, with the greatest decreases when grain was most expensive. These revisions were harsh; even the intendant expressed his disbelief. The new Controller General de Clugny, however, praised the parlement's *tarif* and circulated reports of the outcome. He suggested it would "serve to curb the plans of any bakers in any other city who would like to demand a larger profit," telling words that indicated the minister's desires and his understanding of the effectiveness of this tool.[18]

There were numerous other attempts to use *tarif* revisions to lower the price of bread in cities and towns across the kingdom. Laval's litigation over *tarifs* continued for almost forty years, from 1749 to 1786. Shorter, but no less heated, skirmishes emerged in Tours, Carcassonne, and Marseilles. Following the report by the Royal Academy of the Sciences, Chief of the

[16] BN JF 1161, p. 12; BN JF 1742, pp. 92, 192–3; BN JF 1743, pp. 108–109, 116.

[17] He noted in 1768 that the Rouen *tarif* gave bakers inappropriately high profits and urged Miromesnil to oppose any demands by bakers for higher rates. Miromesnil, *Correspondence*, 5: 127.

[18] "Etat du pain cuit et distribué à Roissy depuis le ler juillet 1775 jusqu'au 11 août suivant," AD Somme C 88; Turgot to the Intendant of Amiens, 17 September 1775, AD Somme C 91; AD Calvados C 245, pp. 184–91; CG to the Intendant of Caen, June 19, 1776, AD Calvados C 245; "Requête des boulangers de Caen," 31 July 1775, AD Calvados 1B 1665 B. Turgot, *Oeuvres*, 4: 229–31, 494–95. Further documentation is scattered in AD Calvados 1B 2074 B; AD Calvados 6E 53 (19); and ADSM 1B 290, 291, 293, 1626.

Royal Council of Finance Charles Gravier Comte de Vergennes pressed officials in many bourgs to draw up clearer, more legitimate price schedules. Administrators in Châlons-sur-Marne, Troyes, Montauban, Château-Thierry, Epernay, and Bar-le-Duc heeded those instructions.[19] Lesser towns, including Caudebec in Upper Normandy, and those within the Paris Parlement's jurisdiction, such as Epernay, Saumur, Chaumont, Crécy-en-Brie, Thouans, Vertus Comté, and Chateaurenard (whose bakers numbered all of three), were told to proceed with *tarifs* or more regular pricing procedures.[20]

Bread Prices and the Flour War in Paris

The most famous episode of the Flour War – the riots in Paris on 3 May 1775 – revealed the risks at stake. In general, the capital's bread prices were held lower than those of the surrounding area, despite the bakers' higher rents and other costs of production. Whether or not there was an official price schedule in effect in Paris, there was a tacit understanding that the police and the parlement would oversee the baking trade and hold bread prices in check.[21] Turgot's repudiation of that understanding led to the

[19] The Academy's conclusions, especially those favoring higher prices for bakers, did not please local authorities. The Academy's position was not, however, a repudiation of the officials' position. Instead, the Academy pushed for higher prices for second- and third-quality bread to encourage bakers to use better flour. This would, it reasoned, lead to a better nourished, more productive and less restive urban work force. Many local officials found these recommendations ill-suited to their immediate needs, however laudable in the long run. Tillet to de Vergennes, 11 September 1784, AN F11 203; Tillet, *Expériences et observations sur le poids du pain* . . . (Paris, 1781), BN Collection Le Senne, 8 Z 4275; BN JF 1742, pp. 38–46, 52–3, 55 and 59–60; BN JF 1743, pp. 1–9, 115–23, 126 and 159–60; 14 February 1786, Bar-le-Duc and Arrêt du Parlement de Paris, 30 August 1786, AD I 23 A; "Observations sur une épreuve . . ." (Carcassonne), AN AD XI 68; Marseilles, 9 May 1777, AN AD I 23 A; "Mémoire pour la communauté des maîtres boulangers," 19 February 1789, in *Les Elections et les Cahiers de Paris en 1789*, 2: 553–62.

[20] BN JF 1111, "6ième liasse;" BN JF 1161, pp. 10, 15–21; BN JF 1163, p. 18; BN JF 1742, pp. 11–12 and 20; BN JF 1743, pp. 108–114; ADSM C 102, pp. 14, 24–5; Miromesnil, *Correspondance*, 5: 80.

[21] Because the records of the municipal administration were destroyed in 1871, more conclusive evidence about Parisian practices can only be pieced together from fragmentary sources. The records of a 1757 assay (but not the *tarif* itself) are in a report by the Lieutenant General of Police of Alençon. A different *tarif* outlined by Turgot and De Clugny yielded very low rates, and Turgot discussed Parisian prices at some length in a 17 April 1775 letter to the Intendant of Champagne. His explanation implied a set of controls was in place in the capital. The city's limited records reveal that actual prices may have been higher and that the bakers were left alone in good years. Even without a formal *tarif*, the Châtelet police indicated appropriate price ranges, and when necessary, imposed a *taxe*. The precise status of the ceilings in 1775 is unclear. Lieutenant of Police Lenoir, who had the misfor-

bloodshed of early May in the capital. It seems that for several months, Louis XVI had supported Lieutenant of Police Lenoir's desire to hold bread prices down. He even may have pledged to maintain bread prices at two sous per pound when crowds attacked flour warehouses in Versailles, or, at least, so the rumors ran. Instead, on 2 May Turgot abruptly ordered the city's bakers to sell at whatever price they desired. He promised to place armed guards outside bakeries and to compensate owners if their wares were pillaged. Overnight, the bakers increased their rates from 13.5 to 14 sols per six-pound loaf. On the morning of 3 May, thousands of enraged buyers roamed from bakery to bakery, in and out of market stalls, forcing doors, shattering windows, seizing loaves, and slitting sacks of flour. The restored Paris Parlement took their side and urged the king to "reduce the price of bread to a rate proportionate to the needs of the people." Turgot requisitioned troops, posted them at bakery doors, arrested instigators, and then fired Lenoir.[22]

The riots in Paris and La Rochelle reveal the tensions surrounding the bread *taxe*. The issue was not solely that prices were high; it was also that the municipal and parlementary officials were clearly willing to use rewritten *tarifs* and the setting of the *taxe* to protect their people from the worst ravages of shortages. This was a responsibility that had not lessened in any way after 1763. Indeed, given the continued reliance of local officials on *tarif* reforms in the late eighteenth century, the message was at some

tune to hold office during the Flour War of 1775, reported simply that the city no longer set the *taxe*. Yet, the riots of 3 May 1775 were triggered by Turgot's instructions to allow bakers to sell for whatever price they wanted. In any event, an anonymous administrative report expressed great concern that low profit margins were forcing the capital's bakers out of the business. (That memoir may have led Turgot to suspend the procedures in 1775.) Whatever the specific practices in a given year, bakers were fined intermittently for selling loaves above the *taxe* during the 1770s and 1780s. *Sentence de police*, 28 June 1731, 4 Z Le Senne 2263; *Ordonnance*, 18 July 1631, BN JF, 1107, p. 126; "Avertissement," 1765, BN Mf 11347, p. 195; BN Mf 21638 (Collection Delamare), especially p. 218; 6 February 1784, AN Y 9497; 14 April 1741, AN Y 9499; AN AD I 23 A, especially the *sentences de police* of 11 June 1773 and 15 December 1775; "Prix du Pain," September 1774, AN F11 265; De Clugny to Intendant Fontette, 11 July 1776, AD Calvados C 245; Turgot to the Intendant of Amiens, 17 September 1775, AD Somme C 91; BN Mf 21638; Turgot, *Oeuvres*, 4: 491–2; Robert Darnton, "Lenoir;" *Les Elections et les Cahiers de Paris*, 2: 551–2, 554; Kaplan, *Bread*, 1: 328–39 and 2: 483–4, 507–9, 522.

[22] The Controller General later struggled with the parlement over the proper venue for prosecuting the wrong-doers, finally succeeding in a *lit de justice* to have them tried in the *prévotal* court, rather than by the more lenient parlement (a measure that was annulled the following November). AN Y 10558; 3 May 1775, BN 4 Le Senne 2263; BN Mf 6877; Turgot, *Oeuvres*, 4: 421; Bouton, *Flour War*, pp. 88–9; Turgot, *Oeuvres*, 4: 421, 474–5; Vladimir S. Ljublinski, *La Guerre des farines*, pp. 305–51; Rudé, "La Taxation populaire de mai 1775 et dans la région parisienne"; Faure, *Turgot*, pp. 250–1; Darnton, "J. P. Lenoir;" Stone, *The Parlement of Paris*, pp. 132–3.

variance from that of the Physiocrats' treatises. By lowering bread prices, urban authorities communicated unequivocally that they were willing to take steps to shelter their communities from grain price rises, whether due to shortages or to free trade. Given the context – that of free trade reforms on the one hand and *tarif* revisions on the other – many consumers drew the very clear conclusion that free trade was to be tempered and adapted to their needs.

The *tarif* revisions and related practices had advantages beyond those of keeping bread prices within afforable ranges. They allowed authorities to avoid public confrontations, whether with bakers or buyers. Bread prices – announced by town criers, posted in shops, and enforced by the police – were a visible demonstration of the public authorities' concern for their people. But, while they were an unmistakable display of official partiality, they also conveyed a sense of routine and regularity, even in the most extreme circumstances. They did not invite raised voices and fists in market squares. Any overwrought protests by either bakers or buyers were to be limited to police bureaus and courtrooms.

Authorities recognized the dangers inherent in marketplace theatrics. They could have predicted the outcome of the ill-considered actions of a hot-tempered Caen police *commissaire* in 1772. The *commissaire*, furious that the bailliage court in Bayeux had yielded to bakers' pleas to raise the *taxe*, canvassed the city's bakeries with a crowd of several hundred angry buyers in his wake. He encouraged his expectant audience to refuse to pay the higher prices. They responded: Riots shook the city for three days.[23] The *commissaire*'s imprudent mission was similar that of the Honfleur magistrate in 1757. Essentially, both men had contrived to use the buyers' wrath to force bakers to lower their prices. Both efforts backfired. These sorts of disturbances, possibly avoidable, were precisely what many authorities feared. They could not maintain order by drawing consumers into the quarrels with bakers. They had to remain behind their desks, penning sentences and summoning bakers to their bureaus. Public order depended on private accords, urban calm on administrative harmony.

The public authorities in smaller towns and cities would have concurred. It fell to the local officials to determine how far free-trade doctrines could be allowed to transform market practices. Paterelle, De Franconville, and Doré had reshaped free trade to accommodate the needs of their towns – finding ways to lower prices, force sales, and guarantee modest protection to their townspeople. Their counterparts in cities – the lieutenants of police and the parlementary presidents – crafted bread *tarifs* to buffer the impact of high prices in cities. These many officials, urban and rural, addressed the

[23] The police *commissaire* was eventually punished. See the dossier in AD Calvados 1B 1665A.

problems as they understood them. If they could not intervene directly in the grain trade – and even before 1763, many of them opposed direct intervention – and if they were barred from simulating sales, they still had to find ways to feed their townspeople. In cities, this realization caused authorities to look to the traditional regulations of the baking trade. There they found a still-viable means for combating hunger and violence. Timely *tarif* revisions and the adept use of price-setting allowed them to cushion their cities from the worst that free trade and shortages might bring.

Free Trade on the Eve of the Revolution

Stepping back from these episodes – from the marketplace harangues of subdelegates, from the midnight discussions of exhausted intendants and parlement *présidents*, from the bakers' incessant complaints of the losses the bread *taxe* cost them, and from the petitions of at least one irate grain merchant forced into hiding – what is the meaning of the activities of the eighteenth-century grain trade? What do they tell about about the nature of the state and the economy as the Revolution approached? There are at least two conclusions to be drawn, first about the nature economic change, especially the forces that drove free trade, and second, about the nature of the French state on the eve of the Revolution.

This analysis of the eighteenth-century grain trade provides a concrete sense of how free trade was defined, and how that definition came into being. While one might argue that humans pursue their interests as they understand them – and admittedly, that is in itself a highly problematic statement – the framework through which they can proceed and the ways in which their efforts were understood are ever changing. The activities and discussions of Upper Norman markets reveal specific circumstances and channels. First, there was an ongoing process of accommodation to market concepts that had probably existed for many centuries, but at the very least from the late seventeenth century. This process continued through the steady refinement of the strategies that local authorities hoped would work to supply their markets. This was an uneven process. Even though Desmarets rejected severe measures in 1709, he eventually yielded to the crisis, and imposed them. Throughout the eighteenth century, more nuanced means were reworked to good effect. The Physiocrats accelerated the process of change, but did not lead it. In some sense, they, and in particular their minister-advocate, Turgot, forced authorities – the intendants, their subdelegates, village magistrates, and an occasional parlementary *président* such as Miromesnil – to make clearer sense of what free trade was to mean and pushed them to tread a finer interpretive line than had their predecessors.

In the years before Turgot, the sense of market mechanisms, and of the need to shape (not thwart) them, had gained credence. While there was no coherent concept of free trade in and of itself, most authorities evinced a general faith in *libre circulation* and in some of the more fully developed claims of early liberal theorists such as Cantillon, Herbert, and (finally) Quesnay and Mirabeau. Through these authorities' attempts to amass and distribute grain supplies, and to encourage merchants to do the same, they had been involved in an extended, meaningful, and probably conscious series of experiments to test the nature of the market and their own capacities to stabilize it. They had concluded, even in 1757, that the owner's perceived right to name a price could only be violated at great risk. Thus, their efforts centered on simulated sales – a means to influence prices, without attacking them directly.

Their definition of free trade relied on ministerial language and directives, a smattering of economic theory, and on their own knowledge of what brought buyers and sellers to their market halls. Perhaps some of them had heard of Quesnay, or had skimmed pages of Dupont de Nemours' *Ephémérides du citoyen*. Certainly Intendants Feydeau de Brou and De Crosne and *Premier Président* Miromesnil had more than a passing familiarity with enlightenment thought and ministerial battles. One would search in vain for any mention of the Physiocrats in the correspondence of the Upper Norman subdelegates, however. The letters of these lower-level officials make little mention of even the free-trade laws themselves. The anguished questions of a Gournay subdelegate, Paterelle, though, bespeak at least one uneasy confrontation between a small-town official, buyers, and liberal theories. Assembling the price records from 1770 to 1775, Paterelle demanded to know what supply, at what moment in time, served as the basis for a market price. "If one attributes the cause of the steep price rises to the poor harvest of 1774," he reasoned, "the public answers that the state of the harvest was known in October, during which the average price of the four markets was 4 livres 10 sols per boisseau." By April, it had reached 6 livres 2 sols, prompting his question: "If it is the same harvest, why is it not the same price, more or less?"[24] Turgot had argued that supplies and prices moved inversely, and Paterelle's query implies some contact with the vocabulary and content of free-trade theories. But to this subdelegate, struggling to assess whether his own market conformed to expectations, the problem was not merely speculative. It concerned more than proposals to promote long-term agricultural productivity. It had immediate material consequences for the daily lives of the impatient villagers he faced each week as he rode across town to the marketplace in the midst of the Flour War. The price, and the "master's" right to set it, had become the

[24] Paterelle to intendant, 4 June 1775, ADSM C 107.

standard by which his market, and his every action, were to be judged, and he knew it.

How had the sanctity of the price become the standard of eighteenth-century free trade? The answer lay in the process underway over the long course of the eighteenth century, culminating in simulated sales. As early and mid-eighteenth century authorities pondered how to secure grain and flour for their markets, they winnowed the list of acceptable and effective measures. The broad range of 1709 strategies narrowed. Increasingly, authorities used methods that imitated the normal practices of commerce. Simulated sales, if well-concealed, could supplement purely commercial undertakings. Owners would be none the wiser, prices would fall, and tranquility would be assured. Authorities allowed the augmented supply to act in their stead, and thus they could refrain from interferring directly with prices. That strategy – clear in 1757, and fine-tuned in 1768 – placed enormous emphasis on restraint with regard to any intervention in prices. As the implications of free-trade policies, and especially Turgot's stricter guidelines, became known, the process proceeded further along that path. By 1775, free trade had come to have two related rules: Prices could not be set and owners could not be forced to supply markets. It is reasonable to conclude that the processes developing before 1763 – the rejection of aggressive measures, and the focus on prices that underlay simulated sales – prepared the way for the definition of free trade that developed under the dual pressures of shortage and Physiocracy.

Not surprisingly, simulated sales received a second wind after Turgot's dismissal, although the authorities exhibited less skill and coordination than they had in 1768 and 1772. In 1784, prices once again soared in northern France, the result of a midsummer hailstorm that destroyed the region's entire crop in a few hours. Grain from the government's *régie des vivres* was dispatched to towns near the Channel, with the halls of Montivilliers and Rouen as its distribution points. To De Crosne's distress, the merchants involved unwisely "took a great pleasure in announcing [the grain's arrival] publicly" wherever they went, thus endangering the shipments and the success of the sales. As more grain arrived in Channel ports, the intendant demanded that it be sent "as safely and as secretly" as possible to its many destinations.[25] The distribution of the grain did not follow earlier clandestine strategies either. In Doudeville, the subdelegate and a *curé* sold off their allotted wagon-load directly to the poor. As De Crosne could have predicted, the gathering crowd soon imposed lower prices.[26] The actions of the

[25] Oursel to intendant, 23 May 1783; and intendant's reply; Gilbert, curé to intendant, 23 May 1784, ADSM C 106, pp. 206–8.

[26] The subdelegate and the *curé* requested troops for the next distributions. Theroulde to intendant, 15 May 1784, ADSM C 106, p. 299.

subdelegate at Les Andleys, however, were somewhat more successful. When he told local grain owners that low-priced government imports would soon rob them of profits, they hurried their supplies to market.[27] A Saint-Lô official in the far western area of Lower Normandy, too, could claim some victories. Explaining to the intendant that he did not want to drive away local farmers, he used intermediaries to sell small amounts of rye at the asking price through the hall.[28] Overall, however, the measures during the 1780s did not work as well as De Crosne had wanted. Perhaps the experience of that preceding generation had been lost in the intervening decade and a half – although Le Chevallier was still present at Honfleur. The names of many of the other officials were new, and so their experience is hard to assess.

The activities of the eighteenth-century grain trade in Upper Normandy give specificity to any conclusions about the evolution of free trade. The course of liberalization across France was complex and took many decades; there was no single moment when free trade came head to head with outdated and obstructive regulations. But that does not mean there were no important waystations in the process. Desmarets' frustrated 1709 efforts held within them the possibilities for change. By 1757, those possibilities had emerged as simulated sales. By 1775, simulated sales had prepared the way for a definition of free trade based on the sanctity of the owner's price. There would be other moments of consensus and illumination in the Revolution and the nineteenth century, as the wisdom of the Old Regime was discarded and then taken up again. Nevertheless, the process of transforming unworkable policies into ones with the possibility of success was underway.

There is then the question of the French state, a matter that is presently very important for scholars of the Old Regime and the Revolution. The monarchy had set the grain trade as one of the public tests of its commitment to enlightenment, progress, and fiscal reform. Its apparent failure placed one more black mark in the Crown's column of losses, and Steven Kaplan's exploration of the impact of the liberal reforms made a substantial and early contribution to our understanding of the monarchy's struggles. Had not the Physiocrats and Turgot proclaimed with great confidence that free trade would feed the starving, while simultaneously restoring the monarchy's treasury and thus its legitimacy? Had not Calonne returned to the problem of the grain trade for one last time in 1787, vanquishing doubts – *fausses lumières*, he called them – siding with "experience" and "truth" and deciding "that the time had come to establish the principles on this matter?" Louis XVI's resulting legislation reinstated export and declared

[27] Subdelegate Les Andelys to intendant, 26 May 1784, ADSM C 106, p. 215.
[28] Lantier, "Saint-Lô," p. 22.

"that the freedom of the grain trade should be regarded as the usual and normal state in our kindgom."[29] Within months, of course, that "normal and usual state" was abandoned and export was again suspended, another illustration of the failures of eighteenth-century provisioning policies.

The wavering policies for export could be added to the list of other such defeats and confusion: the parlementary repudiation of free trade in 1768, Terray's eventual return to regulation in 1770, Turgot's humiliating dismissal in 1776, and Necker's reinstitution of the guilds after 1778. All of those vacillations became useful ammunition for the Crown's many enemies, the Parlement of Rouen among them. Thus, royal indecision and ministerial missteps could be employed to confirm the prevailing scholarly paradigm of the monarchy's painful ideological collapse in the eighteenth century. Yet local authorities evinced no such sense of failure – intermittent confusion and frustration perhaps, but not failure. They heeded the orders of controllers general, but found their own way to impose a form of free trade they deemed appropriate. They had been doing so before Bertin's 1763 policies appeared, and continued after Quesnay, Turgot, and others had departed in disgrace. Many of them had been skeptical of heavy controls, and would not have undertaken any aggressive measures; a few were well-versed in liberal doctrines and could argue deftly for *liberté*. They took the changes in stride. Each time new legislation found its way to their desks, they evaluated it in light of their understanding of their town and its place in surrounding market networks. Like the *inspecteurs des manufactures*, like the guild officers of Lille and probably elsewhere, the authorities who supervised the Old Regime grain trade were open to innovation.[30] They took what they knew of the marketplace, of self-interest, and of hunger and created a circumscribed definition of free trade that they could carry into grain halls and bakeries.

One might object that these authorities' measured responses did not work. After all, riots and shortage continued, and the Old Regime expired in the midst of the some of the worst shortages of the century. Yet, the strategies worked better than the plans of local officials who did not intervene or who waited until late in the game. De Fraconville, for instance, credited his efforts and those of his *hallier* with sparing further agony in 1775. At the very height of the Flour War, they had enabled sales to continue "without tumult, and without fanfare."[31] To have relinquished the

[29] "Déclaration du Roi pour la liberté du commerce des grain," 17 June 1787, ADSM C 103, p. 3.

[30] Philippe Minard, "L'inspection des manufactures en France, de Colbert à la Révolution," Doctoral Dissertation, Université de Paris-I, 1994; Gail Bossenga, "Protecting Merchants: Guilds and Commercial Capitalism in Eighteenth-Century France," *FHS* 15 (Fall, 1988): 693–703.

[31] De Franconville to intendant, 6 May 1775, ADSM C 110.

marketplace to the workings of supply and demand alone would have increased the violence and suffering that were chracteristic of any shortage. Authorities developed means to ease their cities and towns toward freer trade, while refining the protection they could offer. Their choices reflected those conflicting pulls, and were often made under duress. There is nevertheless a consistency in their responses even in the most turbulent situations. They moved from a wider to narrower range of controls – from visible to clandestine means, from erratic to more routine methods, and from those that the interfered with market forces to those that simulated them. This was not a perfectly smooth process, of course, and was punctuated by policy reversals and an occasional abrupt return to harsh regulations and riots.

Were the officials' efforts ultimately satisfactory? Not completely and not without effective enforcement. The lack of troops played an essential role, although there is some evidence that there too, strategies were being refined. At some points, the distribution of troops may have been more precisely organized than that of grain. The lists of the *frais de justice* connected with food riots suggest more aggressive – or at least more costly – efforts by the courts to round up and try the alleged guilty.[32] The police official, Des Ervolus – "*homme aussi actif que sage*," wrote Miromesnil – repeatedly came to the fore in Upper Norman frays. In 1768, realizing that his men were needed simultaneously in Rouen and in nearby Darnétal, he cleverly dispatched half of his detachment to Darnétal, while ordering the others to make a display of marching around Rouen all night. Hearing the constant thud of boots in the mud and the call of sentries, the Rouennais did not suspect that the city was only partially manned, and they halted their assaults on grain and cotton warehouses.[33] In La Roche-Guyon, site of the 1775 attack on Planter's barges, Des Ervolus managed to extort restitutions of 500 livres from the townspeople, and continued his mission of intimidation throughout the summer. He rounded up his suspects, dragged them before the *curés*, and inspired the Old Regime's "salutary fear" in them. Those efforts earned him a nomination by the governor, Maréchal d'Harcourt, for letters patent of nobility.[34]

During the height of the Flour War, Neufchâtel Subdelegate Bequel sent soldiers to the market in advance of its opening, had them hide their rifles in nearby buildings, and then blend in with the crowd. The detachment's drummer remained alert and ready to beat the alarm. Bequel also received

[32] Even before the Flour War, De Crosne blamed "the dearness of grain" and unrest for the rising price of justice. In 1754, the total had been 25,328 livres; in 1773, 69,597 livres. See the correspondence in ADSM C 945.

[33] Draft, Miromesnil to the Chancellor et al., 24 March 1768, ADSM C 107.

[34] Des Ervolus to De Crosne, 24 June 1775, ADSM C 108 p. 310; "*Etat* of grain stolen, lost and of restitutions," folio in ADSM C 107; "*Etat* of requests that the maréchal d'Harcourt proposes to make," [n.d. 1775], ADSM C 107.

Harcourt's recommendation for a reward.[35] In 1784, the Dieppe municipal officers stationed twelve riflemen strategically in the swarming hall. At the first sign of trouble, a group of police arrived, swiftly grabbed one buyer armed with a knife and turned her over to riflemen, who led her directly to the town hall. The day's market proceeded peacefully, although the city officials begged for more grain from ongoing operations.[36] Of course, there were not enough soldiers to organize such tactics at every market. In the absence of adequate troops – and in any event, no official would have wanted free trade to be enforced at musket point – local authorities were substituting shrewdness for manpower. They did what they could to mobilize police *commissaires*, the unwilling *milice bourgeoise*, and soldiers effectively, and to link them in the strategies they hoped would prevent the evils that hunger incited.

Thus, even as the monarchy seemed to collapse, local authorities shouldered the enormous burden of confronting contradictory, even unworkable, policies for many decades, and had created some hope for order in their cities and towns. The Physiocratic reforms were abstract and perhaps impracticable on the ground, and local administrators felt this most keenly. Even the most enthusiastic proponents, such as Intendant de Crosne, were without clear answers to many practical questions. Nonetheless, these officials were not without cunning and expertise, and so the story cannot end with the inconsistencies of royal policy at the end of the Old Regime. Even as the monarchy's disintegration began, its representatives were looking skillfully for ways to carry out its bidding. Where the Crown failed to provide guidance, these men marked off a path. When the Crown faltered, they carried on.

[35] Bequel to intendant, 6 May 1776, ADSM C 110; "*Etat* of requests that the maréchal d'Harcourt proposes to make," [n.d. 1775], ADSM C 107.

[36] Maire and Echevins of Dieppe, 22 May 1784, ADSM C 106, p. 22.

PART TWO

Maximum: Feeding France in Revolution and War

To write of the Revolutionary economy is to invoke the controls and animosities of Year II (1793–4). Restless lines outside bakeries, price lists for the General Maximum, Parisian sansculottes jeering at the finer coat of a more fortunate *citoyen*: These are the stock images of the Revolution's economic iconography. By 1793, the early Revolutionaries' vehement insistence on free trade had been brushed aside, and louder, more radical, voices prevailed. In that year, the Jacobins of the Convention seized the opportunity to transform popular calls for economic justice into a political platform that helped propel them to the leadership of the Convention. Within months, that body established the heavily regulated economy of Year II, with its price ceilings and requisitions, and bent them to the needs of cities and armies. The language of these years combined mythic republicanism with the melodrama of the Old Regime: starving families at the mercy of an assortment of tyrants – *vils affâmeurs*, *vampires*, *aristocrates*, *capitalistes*. The words even defy translation, so charged were they with meaning. The language was one of oppositions and absolutes. Roughly, the selfish had allied against the poor and powerless. The rhetoric of the Revolution admitted little discussion about which of the needs of merchants should be addressed, and which could be challenged or ignored, nor how to evaluate the demands of hungry populaces for food. The careful balancing of strategies, interests, and laws that had been the center of Old Regime provisioning practices disappeared. The circumstances themselves allowed for little reflection. The avalanche of crises – shortages, inflation, political fissures, war, and unrest – demanded constant action. The men of '89 and of '92 and '95 would struggle with the burdens that the accumulation of adversities had left them. They would not have the experience of the Old Regime – for better or worse – to guide them.

CHAPTER FIVE

1789: Municipal Revolutions and the Origins of Radicalism

"... le vertige, la force, l'injustice prennent la place de l'ordre ..."[1]

What had become of the carefully calculated measures of the Old Regime? By July of 1789, they seem to have vanished. The opening months of the Revolution witnessed the rapid-fire radicalization of provisioning policies among the new administrators and their restive citizenry. There were several elements that came into play in the space of only a few weeks, and that determined much later provisioning policy. In brief, these elements eliminated the structures, practices, and personnel that had guided the grain trade during the Old Regime. They included the shortage of 1788–9 itself, Controller General Necker's return to aggressive intervention in the autumn of 1788, and the political upheaval that began with the Assembly of Notables and the Provincial Assemblies and extended throughout the municipal revolutions of the summer of 1789. That process not only vanquished the Old Regime authorities, but it split townspeople into numerous warring factions, and essentially closed off possibilities for the moderation of the Old Regime's provisioning tactics and for it to continue in the new regime.

In the background was hardened frustration over the free-trade policies of the late 1780s, the result of efforts by Calonne, the Provincial Assemblies, and the Assembly of Notables to enact fiscal and economic reforms. The Eden Treaty of 1786, for instance, had exposed French manufacturing to competition from British goods. It had made working people, particularly those with any connection to the textile trades, hostile to economic liberalism. That animosity had only increased with the king's promulgation of legislation that once again permitted the export of grain in June 1787. The

[1] The former municipal officers of Torigni, cited in Gabriel Desert, *La Révolution française en Normandie* (Toulouse, 1989), p. 85.

harvests of the mid-1780s, with the exception of 1784, had been generally good, and many judged that circumstances boded well for reforms. The Provincial Assembly of the Generality of Rouen, dominated by merchants and businessmen, for example, had concluded optimistically that "grain is not expensive, and . . . one cannot fear a dangerous revolution" in 1787.[2] By the spring of 1789, however, a revolution of a wholly different nature had arrived, and it swept away the nuanced understandings and policies of the Old Regime's intendancies, subdelegations, and police bureaus.

The Growing Crisis, 1788

The initial steps that led to the rejection of moderate practices came on the eve of the Revolution. Two bad harvests in 1787 and 1788, the last further damaged by a hailstorm in northern France, drove prices up sharply. In the autumn of 1788, Necker, as the newly appointed Controller General, swiftly closed borders and restricted sales to markets. He had never supported the full blown free trade of the Physiocrats, and had instead argued for policies that addressed circumstances. The 1788 shortage, he reasoned, demanded controls. The *arrêt* of 23 April 1789 permitted local authorities to "use the power they held" to bring grain to market. Admitting that this violated "the most perfect freedom which each proprietor . . . should enjoy," Louis XVI nonetheless allowed grain censuses and, if necessary, allowed officials to force "those with grain stored to supply markets."[3] This was a profound reversal of the definition of free trade that had been hammered out at the local level.

The 23 April *arrêt* not only encouraged authorities to scour the countryside in search of grain, it also signaled to them that the time for moderation had ended. The parlements, once again returned from exile, were anxious to take on the monarchy, and used provisioning policies as one means to that end. Among their traditional roles had been the supervision of the grain trade, and they were not going to renounce those claims in the tense spring of 1789. Rival courts faced off: The Paris *présidents* insisted that *libre circulation* be maintained, and reimposed the regulations that had been lifted during their exile.[4] Resisting such pressure, the Parlement of Burgundy, for example, issued an *arrêt* on 31 March 1789 that halted any

[2] "Procès verbal des séances de l'Assemblée provinciale de la généralité de Rouen," November–December 1787, ADSM C 2111.

[3] "Arrest . . . concernant les grains et l'approvisionnement des marchés," 23 April 1789, Archives de la Seine et de la Ville de Paris (ASVP) D^5 Z 9; Clay Ramsay, *The Ideology of the Great Fear: The Soissonais in 1789* (Baltimore and London, 1992), pp. 60–3.

[4] "Arrêt de la Cour de Parlement," 18 December 1788, ASVP D^5 Z 9.

grain shipments leaving its jurisdiction.[5] As food riots recommenced that spring, other parlements – such as those in Rouen and Lille – refused to punish unrest severely, seeming to blame the monarchy for the shortage.[6] The enormous pressure of the needs of metropolises heightened concerns, and every incursion from Paris, Lyon, or Rouen brought immediate reactions from local bodies intent on safeguarding their dwindling supplies. These parlementary tacks served multiple purposes, not the least of which was to check royal policies and authority, while holding onto scarce resources. The cutting off of shipments between regions had a domino-like effect across France. Each border closing triggered other ones, and each act of intervention encouraged others in neighboring towns.[7]

Necker's return to regulation – and its implied consent for harsher measures at the local level – took place in the context of deepening antagonisms between local elites and the widespread politicization of 1788–9. The events of the late 1780s dissolved any remaining goodwill among elites and between the various Old Regime institutions. The creation of the provincial assemblies in 1787 elevated traditional rivalries to a new level. These assemblies were organized by the Crown to assist with much-needed fiscal reform. Quickly, they turned to consider a wide range of economic questions, and generally supported liberal measures.[8] In Rouen, for instance, the Assembly drew up plans for standardizing weights and measures, and pondered eliminating the *tarifs* of the baking trades.[9] Their schemes generated immediate opposition from the parlements. The Rouen *présidents* found it unbearable that members of the Third Estate, such as the lawyer Thouret, the merchant Planter, or the Protestant *échevin* Le Couteulx de Canteleu, dared to pronounce upon matters of such import.

That the assembly's activities had been greeted warmly in the region, and had elicited much comment, especially in the local press, caused the *présidents* even greater alarm. At the same time, of course, sovereign courts across the kingdom, Rouen's noteworthy among them, were preoccupied by their own lengthy struggles with the Crown over loans and tax reform. Those battles had brought a new run of transfers and exiles, and finally Keeper of the Seals Lamoignon's attempted coup, which stripped the Parlements of much of their power in May 1788. The intendants were soon deeply involved. In Rouen, Intendant Maussion – from a dynasty of Parisian

[5] Bouvet, *Maconnais*, pp. 50–1.

[6] Lemarchand, *La fin du féodalisme*, p. 417.

[7] Ramsay, *Great Fear*, pp. 26–9; Bouvet, *Maconnais*, pp. 54–5; *Documents relatifs à l'histoire des subsistances dans le District de Bergues pendant la Révolution (1788-An V)*, 2 vols., Georges Lefebvre, ed. (Lille, 1914), 1: 39.

[8] Jones, *Reform and Revolution*, pp. 119–20, 139–56.

[9] ADSM C 2111 and 2120; *Procès-verbal des séances de l'Assemblée provinciale de la généralité de Rouen . . . aux mois de Novembre et Décembre 1787* (Rouen, 1788).

magistrates, and who had only arrived in 1787 – became the target of local jokes when he tried to force the *procureur du roi* of the bailliage of Cany, Cherfils, to side with the Crown over the parlement. By the autumn of 1788, a comedy "Le Grand Bailliage," regaled Rouen audiences with its depiction of the crafty Cherfils' victory over the blustering Maussion.[10]

Thus, by late 1788, the region's many administrative and judicial elites were at loggerheads, and the wealthier bourgeois, especially those involved in the provincial assembly, had been wounded by the parlement's contempt for their efforts. These fissures deepened in the pre-Revolutionary debates over voting rights and the preparations for the Estates General. The broad framework of these conflicts is familiar to any historian of the period. Louis XVI's calling of the Estates General, the elections of the winter and spring of 1788–9, and the drafting of the *cahiers* brought about the collapse of royal, and finally local, government. Pamphlet and press wars built on these battles and further politicized, even polarized, the nascent polity. No sooner had the deputies to the Estates General arrived at Versailles in May 1789 than the entire proceedings disintegrated over procedural issues. The factions retired to the inner chambers of the palace, or charged into the clubs and electoral politics of the capital. Thus, the body that the monarch had convened as a solution to fiscal crisis became the site of fierce debate over representation, constitutionality, and the needs of the newly envisioned nation.

The Summer of '89: Elections and Desperation

The dynamic that emerged in these months essentially destroyed any possibility for the provisioning practices of the Old Regime to prevail. The essential fact was that the men who had created and refined those tactics were rapidly pushed aside, leaving less administratively experienced authorities to confront extreme shortage and violence in those months.[11] Some of the Old Regime officials retired in the first days; others were ousted in tumultuous elections later that summer and autumn. And, in general, any late elections were the result of demands for a "regenerated" government – often termed *comités permanents nationaux* – that reflected a broader

[10] J.F. Bosher, *The French Revolution* (New York and London, 1988), pp. 105–16; Léopold Soublin, *Le premier vote des normands (1789)* (Fécamp, 1981), pp. 27–38; Richard Mowery Andrews, *Law, Magistracy and Crime in Old Regime Paris, 1735–1789* (Cambridge, 1994), 1: 206–40.

[11] Lynn Avery Hunt gives a detailed analysis of the municipal revolution. "Committees and Communes: Local Politics and National Revolution in 1789," *Comparative Studies in Society and History* 18 (July, 1976): 312–46; *Revolution and Urban Politics in Provincial France: Troyes and Reims, 1786–90* (Stanford, 1978).

suffrage. By the autumn, there were few familiar faces in the city halls of France. And those that had been held over had to coexist with the newcomers. Adding to the instability, volunteer militias had formed, and were launching out on independent missions to secure provisions.[12]

In Upper Normandy, as was often the case elsewhere, the new cadres of men came from the moderate and progressive nobility, the liberal professions, and commerce. The electors from the Third Estate from Rouen and Le Havre, for instance, predominantly chose wealthy merchants and lawyers, reflecting the heavy parlementary and commercial interests of those cities.[13] These men of '89 may have had lengthy experience in the courtroom or at a *comptoir*. Some even – as in the case of Rouen *négociant* and manufacturer, Pierre-Nicolas de Fontenay, a deputy to the Provincial Assembly and then to the Estates General, appointed to the National Assembly's Committee of Agriculture and Commerce, elected mayor of Rouen in late 1791 and to the presidency of the departmental administration in 1792 – had a broad knowledge of markets and business practices across Europe. Joining De Fontenay were other equally distinguished members of the Third Estate: Rouen financier Le Couteulx de Canteleu, Le Havre merchant J.F. Begouën, and the lawyer Thouret, whose speciality was the Normandy Coutume. (Thouret would play an important role in the National Assembly's Constitutional Committee.) These men, however, did not have much experience with the complicated balancing acts of the many Old Regime officials. They had not spent hours contending with bakers; they had not ridden from farm to farm to pressure estate agents to load up wagons; they had not developed a network of intermediaries on whom they could rely for extra sacks of grain.[14]

The push toward radical measures came in large part from the violence that the combination of dearth and municipal revolutions had provoked. This upheaval generated not only rival groups of municipal officials, but also competing militias, and townspeople intent on making their needs

[12] An excellent overview of the formation of local militias in the Soissonais is found in Ramsay, *Great Fear*, pp. 215–40.

[13] Délibérations, Municipalité de Rouen, 18 July 1789, AM A 40; Lemarchand, *La fin du féodalisme*, p. 430; Guy Lemarchand and Claude Mazauric, "Le concepte de la liberté d'entreprise dans une région de haut développement économique: la Haute-Normandie, 1787–1800," in *La Révolution française et le développement du capitalisme*, pp. 135–53; Soublin, *Le premier vote*, pp. 237–40, 310–1.

[14] The lack of administrative experience was common in the 1789 municipal governments of Upper Normandy, regardless of whether the city's officials were drawn from the ranks of influential liberal nobles and businessmen, or from *corporations* and shopkeepers. *La Révolution en Haute-Normandie, 1789–1802*, 2nd ed. (n.p., 1989), pp. 35–46; Jacques Dimet, *1789: Evreux, La Révolution* (Paris, 1988), pp. 74–81; *Histoire du Havre et de l'estuaire de la Seine*, André Corvisier, ed. (Toulouse, 1987), pp. 137–40.

heard. The events in Rouen in 1789 illuminate the complicated dialectic between the provisioning crisis and municipal revolution as it developed over several months. Significantly, this instability resulted in a recourse to extreme measures by the autumn of 1789. The problems began in the spring with intense struggles over voting rights and procedures in the bailliage of Rouen. To a certain extent, they built on the earlier rivalry between the parlement and the provincial assembly.

These conflicts split the region's political classes. First, the judicial and administrative institutions battled over who had the authority to organize the electoral assemblies – all of which would determine who could vote and how that vote would be tallied. Second, a schism divided the wealthier bourgeois from the guilds. The Rouen bourgeois, anxious to silence the city's outspoken working classes, took numerous illegal measures to limit the guilds' representation in the elections for the Estates General – some even rejected formally by the king. The same bourgeois seized control of the drafting of the *cahiers*, further antagonizing the city's poor and working classes. The tensions between town and countryside were worsened by similar measures – the wealthy of Rouen marginalized the votes and viewpoints of their rural counterparts. The delegation that rode to Versailles left a city deeply damaged by these confrontations. The city's guilds considered the *échevins* to be their enemies, Intendant Maussion had been helpless to intervene, and a deep chasm had opened between the parlement and the city hall. This did not augur well in the midst of a shortage that demanded coordination and agreement between the oft-rival authorities.[15]

The city's plight worsened that summer. On 13 and 14 July, hundreds of men and women attacked textile workshops, smashed machinery, and marched to the docks, where they boarded ships holding grain. Disorder in Rouen and nearby suburbs did not abate for nearly a week.[16] The *troupes bourgeoises*, commanded by the parlement, refused to obey the the city's municipal corps and would not confront the crowds. The situation was not improved by the arrival of Marquis François-Henri d'Harcourt, the governor-general since 1783 and now the *grand-bailli*, who was also the tutor of the royal dauphin. He and his father – the former governor – had had an antagonistic relationship for many decades with Rouen officials. Harcourt, hearing of the unrest, rushed home to Rouen to help restore

[15] *Vivre en Normandie sous la Révolution*, Olivier Chaline and Gérard Hurpin, eds. (Rouen, 1989), 2: 327–8; Soublin, *Le premier vote*, pp. 138–46; Desert, *La Révolution française*, pp. 30–75.

[16] This story is pieced together from numerous, sometimes contradictory, sources: AN DXXIX bis 1, doss. 6; AM A 40; AM Rouen I5 6; E. Gosselin, *Journal des principaux épisodes de l'époque révolutionnaire à Rouen et dans les environs, de 1789 à 1795* (Rouen, 1867), pp. 10–8; *Vivre en Normandie*, 2: 329–32; Soublin, *Le premier vote*, pp. 50, 138–40; *La Révolution en Haute-Normandie*, pp. 35–44.

order. Unfortunately, he only added to the city's tensions by insisting that the troops follow his orders. The *échevins* refused to recognize his authority, and told him to give up his arms. Meanwhile, the guilds demanded that the parlement cede its control over the militia to the municipal officials. Corps of *volontaires* sprung up. At first, these were groups of young men from some of the city's weathier families. (By autumn, rival militias from poor neighborhoods had formed.) The *Volontaires du Tiers-Etat* joined the *garde bourgeoise*, entered the Vieux-Palais where the parlement sat, and claimed the sole right to patrol the city. The attacks on the parlement pleased the municipal officials, but soon their satisfaction turned to horror. Other angry citizens stormed into the intendant's *hôtel* and then the Cour des Aides, where they rifled through papers and destroyed tax rolls. On July 17, after much wrangling, the *présidents* relinquished not only their command of the militia, but also their responsibilities for provisioning the city. Two days later, they tried to reclaim their powers, and, clearly unaware of the meaning of that summer's events, they nullified the city's *cahiers* and the deputies' authority. These deeds only stirred further unrest in the city. By then, the parlement's position was much contested by the municipal officers. The *présidents*' activities were limited to the task of prosecuting those who had led the riots. Asserting their remaining authority, the *présidents* organized the punishments with great ceremony, hanging some of the guilty at a double gallows the court erected near the Seine.

The situation continued to unravel with the arrival of Bordier, an actor, and Jourdain, an *avocat* of the Paris Parlement, the "captain of a company of volunteer *patriotes*." The two had been sent by the Commune of Paris to supervise the safe transit of convoys for Paris and Rouen. The news of their mission to secure food "for everyone" had spread along their path, and the city's poorer inhabitants greeted them enthusiastically.[17] A group of soldiers joined them, and Bordier promised to force Intendant Maussion to raise their wages. Eventually, Bordier led the people in an attack on Maussion's *hôtel*, searching for "a victim to immolate." Facing the intendant himself, Bordier drew his saber and announced that "he was from Paris and that when we wanted to punish Berthier [de Sauvigny] we did not wait for someone to come to his aid." (Another account claimed that the intendant was able to escape through the crowd because no one recognized him.) Meanwhile, the volunteer corps refused the mayor's orders to protect the grain in the marketplace. One soldier instead marched into the market and "imprudently announced that the grain price would be lowered, and the population gathered." Eventually, the parlement and the *maréchausée* did restore order, and Bordier and Jourdain were tried and hanged.

[17] AN DXXIX bis 1, doss. 6.

The First Subsistence Committees: Lawyers, Merchants, and Other Notables

Given the depths of the provisioning crisis and its connection to much of the unrest, securing food was foremost among the duties of these new authorities in Rouen and elsewhere. The National Assembly, for instance, named a Subsistence Committee in June and charged it with coordinating imports for the country and maintaining Parisian supplies. Its thirty-two members were from the Third Estate, with many merchants such as Le Havre's Begouën, and, later, bishops and nobles.[18] By the autumn, the Assembly had disbanded the committee and renamed it the Committee of Agriculture and Commerce.[19] In Rouen, the political and administrative reorganization followed a similar path. On July 19, the Rouen government – the *comité permanent* – was formed. It consisted of an oligarchy of the city's former largely ceremonial officers and the electors. Many were local nobles whose liberal views had distinguished them during the elections, or influential local merchants and manufacturers. Among them were the mayor, the Marquis de Martainville, and the Marquis d'Herbouville, who led the *Volontaires du Tiers-Etat*. The municipal council set up its Subsistence Bureau the following day, part of De Martainville's plan for four new committees to help govern the unruly city. Most of the twelve members of the Subsistence Bureau were drawn from the city's merchant community, primarily textile manufacturers.[20]

They took their responsibilities seriously – and the politicized inhabitants of the city would not have allowed it to be otherwise. As was the case in other cities and towns throughout the new nation, the Rouen committee sought to address the unrest as quickly as possible with highly visible means. Many of these new municipal officials judged that their handling of provisioning would determine how well their communities and their own claims to authority weathered the revolutionary storm. The Rouen bureau's immediate plan was to collect "voluntary donations" to pay for subsidized bread sales to the poor. It organized this charity through the parishes. *Curés* were

[18] Pierre Caron, *Le commerce des céréales: Instruction, Recueil de textes et notes* (Paris, 1907), pp. 13–14.

[19] *Procès-verbaux des comités d'agriculture et du commerce de la Constituante, de la Législative et de la Convention*, 4 vols., Fernand Gerbaux and Charle Schmidt, eds. (Paris, 1906), 1: 7–30. Hereafter *CAC*. Caron, *Le commerce des céréales*, pp. 13–14.

[20] 18, [20] July 1789, AM Rouen A 40. The Committee's records are in AM Rouen F3 4. The members of the Rouen subsistence commission included Levavasseur and Lefebvre l'ainé, very prominent *négociants*. For other regions, see Bouvet, *Maconnais*, pp. 69–70; Hunt, *Revolution and Urban Politics*, p. 85; Bryant Timmons Ragan, Jr., "Rural Political Culture in the Department of the Somme during the French Revolution," Doctoral Dissertation (University of California, Berkeley, 1988), p. 116.

to identify the poor and give them certificates for the cheap bread. The system did not please Rouen's working classes, however. Within hours, posters explaining the plan had been ridiculed, and rumors ran that the townspeople had announced that they "would not subject themselves to asking the *curés* for certificates." Instead, they declared they would set the Hôtel de Ville on fire, which they promptly did.[21] In response, the Subsistence Bureau slashed bread prices and promised to reimburse bakers. To keep the countryside quiet, the Bureau arranged for government grain to be distributed throughout the generality. The members had not anticipated the reaction of buyers accustomed to less burdensome initiatives such as the quiet sales of low-priced supplies to bakers or other such subsidies. Thus, the Bureau's first attempt at maintaining order had failed, and it had been forced to use more drastic methods. The members' actions had reflected their best intentions for both order and food – and also some level of desperation and inexperience.

As flames rose from the Hôtel de Ville, supply lines fell apart. Transporting grain between ships arriving in Le Havre and Rouen and Paris proved to be extraordinarily difficult.[22] As many as forty wagons left Rouen each day for Paris, and Harcourt's *militia* could not protect them. The Subsistence Bureau sent the city's *volontaires* to his aid, but found them only too willing to lead the attacks. The municipal officials had even advised sending no more shipments by land, and had been dismayed to hear that in early August, 4,000–5,000 people had captured a grain boat at Oissel. Other reports recounted the exploits of bands armed with sticks and farm tools that plundered whatever cereals they could find at farms around Rouen. Given the level of violence, neither government grain nor private stores could be depended upon to supply the city.

Rapidly, the city's provisioning situation disintegrated, and the Subsistence Bureau careened toward more militant measures. Already, it had set bread at a very low price – so low that Necker later blamed the Bureau for discouraging sellers from supplying Rouen.[23] As fewer sacks of grain arrived in Rouen's hall in August, the city begged permission from Necker to "force *laboureurs* to bring regular amounts of grain to market, week by week, in proportion with their harvests." The officers argued that these measures were "former regulations," employed the previous spring, and now merited reconsideration. Perhaps they did not know (or conve-

[21] 19, 20 July 1789, AM Rouen A 40.

[22] This account is drawn from AM Rouen A 40; AM Rouen F3 4; and *Assemblée municipale et électorale et Conseil général de la commune. Analyses des déliberations* (Rouen, 1905); Gosselin, *Principaux épisodes*, pp. 2–8.

[23] Necker to the Rouen Municipal Corps, 10 October 1789, AM Rouen F3 4.

niently had forgotten) that such solutions generally had been deemed inappropriate for many years. In any event, permission was refused.[24] They appealed even to the parlement, "respectfully" – certainly an unexpected formulation given the events of July – asking that body to act.

In a few short months, the city had been turned upside down. The parlement, the intendant, the governor, and the municipal officers had become enemies. Granted, there had been few situations where the personnel of the Old Regime had cooperated fully (perhaps only De Crosne and Miromesnil had been able to do so in Upper Normandy), but here the city's many authorities had been pushed into enemy camps. Moreover, they were fearful of the violence that many of the city's poor had directed at both new and old officials. The townspeople had enacted their own revolution, and the militias had been at their sides. The new political elites – liberal and moderate nobles, *négociants*, and textile manufacturers – were confronted with daily reports of disturbances and shortage, and had few tools with which to restore public order. The combination of municipal revolution and hunger was rapidly bringing recourse to extreme measures.

The End of Moderation: The Law of 29 August 1789

At this point, the National Assembly provided the final element that destabilized provisioning in the autumn of 1789 – its free trade legislation of 29 August.[25] The decree effectively undermined any remaining openings for moderation in the cities and towns of Upper Normandy, and probably across France. The Assembly officially halted exports, which was applauded widely. Its orders to maintain *libre circulation*, however, increased ill will in many regions subject to the pull of urban demands.[26] Its most damaging articles reestablished off-market sales. This was a sharp reversal of the policies of the preceding year, and one that immediately diminished supplies at markets and increased fears of hoarding and famine plots. Grain owners, frightened by months of marketplace agitation and encouraged by the Assembly's stand, understandably deserted the retail trade. Cities lurched toward starvation.[27]

For months, local officials had bombarded the National Assembly with lengthy pleas for price ceilings, emergency granaries, requisitions – which would have continued Necker's policies of the previous April – and any

[24] Rouen Subsistence Committee to Necker, 24 August 1789, AM Rouen F3 4.

[25] Their early economic decisions reflects a clear desire to uphold property rights and encourage trade and innovation. *CAC*, v. 1; Caron, *Le commerce des céréales*, pp. 29–31.

[26] Rouen Subsistence Committee to Necker, 3 September, 1789, AM Rouen F3 4.

[27] Rose, "18th-Century Price-Riots."

other possible means to supply their cities.[28] Their disappointment with the Assembly's 29 August law was acute, and led them swiftly to even more aggressive actions. The Rouen Subsistence Committee – as the Bureau was renamed – at first hoped to appease both its citizens and the National Assembly. It called an unsuccessful meeting of the lieutenants of police of almost twenty surrounding towns, asking them to coordinate their supplies and send surpluses to Rouen. When those efforts yielded little grain, the Committee sent deputies to Paris to demand greater powers and grain from the National Assembly, Necker, and other ministers. They were told that the government had no more help to offer and that any grain arriving along the Channel had to be forwarded to Paris.[29]

The city officials, judging they had only a few days of grain left and that their own powers had been clipped too closely, decided to hold the capital's supply hostage. On 2 October, they blocked the Seine. They confiscated ships and threatened to starve Paris if Necker would not "inform them what quantities of grain were being turned over to the city's disposition."[30] This got the controller general's attention. First, Necker rerouted the grain arriving in Upper Norman ports instead through Saint Valéry and Amiens to Paris. Second, he promised the city some portion of incoming grain if it would protect other shipments to Paris. Third, he encouraged the city to purchase up to 60,000 setiers of grain – although he would provide no funds for this. The city appealed to deputy Le Couteulx's Parisian banking offices for a line of credit, and then wrote preliminary letters to several contacts in an attempt to find surpluses.[31]

The nadir came in the next weeks. The city received only paltry amounts of grain from Necker. Rival militias were springing up throughout the city and in its densely packed suburbs. Independently, the militias surveyed the countryside and drew up a list of over 150 clandestine grain stores ready for impounding. The Rouen municipal government branded the list "notoriously false," but shared the militias' concern. In mid-October, those militias journeyed to a nearby chateau and apprehended the mayor along with the man they wished to head the reorganized *garde bourgeoise*, the Marquis d'Herbouville. They marched back to the city with the unwilling Herbouville, captured the Vieux-Palais, and moved the city government into the new headquarters. In the midst of those rebellions, the municipal

[28] See the many petitions and reports sent to committees of the National Assembly, AN DXXIX bis 1; AN DXXIX bis 25; AN DXXIX bis 26; AN DXLI 2; *CAC*, 1: 30–64.

[29] Rouen Subsistence Committee to the lieutenants de police and procureurs du roi of Elbeuf, Pont de l'Arche etc., 4 September 1789, AM Rouen F3 4. See the other correspondence in that carton.

[30] "Arrêté qu'on suspendrait l'expédition d'un bateau . . ." 2 October 1789, *Déliberations*; AM Rouen F3 4.

[31] October 1789 correspondence, AM Rouen F3 4.

officials were hard pressed to maintain order in markets and protect grain convoys going to Paris. Worse, purchasing agents for the capital had made deep inroads into Rouen's zones in the Vexin, and the shiploads of grain arriving in Channel ports were destined primarily for Paris. The administrators of Upper Normandy would have to protect those supplies as they passed through starving towns.[32]

Rouen's seizure of grain traveling up the Seine was not an isolated act of insubordination. Since the mid-summer, municipal authorities throughout Normandy had been engaged in similar exploits. Evreux's government, consisting of numerous representatives from corporations, had confiscated grain in local warehouses and ordered bakers to produce only low-cost whole wheat loaves.[33] In Louviers, plans were more organized: The "municipal officers and electors" commanded forty-seven neighboring villages to supply its hall.[34] Once the National Assembly's 29 August *arrêt* became known, the public authorities' defiance mounted. The "comité national et permanent" of Caen, for instance, provisionally set price ceilings on grain on 20 October. These orders threw commercial networks throughout Normandy into chaos, and further emptied markets.[35] Les Andelys levied requisitions on local farmers.[36] At some of the major grain distribution centers in the Paris zone, among them Vernon, Gisors, and Etrepagny, local authorities prohibited off-market sales, and even told an agent buying for the capital that he could make purchases only in public halls. According to the agent's report, the Vexin authorities seized grain shipments, imprisoned grain owners, imposed requisitions, and set prices. Morever, the agent claimed that their example was inciting their citizens to demand more extreme measures.[37] As late as the spring of 1790, towns in the Pays de Caux maintained they had the right to set grain price ceilings and organize requisitions, all in clear defiance of the National Assembly's admonitions. The authorities had been forced by their townspeople – or so they claimed

[32] "Reglé le 15 octobre 1789," AM Rouen I5 6; *Vivre en Normandie*, 2: 336. See also the correspondence in AM Rouen F3 4.

[33] Dimet, *Evreux*, pp. 74–9.

[34] They did so on the dubious grounds that the towns fell within the jurisdiction of its *grenier à sel*, according to letters patent of October 1725, and the Rouen Subsistence Committee quickly reprimanded them. AN DXXIX bis 2, doss. 17, p. 52; Rouen Subsistence Committee to the Municipal Officers of Louviers, 4 September 1789; Rouen Subsistence Committee to Necker, n.d. [ca. 5 September 1789], AM Rouen F3 4.

[35] Rouen Subsistence Committee to the Comité national et permanent of Caen, 4 November 1789, AM Rouen F3 4.

[36] J.J. Rousseau and Grandin [deputies to the National Assembly on mission to Vernon] to the National Assembly, 2 November 1789, AN DXXIX bis 2, doss. 17, p. 6.

[37] The people of Richeville had held an innkeeper hostage and had made the rounds of local farms. The agent alleged that the municipal officials approved of such illegalities. Réal to the National Assembly, 4 November 1789, AN DXXIX bis 2, doss. 17, p. 8.

– and only the arrival of 150 soldiers from Rouen on 1 May 1790 brought calm.[38] Despite the National Assembly's policies, or, precisely because of the few alternatives they offered local officials, the authorities believed themselves forced to embark on such missions.

Municipal Rivalries and Radicalism

A brutal uprising in Vernon provides an additional opportunity to analyze the relationship between municipal revolution and more extreme measures. Two city governments – one representing the city's elite, the other representing roughly the "artisans and the middle class" – vied for control. The latter, the "provisional government," won the mid-summer rounds, and then joined with other municipalities in the region to secure food. It requisitioned grain from farmers, grabbed shipments to Paris, and set cereal prices in the early autumn. In late October, the people of Vernon, encouraged by the provisional authorities' apparent support, stormed Planter's infamous tower granaries and attacked the merchant himself. The bourgeois militia dragged Planter from his home, presumably for his protection, and "abandoned him to the fury of the people." Planter was beaten severely, and the crowd mockingly dangled a noose above his head before he managed to escape, aided in part by the members of the provisional government, who suddenly feared things had gone too far.[39] The "legitimate" former officials pleaded with the National Assembly to send troops to restore them to their rightful positions. The Assembly dispatched 500 men under the guidance of two deputies with orders to disband the provisional government and oversee "free elections" for a committee to aid the "legitimate" municipal officers.[40] As for Planter, the National Assembly wanted

[38] "Proclamation du roi," 2 May 1790, ADSM LP 6850; Lemarchand, *La fin du féodalisme*, p. 419.

[39] It took a number of days to ascertain that the merchant had survived his numerous attacks. The members of the protest government disavowed any knowledge of the identity of his assaillants. Des Ervolus, the Lieutenant of Police of Evreux who had been so industrious in rounding up the guilty near La Roche-Guyon during the Flour War, now a commandant of the *maréchausée*, was called in to search for Planter's attackers. AN DXXIX bis 2, doss. 17; *CAC*, 1: 374; and René Rouault de la Vigne, "Les débuts des troubles de Vernon en 1789 d'après une lettre inédite," in *Actes du 81e Congrès national des Sociétés savantes*, pp. 437–42.

[40] The election resulted in a committee evenly divided between the legitmate and the provisional officers. Eventually, Grandin and Rousseau urged three members of the provisional government, whom they held responsible for the violence, to leave town, and ordered the militia members who had sided with the provisional government to resign. AN DXXIX bis 2, doss. 17, p. 10.

his assaillants punished, his safety guaranteed, and his grain and flour rushed to Paris.

While this series of events was complex, the salient point is that deep political fissures, unrest in the militia, popular pressures, and shortages led to extreme measures. Perhaps the beliefs of the provisional officials ran toward stronger intervention in the trade; perhaps it was their ambitions that led them to seize the political opportunities local hunger – and Planter's notoriety – offered them. Perhaps they had had little experience with these matters before the Revolution and were grasping for any solution that would appease their *citadins*. Regardless of their motives, the political ferment of 1789 in Vernon spawned local officials willing to challenge the National Assembly's wish to establish free trade.[41]

Why had the Subsistence Committee and its counterparts across Normandy chosen these strong-arm tactics? Certainly a number of the Rouen merchants and *fabicants* who staffed the Subsistence Bureau had winced at the thought of imposing price controls or seizing barges bound for Paris. They may have been uneasy all along with such methods. When the Subsistence Committee eventually confronted the Caen municipality's use of ceilings, for instance, it argued that high prices alone would draw grain. "The inconvenience of paying higher prices for grain is infinitely less to be feared than not having any grain at all," it wrote Caen. (This letter was written in early November, however, after *la loi martiale* had been declared, the militias calmed, and the National Assembly's power reinforced.[42]) Overall, the committee's correspondence suggests only scant reflection on the breakdown of the grain trade, and only slight knowledge of previous methods for addressing shortages.

The first missteps in confronting high bread prices in July and August 1789 revealed precisely those gaps in their understanding, and had led to the fires in the Hôtel de Ville. Later, the committee seemed to have no information on when harvests would appear at surrounding markets. The members expressed surprise at finding they could expect little grain before November. This would have been rudimentary knowledge for any Old

[41] Events elsewhere in Normandy – such as in Fécamp and Elbeuf – suggest that other combinations of municipal unrest, militia involvement, and desperation brought strong support for greater controls on the trade. Catherine Rabat, "Les Jacobins de Fécamp (1789–95)," in *A travers la Haute-Normandie en Révolution 1789–1800* (n.p., 1992), p. 114; Kaplow, *Elbeuf*, pp. 128–30. Events in Upper Normandy had their counterparts across France, and the various militias and guards played significant roles. See the problems the Paris zone, Bergues, Cambrai reported in: n.d., AN DXXIX bis 26, p. 3; n.d. and 20 October 1789, AN DXXIX bis 2, doss. 17, pp. 8, 16; *Bergues*, 1: 77–80, 130–1.

[42] Subsistence Committee to the Comité national et permanent à Caen, 4 November 1789, AM Rouen F3 4.

Regime subdelegate or police official, and the query in the correspondence is striking.[43] When it came to organizing grain imports, the committee clearly had few contacts, little experience finding monies, and an insufficient sense of the choices for distributing them. In early September, for instance, the committee had contacted merchant houses and found, as expected, that no *négociant* wished to be involved in the trade. The Committee even agreed with the merchants' widespread fears of having "their life, their honor and their fortune compromised." They seemed unable to press ahead, perhaps lacking contacts, information, or financing – all the elements that Miromesnil and De Crosne had had at their disposal. Instead, the 1789 authorities immediately appealed directly to Necker for grain.[44]

The committee did make two stabs at bringing grain into the city in ways that had some similarity to Old Regime methods. The details were not spelled out fully, making it hard to evaluate how well the members understood their tasks. In the first of these two initiatives, the city's Chamber of Commerce brought 130,000 livres of grain to Rouen. The records give no indication of how it was used. It may have been sold to bakers, exciting the rumors of spoilage and corruption that rocked the city that year.[45] Second, the committee offered bonuses for any grain that merchants brought to the city's hall in October. The plan for bonuses was somewhat successful, bringing about half of what was needed, although again, the records contain little information.[46] In addition to grain from these efforts, the committee oversaw the distribution of supplies granted by Necker. The city sold these wares directly to the bakers and through the hall. It did so publicly and at a very low price, sparking rumors that the grain was spoiled. (In Bolbec, identical sales brought the same accusations and unrest. There, many of the experienced officials who had been elected in August 1789 had declined to accept posts, leaving a collection of *avocats* and *négociants* to run the city.[47])

The consistent lack of description on the means for dispensing the grain was accompanied by reports of resulting disturbances. This combination strongly suggests that the committee of 1789 was operating with the best of intentions, but with little knowledge of the complicated processes it was undertaking. The members did not have a grasp of the complexities of the

[43] Subsistence Committee to Necker, 24 August 1789, AM F3 4.

[44] Subsistence Committee to Necker, 3 September 1789, AM F3 4.

[45] Subsistence Committee to Necker, 3 September 1789, AM F3 4.

[46] The city offered a bonus of 20 sous per mine of grain. In late October, the Committee reported that the *prime* had brought about 1,000 mines per week, but that the city needed twice that much. Necker to Rouen Municipal Corps, 10 October 1789; Committee to Tarbé and Levavasseur, 21 October 1789, AM Rouen F3 4.

[47] Jean Pigout, *La Révolution en Seine-Maritime: Bolbec 1789–1794* (Luneray, n.d.), pp. 43–4.

pricing and distribution of the grain – and perhaps they did not even recognize these as issues that had lengthy Old Regime histories. The voice of Old Regime restraint in the 6 April 1789 instructions of the Provence intendant that "it is not possible to force merchants and *fermiers* to lower [their price]" was being overtaken by less temperate, and less knowledgeable, administrative judgments.[48]

There are two additional points to be made about the efforts of the municipal corps. First, having few other possiblities within their reach, these officials looked immediately to the ministries and the National Assembly for help. They were frantic, and expected the National Assembly to respond with a sympathy and generosity beyond what the monarchy previously had offered. The National Assembly, after all, could not ignore the pleas of its own citizens. At the very outset, the notion of a government attentive to the needs of its people forced the conclusion that the Assembly had an obligation to feed the people who had supported it with their lives. Given that the local officials had few other alternatives before them, they turned expectantly to the National Assembly and ultimately were disappointed.

Second, many of these expectations were expressed in the new language of republicanism. This vocabulary had been present in some of the *cahiers* of the preceding spring and built on customary themes of the Old Regime famine plots – namely, that the corrupt and powerful were hoarding grain in order to starve the poor and unprotected. In Normandy, one community blamed shortages on the existence of "men injust and barbaric enough" to engage in the grain trade, taking supplies out of the village "against any sentiment of humanity."[49] Another asked, "Why then so much misery in the midst of abundance? It is the greed of the individual that leads them [sic] to fatten themselves [sic] at the expense of the public. . . ."[50] The petitions to the National Assembly were riddled with these images. They stressed the general interest over the particular, and argued in terms of fraternal responsibilities. A petition from Brittany began with the seemingly incontestable assertion that "we are all French, we are all brothers" and then urged the National Assembly to send supplies to the region.[51] A Meaux notary wrote that "[t]he poorest citizen has as much right to live as the rich. His indigence gives him the right to the government's protection, which should help him get provisions easily."[52] Using the same charged words, the Rouen Subsistence Committee berated neighboring areas for refusing to

[48] Cubells, "*Les mouvements populaires*," p. 323.
[49] *Les cahiers de doléances . . . du bailliage de Rouen*, 2: 389.
[50] *Cahiers de doléances du Tiers Etats du Bailliage d'Andely*, p. 132.
[51] 19 July 1789, AM Rouen A 40; 31 July 1789, AN DXXIX bis 2, doss. 17, p. 37. See also the petition from the bailliage of Auxerre n.d., AN DXXIX bis 2, doss. 17, p. 30.
[52] n.d., AN DXXIX bis 2, doss. 17, pp. 34–5.

send grain. Writing to the municipal officers of Elbeuf, the committee stressed that "only by . . . the fraternal union, and the mutual aid. . . . between cities, can France hope to see peace reestablished in its breast . . ." The "destructive system" of the monarchy had severed such bonds. Now, however, freed from "tyranny," regions with abundant supplies could yield to their fraternal sympathies and send grain shipments coursing to Rouen. Marat could not have said it better.[53] By the spring of 1790, such language was universal, embittered, and highly politicized. The municipal officers of Tours, for instance, demanded in May 1790 that the National Assembly become much more forceful with "the Revolution's enemies," who were "profiting from the unlimited freedom of the grain trade [by] hoarding [supplies]." A play, "Le Patriote zélé," sent to the National Assembly, denounced "monopolies, usury, aristocrats of every type, [and] hoarding."[54]

These striking oppositions came to be the dominant rhetorical tools of late 1789. They did not leave much room for moderation and deliberation. They expressed instead the frantic nature of the year's circumstances. The National Assembly had left communities with few legitimate means of securing supplies, and in all fairness, it did not have the money and experience to be of much additional help. Thus, ultimately, municipal officials had to rely on words, not force. They made desperate comparisons between a neighboring town that refused to send grain and the actions of a brother who had turned his back on his family. Grain merchants who stored supplies while waiting for higher prices became "vampires" who wished to drain a region of its blood. When words did not bring grain, the authorities had no options save breaking the law and embarking on aggressive and impetuous attempts to find food.

By the first months of the Revolution, the concerns of the previous decades about how one defined free trade and how it could coexist with marketplace realities had disappeared. The careful April 1789 explanation of the intendant of Lille, in northern France, that a grain merchant could be ordered to bring grain to market, but not forced to bring it to any specific hall – a violation of *libre circulation* – is a rare example.[55] Several petitions from towns across the nation to the National Assembly did raise such issues, but generally only in order to reject the possibility that such concerns were either appropriate or relevant. Considering regulations from a practical standpoint, for instance, a Baumes-les-Dames petition argued that censuses and formalities were not "troublesome to commerce," and that given the level of scarcity, "the preferences of commerce merit no consideration."[56]

[53] Rouen Subsistence Committee to the Municipal Officers of Elbeuf, 24 August 1789, AM Rouen F3 4.

[54] *CAC*, 1: 259, 309.

[55] *Bergues*, 1: 62.

[56] n.d., AN DXXIX bis 2, doss. 17, p. 27.

That the "preferences of commerce" were no longer of interest revealed a striking contrast with the previous administrators' policies. A sea change had taken place during 1789. In short, the solutions and concerns of Old Regime officials had vanished. So too had many of the men who had formulated the Old Regime solutions. Many had left public life and were replaced by the more interventionist municipal authorities who emerged from the agitated political circumstances of that summer. These cadres faced a more impassioned and demanding citizenry – and a greater shortage than anyone could remember. The Revolutionary officials built on the exceptional measures of the Old Regime, such as price fixing and requisitions. The desperation of their townspeople, from whom they claimed their legitimacy, seemed to demand such extremes, and they rapidly imposed them despite the National Assembly's opposition. They had to use language – much of it highly inflamed – because the National Assembly would not allow other measures.

The shortage of 1789 had been extraordinary, and the high price of grain had consumed over 80 percent of working-class wages. So great was the suffering that even the most accomplished Old Regime authorities probably would have been stretched beyond their capacities. Moreover, the Revolution had created institutions and activities that answered to a more vocal citizenry. The entry of new personnel, embattled, yet energized by their sense of responsibility, provided a context in which radical ideas came to the fore. Unfortunately, the National Assembly determined that only completely free trade would accomplish its goals. That few of its committee members had any genuine provisioning experience was evident in this choice. The 29 August 1789 *arrêt* polarized matters and left local officials no leeway. From their windows, they heard the cries of their townspeople; on their desks, they saw the instructions from the Assembly. They could not obey both, and so sided, perhaps unwillingly, with their people. A further impetus was provided by the vocabulary of republicanism. Its oppositions and moral categories posited antagonistic relationships between buyers and sellers, rich and poor, those with grain and those without it. Language, desperation, and extreme provisioning strategies fused together in 1789.

CHAPTER SIX

Unity and Interests

"I maintain that we must not only protect the grain trade; we must make it honorable . . . I regard a man who enters the grain trade as *one of the benefactors of the country.*" Representative Le Quinio, from the Committee of Agriculture and Trade, was addressing the Convention in late November 1792. He wished to counter the stark demands for economic controls that were overwhelming the Committee. He urged his audience to hold firm to its earlier commitments to free trade. The young republic had enough grain, he insisted. It was just a matter of moving it swiftly from producers to consumers. Outside in the streets of Paris, the *enragés*, led by Jean Varlet and Jacques Roux – "amibitious hypocrites," Le Quinio called them – had begun their campaign for regulations. Their impassioned speeches attracted crowds who were feeling the impact of another short harvest and the collapse of the grain trade. Le Quinio and his committee members were wearying of the sections' constant interruption of their meetings. The sections blamed farmers, merchants, and even the Minister of the Interior for their hunger, and demanded that the Convention come to their aid.[1] Beyond Paris, similar shouts could be heard. Indeed, petition after petition to Le Quinio's committee spoke in immoderate tones, and delegations arrived to press the members for supplies.[2] By late November, another crisis was underway, and Le Quinio and his fellow committee members sought to ward it off.[3]

Their main line of argument was that the nation had enough grain. Le Quinio's explanation for the shortage was that farmers and merchants had been terrified by the violence of the past years. They either had hidden their grain out of fright – they were not, he insisted, hoarding it – or had decided

[1] *CAC*, 3: 29–31.
[2] *CAC*, 3: 33–4.
[3] *Archives Parlementaires de 1787 à 1860, recueil complet des débats législatifs et politiques des chambres françaises*, première série (Paris, 1862–1988), 15: 337. Hereafter *AP*.

to give up grain production entirely. Merchants, too, had fled the trade. The sole remedy was to protect the lines of transport. Throughout those months, delegate after delegate climbed up to the Convention's podium and pronounced on the matter of *subsistances*. Many from the Committee shared Le Quinio's view and opposed the extreme measures that the Paris Commune and some sections were demanding.[4] The handful of radical representatives to the Convention and on that Committee – some Jacobins, others from the Cordeliers club and sections – launched noisy campaigns against moderate policies.[5] Jean-Marie Roland de la Platière, the embattled Minister of the Interior, arrived several times to defend liberal policies as the sole long-term solution.

The Impasse of 1792: Roland and the Paris Commune

Roland had to defend not only his policies, but his reputation. Under his watch – and those of his predecessors, Claude-Antoine Valdec Delessart and Cahier de Gerville – provisioning had fallen apart. Roland had returned to the post after the Parisian uprising of 10 August 1792. He was confronted by the Commune, more militant than ever. It demanded cheap bread, along with the death of the king. Some of the Commune's most outspoken radicals, such as Danton and Pétion, had been elected to the new legislature and carried on their daily crusade from the Convention's lecturn and hallways. Roland had earned their emnity. Long a partisan of free trade – before the Revolution he had been an inspector of manufactures – he had opposed intervention in the grain trade the previous spring. He had criticized some cities' efforts to hold bread prices below the cost of grain, and urged the Commune to take more responsibility for the city's provisioning. In fact, some of the origins of the crisis of the autumn of 1792 could be found in the tensions between the Ministry of the Interior and the Commune over which one had failed to supply the city.[6] Worse, section radicals were alleging that graft riddled the imports the ministry had arranged using government money.

In fairness to Roland, the situation had been unraveling since the autumn of 1791, when the first shortages had been apparent. Numerous food riots

[4] His Committee had a number of Girondins and their conflicts with the Paris Commune and the sections had broken into the open by the autumn of 1792. Identifying specific factions is problematic, however. M. J. Sydenham, *The Girondins* (London, 1961); Alison Patrick, *The Men of the First French Republic: Political Alignments in the National Convention* (Baltimore, 1792).

[5] 16 November 1792, *AP*, 53: 434–4.

[6] Susanne Petersen, "L'approvisionement de Paris en farine et en pain pendant la Convention," *AHRF* 56 (July–September 1984): 366–85.

had pushed the outgoing National Assembly to take firmer measures against disorder. In particular, it called on the local authorities, who were proving unpredictable, to confront crowds. Authorities were to shout "*Obéissance à la Loi!*" three times and warn that force would be used.[7] The law of 25 September–6 October 1791 set the punishment for pillaging at as much as six years in irons, if committed "with a crowd and violently."[8] Nonetheless, local authorities were either incapable or unwilling to repress disturbances, and the national guard lacked the strength to aid them.[9] The food riots of 1789 had never completely abated, and in 1791, the unrest worsened.[10] The subsequent shortages were intensified by the tax evasion that had characterized the Revolution since its first weeks. Fewer peasants needed to market their grain immediately after the harvest. Thus, the usual pattern of sales and tax payments was interrupted by the autumn of 1792. Lower tax receipts only further restricted the government's ability to purchase grain for the nation, and increased the rate of inflation.

Despite the legislatures' insistence on free trade, there had been a constant flow of petitions pleading for more government imports, a return to controls, and for the creation of municipal or departmental granaries. Worse, some jurisdictions continued to block shipments.[11] The National Assembly's *Comité des recherches* and its Committee of Agriculture and Trade had rejected the immoderate demands.[12] By the spring of 1792, however, the Committee of Agriculture and Trade considered reforms, especially controls.[13] During these months, momentum built for laws against hoarding and for the setting of price maximums. The departments took the lead.[14] In June 1792, the Legislative Assembly yielded, and permitted a grain census, but enjoined local officials to maintain free trade.[15]

[7] 26–7 July 1791, *Code criminel et correctionel*, 2 vols. (Paris, An XIII-1805), 1: 26–36. See also, 2 October and 27 November 1791, ADSM L 8650*.

[8] *Code criminel et correctionel*, 1: 37–76.

[9] In Le Havre, local officials confiscated grain on ships, forced the owners to sell it to townspeople in the summer of 1791, and then obstructed shipments leaving their jurisdictions. MI Delessart to the Directoire S-I, 7 and 15 August 1791, AN F11 220. See also his instructions in AN F11* 3.

[10] *AP*, 15: 337; Kaplow, *Elbeuf*, p. 131; André Endres, "Une émeute de subsistances à Meaux en 1790," *Actes du 86e Congrès national des sociétés savantes*, pp. 475–83; F. Evrard, "Les subsistances en céréales dans le département de l'Eure de 1788 à l'an V," *Bulletin d'histoire économique de la Révolution* (1909), p. 36.

[11] "Arrêté," Seine-Inférieure, 9 July 1791, ADSM LP 8650*; *CAC*, 1: 166–7, 248, 506, 512; 2: 25–6, 138; Lemarchand, *La fin du féodalisme*, pp. 420–22.

[12] *Bergues*, 1: 185–6.

[13] *CAC*, 2: 536–46.

[14] See the correspondence from Rouen, Yvetot, and Les Andelys, and the incidents reported to the Committee of Agriculture and Commerce: 22 March 1792, AN F11 220; 12 May 1792, AN F11 213; *CAC*, 2: 661–815, 3: 612–55; Rose, "18th-Century Price-Riots."

[15] *Bergues*, 1: 155.

Provisioning Efforts Since 1789

What measures had been tried, and which ones had failed? There had been three forms of provisioning initiatives since 1789. First, modest bonuses were once again offered to merchants. Unfortunately, the unrest and then the war had dampened commercial enthusiasm. Second, the government advanced money for grain purchases, and also permitted some cities to use the grain in military warehouses. It then struggled with the cities that claimed later to be unable to repay either in kind or in cash – suggesting confusion and inexperience with such plans, along with genuine poverty and hunger.[16] In addition, the Legislative Assembly had given Roland and his predecessors, Delessart and Cahier de Gerville, several grants of money with which to buy grain since October 1791. The funds had been advanced in several blocks of assignats, generally 12 million livres at a time. There had been trouble finding merchants willing to accept the paper money or to operate on credit, and there were continuing questions about what branch of the treasury was to disburse the money.[17] The Ministry of the Interior used the funds for purchases and also loaned it to departments at 5 percent interest. Rapid prices rises followed. Both the national government and a multitude of departments, districts, and cities sent agents scattering across France – and occasionally to Great Britain, Amsterdam, and Genoa – in search of grain. Delessart had hoped to arrange all purchases through his office, but was denied that power in September 1791. Le Quinio had tried unsuccessfully to reintroduce a plan for ministerial supervision of the purchases four months later.[18] The agents' activities had been chaotic and had further angered merchants. Parisian purchasing agents had been especially quarrelsome and had antagonized many surrounding towns. In addition, war suppliers, struggling to feed a rapidly expanding army, were criss-crossing the same paths in search of victuals.[19] Moreover, the funds that the Ministry had to spend were being eroded by inflation – owners

[16] There were numerous such distributions of grain or money, and local communities frequently defaulted. In some cases, the municipal officers alleged that they could not find buyers for the grain, given its poor quality and high price, and so could not recoup what had been advanced. In the Seine-Inférieure, Neufchâtel, Bolbec, and Gournay were among the municipalities that could not repay the arrears. AN F11* 3; F11 220.

[17] See, in particular, Minister of the Interior de Gerville's March 1792 defense of these operations, *AP*, 40: 435–8. Some of this grain came from England. AN F11 225–6; ADSM L 383; *AP*, 39: 383–5; *Le District de Bergues*, 1: 158–60.

[18] *CAC*, 2: 417–8; *AP*, 34: 580; 37: 106–7.

[19] That the Ministry of War had been engaged in its own purges and a battle with the Convention added to the tensions over military supplies and government purchasing in general. Howard G. Brown, *War, Revolution, and the Bureaucratic State: Politics and the Army Administration in France 1791–1799* (Oxford, 1995), pp. 38–71.

raised their prices in advance of the currency's depreciation – and the Convention could not spare more money.[20]

Roland promised to restore order, but even his own agents were hard to tame.[21] When grain did arrive, merchants feared pillaging. They refused to transport the cereals inland or to other ports. In addition, some of the grain had spoiled already or was of very poor quality.[22] Thus, by the time Roland or the members of the Committee on Agriculture and Trade addressed the Convention in the autumn of 1792, the situation was largely out of control and the funds available for imports were insufficient.

The third provisioning initiative concerned bread pricing. In the confusion of the early months of the Revolution, many communities had been uncertain whether the practice of setting the bread *taxe* had lapsed. Some municipalities had continued to announce the ceilings, while in others there had been questions about which bureau, if any, had the responsibility.[23] The municipal laws of 19–22 July 1791 finally outlined the responsibilities of the municipalities, among them the authority to set prices on bread and meat if they wished. Even then, there were queries about what would happen if the officials did not wish to set the *taxe*. Could they be forced to do so, some asked, and if so, by whom?[24] During the summer of 1792, many municipalities, including Paris and Rouen, sold heavily subsidized grain to bakers and held the price of bread very low. The Paris and Rouen prices were well below what the Old Regime officials had sanctioned, and some city officials, such as Rouen's Mayor de Fontenay, evinced great discomfort. These measures drove more merchants and millers from the trade and unintentionally allowed bakers stunning profits. If accusations at the time are to be believed, the Parisian bakers bought the subsidized flour and resold it outside city boundaries for substantial gains. This aroused rumors of graft while exciting the already strong jealousy of rural areas toward

[20] The value of the assignats was roughly 60 to 75 percent of their nominal value in the fall of 1792. Grain prices were somewhat disproportionately higher. Dominique Margairaz, "Le maximum: une grande illusion libérale ou de la vanité des politiques économiques," in *Etat, finances et économie*, 411–5; S.E. Harris, *The Assignats* (Cambridge, Mass., 1930), pp. 121–32.

[21] *AP*, 40: 435–8; 41: 52, 100–1; 44: 453, 480; 48: 165; *CAC*, 2: 417–8; *Bergues*, 1: 157–8; Charles Poisson, *Les fournisseurs aux armées sous la Révolution française: Le Directoire des achats (1792–1793)* (Paris, 1932), p. 20.

[22] "Situation des magasins des subsistances au Havre," 7 Septembers Year I [1792], AN F11 225–226; *Bergues*, 1: 159.

[23] *Code criminel et correctionel*, 1: 9–20; A. Subtil, "Réglementation municipale de la distribution des grains et de la boulangerie avant et sous le regime des maximum et pendant les disettes de l'an III et IV," in *Actes du 81e Congrès national des Sociétés savantes*, pp. 279–92; Kaplow, *Elbeuf*, pp. 130–1.

[24] *AP*, 17: 214; *CAC*, 2: 257, 698.

cities. The subsidies, of course, cost far more than the government could afford, Roland argued.[25] All in all, none of the undertakings – bonuses, purchases, low bread price ceilings – had worked.

Given the upheaval of the past months, and the news of stoppages and seizures along major supply lines, what did the Committee propose in the autumn of 1792? Its turbulent sessions and the members' addresses to the Convention reveal a growing sense of the workings of markets and of a desire for policies that reflected that understanding. In some instances, such as Creuzé-Latouche's 8 December speech analyzing prices for the previous forty years, extensive information was presented for discussion. Within the chambers of the Committee itself, the complexities of bread pricing received considerable attention. Some in the Committee believed the Paris Commune's policy – holding the price of bread far below that of grain, while providing subsidies to bakers – was a grave error. They discussed instead offering bakers bonuses for any flour they purchased outside the city's market. The bonuses would enable bakers to seek out supplies themselves. That strategy would stretch scarce funds further than the present system. It would also encourage bakers to reknit the commercial networks that had unraveled in the past months. The Committee hoped that the Commune would set the price of bread equal to that of grain, so that grain and flour merchants would return to the city markets.

In many ways, these ideas followed Old Regime strategies. The attempts to subsidize purchases less visibly and to remain close to the going prices would have been praised by De Crosne or Miromesnil. It is not clear from the records whether the Committee was aware of the prior methods, or simply had arrived at similar conclusions. One loud voice – that of the well-known radical, Coupé de l'Oise – was raised in opposition. He blamed speculators who were selling spoiled grain. In the end, the Committee reluctantly concluded that it had no authority over the price of bread in Paris. It turned to other plans to subsidize bread for the poor by selling 1,200–1,500 sacks of imported flour daily – almost the city's entire needs – through the hall, running up deficits of 4 million livres.[26]

[25] He estimated that the Paris bakers lost about 12,000 *livres* per day on their sales. Until his dismissal on 22 January 1793, he argued with the sections that they should allow the price of bread to rise with that of grain. *AP*, 52: 144–8, 53: 476–7; Henry E. Bourne, "Food Control and Price Fixing in Revolutionary France. I," *Journal of Political Economy* 27 (February 1919), p. 82.

[26] The naturalized Swiss banker and grain merchant Bidermann and his colleague, Cousin, ran the *Directoire aux Achats*, established in November 1792. AN F11 209; *CAC*, 3: 24.

Cries for Controls

In general, the rush to regulation that began in late 1792 has been understood as a "*politique de classe*," to which wartime needs and inflation provided the final impetus.[27] For many Marxists, the sansculottes movement, decidedly proto-socialist, had managed to impose its will on the government for a brief moment.[28] Other historians, in particular Florin Aftalion, view the maximums as a final – and incontrovertibly hypocritical – capitulation to a radicalized polity and a betrayal of the liberal principles of 1789.[29] Historians, whether for or against the provisioning policies of 1792–4, have noted the pressure from Parisian working people on the Convention, and the Jacobin's acquiesence, as critical factors in the creation of the controlled economy of the Terror. Others point to the wartime pressure for stable, affordable prices, and protection from inflation as the primary considerations. All of these have merit.

However, the members of the Committee and the Convention had their own ideas, too, and these had an effect on local circumstances. The sources for the representatives' economic knowledge at this juncture are difficult to trace.[30] First, the period was one of great legislative activity, but less economic writing. The great matters of the day were to be solved with votes and speeches, and not with reflection, ink, and paper. One sees evidence of some lingering Physiocratic thought in the representatives' policies – especially those of 1789–91 – and there is occasional reference to a "Monsieur Smith" or an *éloge* for Turgot. In the absence of a well-developed secondary literature on the economic ideologies of the period, the outlines of the convictions that animated the Assemblies or the Convention are hard to discern. That does not mean, however, that no underlying assumptions emerged in the debates and committee meetings. In particular, the repre-

[27] Mathiez first evisioned the Maximums in this way, and later Marxists built on his understanding. Albert Mathiez, *La vie chère et le mouvement social sous la Terreur* (Paris, 1927).

[28] Soboul, *The Sans-culottes*; Mazauric, *Babeuf*; Rose, "The French Revolution and the Grain Supply."

[29] The capitulation had begun with the confiscation of Church lands, and continued through the issuing of the assignats. Florin Aftalion, *The French Revolution: An Economic Interpretation*, trans. Martin Thom (Cambridge, 1990).

[30] See the introductory comments by Gilbert Faccarello and Philippe Steiner to the conference volume, *La pensée économique pendant la Révolution française*. There has been some recent scholarship on the topic by a handful of scholars, most notably François Hincker, Dominique Margairaz, Simone Meyssonnier, Francis Démier, and Eugene White. See their articles in *Etat, finances et économie*; *La Révolution française et le monde rural*; *La Révolution française et la développement du capitalisme*; and *La pensée économique pendant la Révolution française*. One recent collection unfortunately contains no treatment of provisioning policies: *Idées écomiques sous la Révolution 1789–1794* (Lyon, 1989).

sentatives did not doubt that the nation's grain supply was sufficient. Le Quinio had been one of the strongest proponents of this nearly universally held position. All else derived from that fact: Requisitions, the maximums, and even, finally capital punishment for hoarding. The representatives may have been close to correct. On the average, French resources may have been more or less adequate. As long as imports could be found, there was hope of covering intermittent deficits. During the Revolution, however, an exceptionally poor run of harvests fell below the nation's needs. War, inflation, and hoarding added to the distress. The many parties that dominated the legislatures insisted, however, that there was enough grain. The problem, they maintained, was that inadequately protected supply lines had been severed and that owners were hiding their stores.[31] Why did bread cost 6 sous in one town and 2 sous in another?, the *Conseil exécutif provisoire* asked. "You need look no further for the cause than in the numerous obstacles to *circulation.*"[32] In other words, the failings were the result of unrest and greed. Food rioters who would not allow grain to leave their communities, and owners with overflowing granaries – "men who calcalute the price of their *concitoyens*' misfortunes" – were equally at fault.[33]

This belief in the sufficiency of the year's harvest helps to explain the legislators' unwavering insistence on *libre circulation*, and their subsequent development of policies that sought to draw grain from hiding. It also accounts for the harsh punishments, including execution, that the Convention decreed for those who interfered with shipments.[34] The representatives' goal was to restore the public's access to the grain. If free trade had not achieved that end, then requisitions, public markets, and even set prices were entirely legitimate.[35] The law of 9 September 1792 allowed towns to levy requisitions within the areas that generally supplied their halls. This legislation provided the first step in the steady move toward the controls of the Year II.[36] The requisitions only deepened the crisis. Districts dragged their feet in drawing up censuses, while authorities from needy areas rushed across municipal and district boundaries in search of grain. They handed out orders that conflicted with those of the next town up the road, and brought swift protests.[37]

[31] See the speeches of Creuzé-Latouche, in *AP*, 54: 676–88.

[32] *AP*, 53: 84.

[33] *AP*, 53: 131.

[34] François Hincker, "L'idéologie économique à l'oeuvre dans les débats et les décisions concernant la fiscalité pendant la Révolution," in *Etat, finances et économie*, pp. 355–63.

[35] Dominique Margairaz, "Le maximum."

[36] AP, 49: 511–2; *AP*, 50: 62–3; Mathiez, *La vie chère*, p. 89; Marcel Dorigny, "La conception et la rôle de l'Etat dans les théories économiques et politiques des Girondins," in *La Révolution française et le développement du capitalisme*, p. 130.

[37] Part of the problem was that new district and departmental boundaries cut across the zones

By the late autumn 1792, the correspondence from across the nation evinced great skepticism, even cynicism, about the Convention's commitment to free trade. The Committee had not been surprised when the Convention had cheered Fabre d'Eglantine's 8 November speech. The representative, a militant member of the Cordelier Club, had asserted that France "in general, reaps enough grain for the consumption of its inhabitants," and then proposed regulations. He demanded that all owners submit declarations of the grain in their possession, that local authorities assess requisitions, and, among other items, that exporting grain illegally be punished by two years in irons.[38] Publicly, many on the Committee still upheld *libre circulation*. Privately, however, they discussed the possibility of imposing a maximum. That same week, they sketched out a law for more stringent grain censuses and confiscations, punishments for grain riots, and for funding more advances for departmental grain purchases.[39] From the electoral assembly of the Seine-et-Oise came an increasingly popular challenge to the Convention's liberal policies. "Is free trade in grain incompatible with the existence of our republic?," it asked belligerently. It answered that the profits on the trade enriched the few at the expense of the many, and therefore had to cease.[40] Such was the crisis that had brought Le Quinio, Fauchet, and others to the Convention's podium in late autumn 1792.[41]

The representatives made one last attempt to uphold the truth as they understood it: There was enough grain, and obstacles to its transport had to give way. If it took harsh measures, so be it. On 5 December 1792, the Convention decreed the death penalty for anyone exporting grain. The next day it voted the same fate for anyone obstructing grain shipments to Paris. In addition, the Convention offered rewards for anyone denouncing those guilty of such crimes, even if they had been accomplices. The strongest injunctions appeared in the famous law of 8–10 December 1792: death for anyone interfering with any grain shipments on French territory, with strict instructions for local officials to repress and punish the guilty. Any damages were to be repaid by the communal authorities of the place where the riot occurred, and presumably passed on to the citizens themselves. Among the last articles was one formally terminating the requisitions allowed by the law of 9 September. The "most complete freedom of the grain and legume

from which cities had drawn their grain in the past. On the conflicts between cities in the Seine-Inférieure and the areas it wished to requisition in the Somme and the Eure, see ADSM L 451 and 453.

[38] *AP*, 53: 131–2.

[39] Roland had objected, arguing that the departments had sufficient authority to repress riots and maintain trade networks. *CAC*, 3: 17–28.

[40] Cited in Mathiez, *La vie chère*, pp. 101–2.

[41] See the distressed correspondence from the department of the Eure in AN F11 213.

trade" and its indispensable counterpart, *libre circulation*, were to be the source of the nation's well-being.[42]

The response was anguish. Facing famine, departments uttered even sharper cries for help. They demanded government imports. Few imports arrived, of course, especially after the outbreak of the war with Britain cut off shipments to France in January 1793. Barring those, the departments wanted requisitions and price controls.[43] In the district of Vernon, thirty-eight communes protested the law of 8 December.[44] They were seconded by Parisian calls for economic controls for a wide range of goods. For many months, the Paris Commune and the Convention had directed much attention and money toward providing subsidized bread to the city.[45] Since bread prices still were firmly capped, high sugar and coffee prices incited the population's greatest wrath. Together, the Convention and the sections denounced hoarders, their common enemies.[46] Both could agree that a central problem was the lack of food, and so their accusations against *affâmeurs* served their desires to move the opposing party to action.

Mayoral Missteps: Rouen, August 1792

The events in the Seine-Inférieure give us a sense of how the local administrators responded to the provisioning problems, and of their faltering attempts to reinstate market mechanisms. Their strategies did not resemble the simulated sales of the previous decades, however. There were no attempts in 1792–3 to keep the government's supplies a secret, nor to use them to shape prices. The government grain was placed for sale – whether in the hall, to bakers, or to communities as a whole – "at the going rate" as the law of 1790 required. The government's desire had been to cover its costs, and if possible, to avoid annoying merchants.[47] As for the authorities' intervention in riots during these months, there were great differences.

[42] Roland to the Administration of the Department S-I, 14 December 1792, ADSM L 383; Duvergier, *Lois*, 5: 86–7, 90.

[43] *CAC*, 3: 40–94.

[44] Jacques Godechot, *Les Institutions de la France sous le Révolution et l'Empire*, 2nd ed. (Paris, 1968), pp. 410–1.

[45] The Commune increased its role as the capital's provisioner, from supplying 5 percent of the monthly cereal needs in March 1792, to 70 percent in August 1792. From December 1792 through January 1793, bakers were offered further bonuses for any flour they purchased from other sources, and they received other subsidies in February and March 1793. *CAC*, 3: 24; Petersen, "L'approvisionnement de Paris," pp. 366–85.

[46] Those demands also reflected tensions between the sections and the Convention. The Convention wished the city to raise the capital's bread prices. *CAC*, 3: 62–3, 89–90.

[47] 12 November 1790, ADSM LP 8650*.

In 1792–3, there were none of the calculated attempts to define free trade or to negotiate prices. Instead, the authorities' repertoire was particularly stunted. To varying degrees, the 1792–3 methods were generally more blunt, and seem to have created much less analysis.

In De Fontenay's bureaus – whether during his 1791–2 term as mayor of Rouen, or as president of the department's administration in 1792–3 – there were intermittent efforts to restore *libre circulation* and to allow prices to rise. Like the Conventions' representatives, he held that high prices would bring sufficient grain to his territory. None of his policies succeeded, however. The first of these, directed at the baking trade, nearly brought down the city's government in August 1792.[48] The city's officers had frozen bread prices in mid-July. In a few weeks, the market was empty, and what little grain could be found was very dear. De Fontenay blamed the bread price ceilings. On the evening of 28 August, he and the municipal council decided to suspend the city's bread *taxe*, thus permitting bakers to charge whatever they wished. The next morning, the people stormed the city hall. Calls for section meetings echoed throughout the city. Working-class quarters clamored for bread at 2 sols. According to De Fontenay's reading of the 1787 *tarif*, bread should have cost 3 sols 5 deniers, and the bakers were charging over 4. The crowds roundly denounced the mayor's policies as "uncivil and unsuitable." A "subscription" to pay for flour subsidies was drawn up. The wealthy were to contribute; the poor had refused. The outgoing Legislative Assembly allowed Rouen to buy up whatever grain it could in the surrounding areas, and even to purchase grain from military warehouses. De Fontenay had underestimated his citizens' attachment to cheap bread. His actions, by any account, had been far too abrupt and had caught everyone off guard. Within hours, he had declared martial law, returned to controls, and reorganized the city's national guard (Figure 6.1).

Two weeks later, the legislature had voted the 16 September requisitions. Provisioning lines fell apart. Every market in the region reported increasingly bare markets, and many were angered by Rouen's aggressive purchasing.[49] The district of Rouen first tried to use the "path of persuasion," sending its officials to visit farmers and offering troops to protect grain wagons. It then demanded its rightful requisitions, and was uniformly

[48] This account is based on the "Extrait des déliberations du Conseil Gal de la Commune de Rouen . . . ," 23 September 1792, AN F11 220; and the numerous records in ADSM L 407, 2286; ADSM LP 6431, 8452; AM Rouen I5 6; AM Rouen F3 2; Claude Mazauric, *Sur la Révolution française; contributions à l'histoire de la Révolution bourgeoise* (Paris, 1970), pp. 181–91.

[49] See the regular market records in ADSM L 407. Pigout, *Bolbec*, pp. 190–3.

Figure 6.1 Mayor De Fontenay explains bread prices and free trade. This is a sketch for the nineteenth-century painting, *Scène de vie publique pendant la Révolution* by Louis Boilly (1761–1845). The artist depicted De Fontenay, the Mayor of Rouen, as a hero stepping forward to address a crowd during the riots of 29 August 1792. Bibliothèque de Rouen, Fonds Leber. Reproduced with permission from the Bibliothèque de Rouen.

refused.[50] Worse, Le Havre had announced that it would not risk forwarding incoming grain to Rouen.[51] By early October, the department reported that several of the more radical citizens of the city, including a former *avocat*, a wine merchant, and a journalist, "have put everything on fire in the sections." The department tried to justify its opposition to price controls to its citizens. In a lengthy "Instruction," written as it stepped down

[50] Arrêté, District of Rouen, Procureur général syndic, 29 August 1792, ADSM L 2286. On problems with the requisitions, see the records in AN F11 213, 220; ADSM L 383, 2286; and ADSM LP 6431, 6432.

[51] Le Havre's mayor was among the merchants who had brought grain to the port for the government's purposes, and thus had a deep interest in protecting the supplies. Bureau municipal du Havre to MI, 27 September 1792; and Rialle et Cie to MI, 15 October 1792, AN F11 225–6.

before the new elections, it blamed the many market and roadside disturbances for frightening away sellers. It also insisted that the shortages were real: All the attempts to scout out illegal stores had found none. The specific effect of the unrest and requisitions had been to cut bakers off from their normal paths of supply, the department officials alleged. When bakers and *blatiers* went to other towns to buy grain, they risked being attacked as they returned. Thus, the department explained, bakers had no grain, and when their shelves emptied, townspeople crowded into the even more barren market hall in search for food. The department argued that calm and time alone would restore the normal provisioning paths, and in the meantime, it demanded more supplies from the Ministry of the Interior.[52] When the calm did not return, Rouen eyed the military warehouses, and finally was permitted to buy the first of many rounds of supplies from that source – 4,000 sacks of flour and grain – and sold them to bakers.

Hunger and Rebellion: The Disintegration of Local Authority

The Seine-Inférieure administrators were tiring of the deputies' endless harangues about the value of *libre circulation*, judging them to be a ruse by which the legislature would favor Paris's provisions, and which also permitted merchants to reap enormous gains.[53] By December 1792, many public authorities in the Seine-Inférieure and elsewhere were no longer willing to abide by the Convention's policies.[54] The elections of November–December 1792 returned somewhat poorer and more Jacobin administrators to the municipal offices across the department. They decried the ruinous activities of "*capitalistes*," such as the grain merchants and landowners who operated with impunity under the Convention's liberal policies.[55] The municipality of Elbeuf, among many others, sent the Minister of the Interior an extensive list of necessary procedures: prices fixed between 8 and 10 livres per quintal, bread set accordingly, assignments given to farmers to supply specified markets, and corporal punishment for those who did not comply. Not waiting for permission from a minister they knew would oppose their designs, the officers put the plan in motion.[56] Seine-Inférieure officials demanded more forceful laws with which to repress the

[52] "Instruction relative aux subsistances," 14 October 1792; Department S-I to MI, 5 January 1793, ADSM L 383.

[53] AN F11 213; *CAC*, 3: 17–39.

[54] Seine-Inférieure to MI, 22 December 1792, AN F11 220; ADSM LP 6431 and 6432.

[55] Lemarchand, *La fin du féodalisme*, pp. 470–1.

[56] Kaplow, *Elbeuf*, pp. 133–4.

unrest. "We recognize the truth of [free trade]," they argued, but "the principles, however true, are in no way applicable to our situation." They recounted the problems facing the department: municipal riots, farmers under seige, communes that barred outsiders from their halls, especially in the Eure. The laws available were "insufficient," and gave them little support for prosecuting the guilty, they believed. They needed to be able to move against both rioters and farmers.[57] Roland's reply was succint. He reminded the department of the enormous help it had received already – almost 25,000 quintals of grain, 150,000 livres in funds, and a loan of 300,000 livres.[58] Surely it could not ask for more.

The funds had been insufficient, and so Rouen tried once again to rouse commercial undertakings by raising bread prices after early November – each increase provoked disturbances.[59] By January 1793, the requisitions of the previous autumn and the continuing unrest had wrecked the grain trade. De Fontenay outlined the dilemma: If the department increased bread prices, more grain might arrive; if bread prices remained low, the city would be forced to purchase every sack needed.[60] The department persuaded Rouen officials to raise the *taxe*, but still no grain appeared. "Commerce provides almost nothing," the department wrote. The department officials were "taking care not to sell the [government supplies] below the going price for domestic grain." The city, however, was matching bread prices to those of the hall, where sellers generally feared to ask a fair market rate. Thus, bread prices lagged behind what the department believed to be a true commercial price. The city had become "responsible almost exclusively for the city's subsistence." The Rouennais lived from hand to mouth that spring, quickly devouring anything the authorities could scavenge. When additional sharp grain price increases in March and April 1793 led to higher bread prices, the people rebelled. Their "*rassemblements*" surrounded the city hall on 1 May, causing the department to send yet another delegation to Paris to plead for grain. The men arrived on 4 May 1793, as the Convention prepared to vote for the maximum.[61]

[57] Administration of the Department S-I to MI, 4 December 1792, ADSM L 383.

[58] Roland to the Administration of the Department S-I, 14 December 1792, ADSM L 383.

[59] 11 December 1792, AM Rouen I5 6. The following is drawn from the correspondence in ADSM L 383. Le Havre also placed bread prices at those of grain, angering its citizenry. "Adresse," 27 April 1793, AM Le Havre PR F4 148.

[60] That was the case in Paris. The government was providing nearly all the grain consumed in the city, at tremendous cost. In Bolbec, too, bread prices had been kept low. *CAC*, 3: 24; Petersen, "L'approvisionnement de Paris," pp. 366–85; Pigout, *Bolbec*, pp. 206–7.

[61] In early March, interim Interior Minister Paré told the department that he had most recently given it 12,000 quintals and that only the Convention could help it now. See the correspondence between the departmental administration, and Paré and the Convention in ADSM L 383; and *Délibérations*, pp. 118–20.

These discussions of pricing strategies are among the few that appear in the correspondence of the many department, district, and municipal bodies. In most other records, the prevalent theme was attempts to secure government grain or loans. The officials demanded supplies from incoming ships or military warehouses – often breaking promises to pay for them. In January, when the department failed to repay the military what it owed in October, for example, the Minister of War cut off further loans of cereals. The department's letters listed the numbers of jobless workers and described the Seine-Inférieure's poor soil. The administrators thanked the Ministry of the Interior and the Convention when they sent food, and argued tirelessly when they did not. They reported how the grain had been distributed: 3,000 quintals to Ingouville, 6,000 to Le Havre, 3,000 of flour to Fécamp. But other than trying to ascertain the going rate, these operations bore little similarity to Old Regime strategies, and occasioned no deep discussion.[62]

A significant problem was the unending episodes of *taxation populaire* and assaults on grain shipments. Once many of the nation's men had marched off to fight, there were few soldiers able to confront the unrest. If the conditions in the Seine-Inférieure were at all representative, it seems that the forces of order had collapsed.[63] The law required officials to confront disturbances and to prosecute the guilty. The costs the reimbursements or fines for stolen grain were to be passed on to the community in which the disturbance had happened. The authorities of 1792–3 appear a feckless lot, but perhaps it was not their fault, for they often had no means of enforcing their will.

The "*attroupement*" at Le Houlme in October 1792 revealed how little local authorities were able to do. For weeks, people from surrounding towns had laid siege to *blatiers* traveling to Rouen. When judicial procedures finally began, the local authorities were questioned. Several claimed to know nothing of the events. The commune's secretary said he had arrived too late to help. He and a cavalry commander had briefly discussed the people's hunger. The commander then had turned to the crowd and explained that he, like any other official present, would be punished if he did not "oppose" it. "Saddened" to see their desperation, he had left, saying that if they repeated this offense, he would find them "even more guilty because now they knew the law." The *procureur*, too, had told them to obey the law, and that "he had no right to sell the grain to them or set a price." This news caused the frightened *blatier* to lower his price and flee; the *procureur* then went home to write up a report. An armed detachment arrived at some point, but realized that they had only twelve ammunition

[62] ADSM L 383.

[63] "Participation d'attroupement au Houlme," December 1792, ADSM LP 6432.

cartidges between them, and sent for help. As they began making arrests, one of the most angered of the "*révoltés*" almost bit off the finger of a *grenadier*. The *grenadiers* then turned on their commander and told him they would not make any more sorties from the city. The authorities had all departed. Only the cavalry detachment had made any attempt to settle the disturbance by bargaining with the crowd and the *blatier*. One can sympathize with all involved – the townspeople, the *grenadier* almost sans finger, the *blatier*, and the outnumbered officials. Such reports were standard in 1792. Darnétal officers, in another instance, received daily threats and wished to resign. Their responsibilities for finding food and maintaining order had overwhelmed them too.[64]

Troops were few, and they were not disposed – or were utterly unable – to halt the unrest, or to negotiate a price or oversee more peaceful distributions of grain. The lack of available force, and the roadside, as opposed to marketplace, locations of many of the disturbances, placed the officials in a very difficult position. Local authorities were not attempting to reimburse losses, which further frustrated owners.[65] When merchants and producers then refused to supply markets, the problems only escalated. The treasury had little supplementary help to offer, and in some cases, government purchases had contributed to the breakdown of ever-tenuous trade networks.

Beyond those difficulties, there is much that the records do not state. Did the officials know of the methods that pre-Revolutionary authorities had tried when confronting unrest, for instance? Given how sharp a line had been drawn between the Old Regime and the new, since June 1789, and how much the former had become vilified, it might have been difficult to countenance any return to Old Regime practices. Each time a representative spoke of the grain trade in the "former government" – meaning not only the Old Regime, but the Constitutional Monarchy, he conjured up the stock images of the "corrupt [royal] court."[66] To invoke the Old Regime in 1793 was to invoke its wholly demonized image. These dividing lines made it difficult to look with favor on any policy associated with the monarchy, and may account for some part of repudiation of former measures.

How did the Seine-Inférieure of officials envision *libre circulation* and free trade? This seems to have been particularly irrelevant. Were Revolutionary crowds more insistent than they had been during the Flour

[64] "Emeute pour du blé à Darnétal," December 1792, ADSM LP 6432. See also problems in Le Havre, District of Montivilliers to the [Municipal Officers of le Havre], 9 November 1791, AM Le Havre PR F4 158.

[65] "Décret de la Convention nationale," 11 April 1793, AM Le Havre PR F4 158.

[66] *AP*, 53: 83–4.

War? Certainly the political lessons of the pre-revolutionary *tarifs* and of the authorities' actions in 1789, not to mention the Parisian uprisings in 1792, could have given them reason to demand better from their officials. Did the officials see their own responsibilities differently? Given that officials regularly insisted that they had addressed their townspeople "as they should" – by crying "Obéissance à la loi!" three times and ordering them to disperse – it is clear that the laws were guiding their responses, or at least their later recounting of the incident. Having shouted those words, did they believe they had further options or duties? The laws promised swift punishment for any authority who became involved in disturbances, which certainly limited his alternatives.

The correspondence and court records reveal few of the nuanced discussions and justifications that had characterized Old Regime strategies. In comparison, the Revolutionary documents seem rather "flat." The need for grain was acute; inflation and war had added further burdens; government purchases were insufficient and only aggravated the few merchants still willing to be involved in the trade; the authorities were to repress disturbances and round up the guilty. In short, there was plenty of blame to be parceled out. Even the Seine-Inférieure's officials' concerted, if sometimes ill-conceived, attempts to revive supply lines met up with those obstacles. The grim circumstances of 1792–3 seems to have overwhelmed the authorities, along with the potential buyers and sellers they rode out to confront.

The First Maximum: 4 May 1793

The further radicalization of the Convention in the spring of 1793 is familiar to historians. On the one hand, the problems the representatives faced were of a magnitude unimaginable even a year before. On the other, within the Convention itself, there was little agreement, and Parisian politics were more restive still. The Jacobins saw their opportunity in the acute pressure for economic controls from the clubs and sections – among other political elements they could build upon. The belief that the country had enough grain – or could grow enough – did not waver. The means by which the Convention chose to get grain to cities, however, did shift. The Jacobins turned on the "guilty" – the producers and merchants hoarding supplies, the émigrés who had abandoned their land. What freedom had not supplied, the Jacobins would use force to obtain. Whereas the Convention had prohibited calls for egalitarian economic policies on 18 March 1793, the Jacobins increasingly saw the redistribution of property as one means for putting every acre under the plow, and thus remedying the seemingly intractable problem of poverty.[67] Emigré lands, some abandoned, also held promise, and so the decree of 23 March 1793 ordered municipalities to farm them.[68]

Still arguing that there was enough grain, the Agriculture Committee began formulating the radical policies the departments had been demanding for well over a year. As news spread that the Convention might countenance more sweeping measures, the committee was overwhelmed by proposals.[69] Large cities viewed calls for uniform price maximums with some ambivalence. Some, like Rouen, doubted there really was enough food. While ceilings would quiet their populace, they would not encourage owners to undertake the transport of wagonloads from rural to urban areas. When Dieppe demanded "that the price of grain be everywhere the same," the departmental officials objected, knowing Rouen would starve. Instead, they wished to tax the rich and use proceeds to furnish cheap bread for the poor.[70] The Committee members pondered such matters in late April and early May. Their unease with the policies, or with the sharp turns of politics within the Convention itself may explain why only a handful of the Committee's twenty-four members attended the meetings in late April where the Maximum was sketched out.[71]

The first grain Maximum was decreed on 4 May 1793.[72] It reinstated many former regulations, such as limiting transactions to the marketplace and ordering a census of all grain owners. It also allowed local authorities to requisition grain from their regions. The price ceilings were to be set by each district, according to the average of area market prices in the preceding months. That ceiling was to be decreased each month until the Maximum lapsed as the harvest appeared in September. It ordered the death sentence for any hoarder who sought to evade the law. There was tremendous initial confusion about the ceilings, and the correspondence of the Ministry of the Interior and the Committee rapidly doubled.[73] Each department, of course, was desperate for grain, and turned the 4 May law to any benefit it could find. In general, department administrators sought to calculate price ceilings slightly above those in neighboring departments, so as

[67] Hirsch, "Terror and Property," pp. 215–7.

[68] It appears few did. Octave Festy, *L'agriculture pendant la Révolution française: les conditions de production et de récolte des céréales* (Paris, 1947), p. 196.

[69] *CAC*, 3: 111–20.

[70] Conseil général S-I, n.d., ca. March 1793, ADSM L 383.

[71] *CAC*, 3: 117–226.

[72] "Decret . . . du 4 May 1793," ADSM LP 8650*; Caron, *Le Commerce descéréales*, pp. 46–9.

[73] There was universal uncertainty whether ceilings were department-wide or if each district could set a different Maximum. The records of the Maximum are in AN F12 1547C.

to attract as much grain as possible. These tepid efforts rapidly undermined the May Maximum.[74] The Seine-Inférieure, for instance, set a high departmental price on 16 May, and but did little to enforce it. Its only interest was in levying requisitions, although it lacked the manpower to uphold those.[75] The Minister of the Interior's 11 June 1793 circular condemned all such machinations, blaming them on "the greed which has calculated the sum of the people's needs, in order to fatten itself on its provisions." The tendency of local officials to over-requisition their territories had made cities "the prey of horrors." "Thus," he lamented, "the salutory measures of the law of 4 May become illusory and without effect. . . ."[76]

The May Maximum required an unprecedented level of cooperation, especially across department and district boundaries, and it was not forthcoming. Without fail, local authorities refused to send grain to areas that traditionally had drawn grain from their markets, knowing the new political frontiers offered some protection.[77] In theory, departments had the authority to send workers and troops to farms to harvest and thresh grain and then bring it to the hall. In practice, however, they could not muster the necessary hands, especially when the fields were in a rural jurisdiction. Thus, the immediate impact of the maximum was to empty markets.[78] In the Seine-Inférieure, there were sporadic attempts by communal and district authorities to send the national guard or other delegations out to seize requisitions, but these were uncoordinated and generally unsuccessful. The department, preoccupied with Rouen's needs, did little to sort out disputes between smaller villages.[79] To remedy shortages, the Ministry of the Interior had offered the Seine-Inférieure imports arriving in Dieppe and Saint Valéry. Unfortunately, when the communes' agents reached the ports, they found that *sociétés populaires*, and municipal and district officials refused to honor the Ministry's mandates.[80] Thus, communities embarked on lawless,

[74] Only twenty-five departments sent copies of their *arrêtés* imposing price ceilings, for instance. Other strategies included delaying the starting date and concealing price differences by using various volume measures. See the many complaints in AN F12 1547C; and Antoine Richard, "L'application du ler Maximum dans les Basses-Pyrénées," *Annales révolutionnaires*, 13 (1921), pp. 208–10.

[75] AN F12 1547C; Lemarchand, *La fin du féodalisme*, p. 482.

[76] "Resultat de la conference qui a eu lieu entre le Ministre de l'intérieur, des commissaires du Département de Paris . . . ," 24 May 1793, AN F12 1547C; *CAC*, 4: 179.

[77] Municipality of Perenas to the Department of the Hérault, 15 July 1793, AN F12 1547C; Richard Cobb, *The People's Armies: The armées révolutionnaires: Instrument of the Terror in the Departments April 1793 to Floréal Year II*, trans. Marianne Elliot (New Haven and London, 1987), p. 262.

[78] AN F12 1547C.

[79] It replied only that the commune or the district would have to send armed force into areas that did not heed the requisition orders. ADSM L 1387.

[80] Correspondence of the District of Cany, ADSM L 1387.

but wholly necessary, forays. Having no mandates to claim grain in port, for instance, the district of Cany seized 1,400 quintals of Parisian grain at Saint Valéry-en-Caux. The district explained only that "hunger silences every law."[81] There were additional problems with Parisian supply lines. Its traditional provisioning zones were angry that the capital had set its maximum so high. They retaliated by cutting off shipments entirely.[82] Illegalities, jealousies, and empty markets marked the first Maximum. In sum, one department concluded that "this law, instead of bringing our people food at a moderate price has made grain completely disappear from our public markets."[83] As the summer's harvest proceeded, one department after another announced an early end to its ceilings.[84] In part, this was occasioned by the vague wording of the May 4 law, which outlined the calculations for price ceilings that steadily decreased month by month during the summer, finally lapsing on September 1 when the harvest was expected to be finished.[85] The incoming harvest was just a pretext, though, for many departments' decision to lift the 4 May maximum. In general, they complained about the unevenness of its application, especially the wildly varying price ceilings. Ironically, the jurisdictions that had observed it most faithfully had been the most likely to suffer.[86]

Their overall assessment, however, was not that the policies should be abandoned. Rather, they argued that greater enforcement was needed.[87] Most significant, many petitions demanded a uniform national maximum so as to avoid precisely the schemes that had undermined the May measures.[88] In the Seine-Inférieure, for instance, the department administrators concluded that the May requisitions had been too disorganized. On 14 August, the department stepped in to construct more effective levies based on a system of provisioning *arrondissements*, thereby designating the zones for the four most restive districts, Rouen, Caudebec, Dieppe, and Montivilliers, which included Le Havre.[89] These orders came at the

[81] 7 August 1793, ADSM L 1387.

[82] *CAC*, 3: 140–2.

[83] Extrait du proces-verbal, Conseil général du Département du Cantal, 20 June 1793, AN F12 1547C.

[84] AN F12 1547C.

[85] AN F12 1547C.

[86] See the numerous criticisms offered, especially from the Cantal, the Hérault, and the Lot et Garonne in AN F12 1547C.

[87] Some regions criticized the Maximum for violating property rights, or for being impossible to coordinate. See the many complaints in AN F12 1547C.

[88] For example "Extrait du registre des délibérations du Conseil général du Département du Sarthe," 16 August 1793, AN F12 1547C.

[89] The department imposed a weekly levy of 67,200 quintals, estimated at 4 quintals per *charrue* farmed from each district: Rouen (5,300), Caudebec (14,700), Montivilliers (11,350), Cany (8,900), Dieppe (11,500), Neufchâtel (9,400), Gournay (6,000). "Etat des

worst time in the year, however. The harvest was not yet available and old supplies were exhausted, so despite the new impositions, the markets were barren. Crowds massed in front of municipal and district offices throughout the department, and once again some jurisdictions proceeded to employ desperate but illegal means to wring grain from other towns.[90]

To counter the summer's unrest, the Convention had decreed stronger measures, preparing the way for the controlled economy of 1793–4. On 26 July, hearing reports that insurrections along the Seine in the Calvados threatened Parisian shipments, it reiterated the death penalty for hoarding.[91] On 9 August, districts were ordered to establish public granaries, using émigré property, and towns that wished could establish public bakeries.[92] To speed grain to the army, the Convention pressed departments to harvest and thresh their crops.[93] The issue of farm labor was critical: Conscription had taken many men, and the landowners' demands for high cereal prices had emboldened others to seek more money. Farm workers joined together, announcing their unwillingness to work for low wages.[94] That raised the ever-present issue of maximums on other goods and labor.

The Minister of the Interior and the Convention's committees, considering the situation, concluded that what was needed was a single price for the nation – the May 4 ceilings had ranged from 13 livres 17 sols in the Loiret to 28 livres in the Cantal. The committee recognized the many problems that might result – not the least of which was how to calculate transport costs – but believed that price ceilings offered greater hope of success than free trade. And it did take to heart the concerns that had emerged in the meeting with the representatives from the Seine and the Seine-et-Oise about Parisian supplies, that a single maximum inevitably hurt a farmer who had to sell his grain for one price, yet buy the other items at "the price that it pleases the seller to set."[95] The requisitions and ceilings had been a chaotic undertaking and needed greater coordination and supervision if they were ever to be used again.

réquisitions qui ont été faites . . . ," 26 Pluviôse [Year II] (14 February 1794), ADSM L 2381; ADSM L 1387; Lemarchand, *La fin du féodalisme*, p. 482.

[90] ADSM L 1387.

[91] 26 July 1793, ADSM LP 8650*.

[92] 9 August 1793, ADSM LP 8650*.

[93] 14 August 1793, ADSM L 8650*.

[94] Ragan, "Rural Political Culture," pp. 133–5.

[95] "Resultat de la conference qui a eu lieu entre le Ministre de l'intérieur, des commissaires du Département de Paris . . . ," 24 May 1793, AN F12 1547C. See, too, Conseil Général of Beziers to the Administration du Département [de l'Hérault], 9 July 1793, AN F12 1547C.

The Controls of Year II in the Seine-Inférieure

By early September 1793, nearly two years of unrest and confusion over provisioning policies had taken their toll. The Federalist movements and counterrevolution had added further burdens to the young republic. Market disturbances were widespread, grain owners feared for their lives, and no clear path was being charted by the Convention. Pillaging and high prices threatened the supplies for the military. The Convention sent representatives to departments to restore order, but these beleaguered deputies found supplies so short that crowds turned on them.[96] In Paris, a war between the Subsistence Administration and the Minister of the Interior erupted. In mid-August, the Subsistence Administration, which was responsible for the capital's provisioning, plastered the city with posters accusing the Minister of trying to starve the city. The Administration claimed he had been far too moderate in his dealings with the outlying departments supplying the city. Angered, it had addressed the departments independently, demanding they send more grain, to little avail.[97] On 5 September, Chaumette, president and *procureur* of the Paris Commune, issued his much-repeated call, "Subsistence, and to get it, force for the law," demanding greater use of troops to supply his city. The Convention's creation of a vigilante force, the Revolutionary Army, to supervise the capital's supply soon followed. Terror was declared the order of the day, and the Commerce Committee was charged to render an opinion on maximums for all "items of first necessity."[98] The September Maximums were underway.

The decision to craft more sweeping controls was supported by the Convention's belief that the country truly had enough grain. The initial news about the harvest was reassuring. Thus, the goal became once again to force the grain out of the hands of owners. To that end, the Convention established the economy of the Terror. On 11 September, it decreed national price ceilings for grain and flour, setting transport costs and instructing departments to coordinate grain censuses and requisitions for public markets and granaries. The best quality wheat was not to exceed 14 livres

[96] See the fate of Chabot and Dumont, dispatched to the Somme, Ragan, "Rural Political Culture," p. 130.

[97] The minister replied to the charges through a meeting with the Convention's Commerce Committee, explaining that he had followed every legal channel available and arguing further that the Subsistence Administration had to work in concert with the Commune, the Ministry, and the Convention, and could not launch out independently. *CAC*, 3: 141–2, 4: 173–9.

[98] *CAC*, 4: 200–1; Sewell, "Rhetoric of Subsistence," pp. 257–8; Cobb, *The People's Armies*, pp. 34–6.

per quintal and the best flour was set at 20 livres, with the appropriate additions for shipping. The results of the grain censuses were to be reported to the Ministry of the Interior so that plans for the nation and its soldiers could be coordinated and fraud prevented.[99] On 29 September, the deputies extended the ceilings to all "goods of first necessity," from cloth to candlewax, to metal and shoe leather, and also to wages. Requisitioned "first level goods," such as grain, were to go to specific markets or granaries, while lesser foodstuffs, dairy products, for instance, could be sent to any market the owner wished. The Parisian Revolutionary Army was sent in particular to the twelve departments that generally supplied the capital.[100]

The activities in the Seine-Inférieure, a department that was incapable of growing all the grain it needed in the best of times, suggest that the harvest was much poorer than the Convention wished to believe, and the supply networks had been far more damaged than was understood (Figure 6.2). The department's additional August levies had been fruitless, and in mid-September, the commune of Rouen sent fourteen agents to the seven districts. They too returned emptyhanded. The commune's appeals to the Eure and to Verneuil had garnered only the information that Paris and the army had laid claim already to those supplies, and that the detachments of the capital's Revolutionary Army would brook no contradictory orders.[101] The Convention, surmising that unrest in the Seine-Inférieure, especially the District of Cany, threatened Paris, sent representatives on mission to the troubled district.[102] They found conflicting orders throughout the region.[103] Moreover, the increasingly powerful districts had taken it upon themselves to step into the vacuum, creating more confusion.[104] The owners, using the chaos to their benefit, only shipped 32 percent of the grain requisitioned for the District of Rouen in September.[105]

[99] "Décret de la Convention Nationale," 11 September 1793, ADSM LP 8650*.

[100] Cobb, *The People's Armies*, pp. 286, 311.

[101] 19 September 1793, AM Rouen F4 16; Cobb, *The People's Armies*, pp. 262–4.

[102] The representatives drew up a painstaking plan for the supply of the District of Cany. "Arrête," 18 September 1793, ADSM L 1387.

[103] ADSM L 453.

[104] ADSM L 1387. From their creation in 1790, the districts were charged with a number of fiscal areas. The Convention gave the districts greater power in its reorganization of local government in November 1792. On 4 December 1793 (4 Frimaire Year II), it brought the districts directly under its own authority and made the districts' *procureur syndics* into *agents nationaux*, the official representatives of the Convention. The goal of such changes was to bring more direct supervision by the Convention to bear on the local population. As the Jacobin policies veered toward greater regulation, the districts received more power to carry out those decisions. The districts came to oversee requisitions for the military and the provisioning of the local marketplaces. Godechot, *Institutions*, pp. 106–7.

[105] 23 and 26 September 1793, ADSM L 1387.

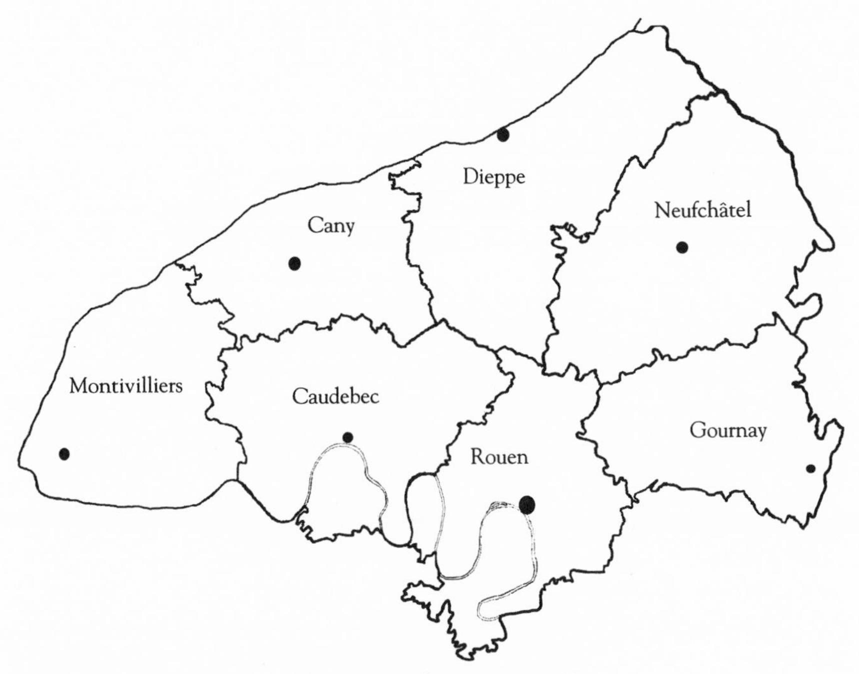

Figure 6.2 The districts of the Seine-Inférieure in 1792.

Such disorder, hardly unique to the Seine-Inférieure, could not continue. Within a month, the Convention restructured and nationalized provisioning. On 27 October 1793, the Convention sought to counter the confusion by placing all the nation's provisioning activities under the purview of the Republic's Subsistence Commission, headed by Robert Lindet.[106] Henceforth, it would issue and approve all requisitions, review censuses, and weigh the needs of its citizens and its armies. Subsidiary committees were to be created at the local level, responsible to the national one. On 10 Brumaire Year II (31 October 1793), the Convention authorized a new levy for the District of Rouen: 10,500 quintals of grain per week to be sent from the four best-supplied disticts, Montivilliers (3,800), Cany (3,700), Caudebec (1,870) and Dieppe (1,130).[107] These amounts, if they could even

[106] On 22 Pluviôse Year II (10 February 1794), the Commission was divided into two bodies, the Commerce and Provisioning Commission, and the Agriculture and Arts Commission. Godechot, *Institutions*, p. 415.

[107] "Arrêté du Départment," 10 Brumaire Year II, ADSM L 379; "Resultant des grains et farines existants . . . ," 22 Frimaire Year II, ADSM L 390.

Table 6.1 *Grain census in quintals according to the law of 11 September 1793*

District	Wheat	Méteil
Rouen	146,366	5,406
Caudebec	383,893	4,107
Montivilliers	510,510	3,986
Cany	422,269	15,943
Dieppe	347,423	8,325
Neufchâtel	155,382	18,065
Gournay	98,770	11,931
Total	2,064,613	67,763

Source: Report of 22 Frimaire Year II, ADSM L 390. See also Ministère de l'instruction publique, *La Commission des subsistances de l'An II: Procès verbaux et actes*, Pierre Caron, ed. (Paris, 1924), pp. 12–3.

be found, would not be sufficient. The city of Rouen alone required 9,800 quintals per week (Table 6.1).[108]

Despite the Herculean intentions of the September Maximums, the execution and results were mixed at best. Owners refused to obey the orders, and market supplies fell further. The censuses, done with little accuracy, whether due to the administrative inabilities or recalcitrance, rendered little sure information. It is in fact difficult to tell how far short the harvest of 1793 fell, although the Seine-Inférieure reported 2,496,907 quintals of wheat and méteil for its population of 631,515 (Table 6.2).[109] If one deducted the minimum one-fifth of the crop needed for seed grain and some livestock, there remained only 1,977,525 quintals, or 3.16 quintals per person. That was less than a pound of wheat per day per person, and meant starvation. The department estimated its yearly per capita needs at about 5.5 quintals, and justified that figure by stating it was well below the 7.2

[108] A 7 Brumaire levy by the department imposed the following requisitions per week on the districts: Rouen (4,000 quintals), Caudebec (9,500), Montivilliers (12,500), Cany (10,500), Dieppe (9,500), Neufchâtel (5,000), Gournay (3,000). These decreased levies on the districts of Rouen, Caudebec, Dieppe, Neufchâtel, and Gournay, while increasing slightly those of Montivilliers and Cany, probably in response to more accurate censuses. "Etat des réquisitions qui ont été faites . . . ," 26 Pluviôse [Year II] (14 February 1794), ADSM L 2381; 19 September 1793, AM Rouen F4 16.

[109] 22 Frimaire Year II (13 December 1793), ADSM L 390.

Table 6.2 *Grain harvest of 1793 in the Seine-Inférieure*

District	Quintals of wheat	Quintals of méteil	Population
Rouen	181,850	7,025	174,742
Caudebec	499,744	4,668	88,043
Montivilliers	554,011	4,014	97,260
Cany	461,278	16,119	74,657
Dieppe	425,240	9,132	98,557
Neufchâtel	221,655	2,439	60,218
Gournay	123,647	14,126	38,038
Total	2,417,427	79,480	631,515

Source: Report of 22 Frimaire Year II, ADSM L 390.

quintals the Old Regime had deemed necessary for adult consumption.[110] In the past, it had looked to the Vexin, Picardy, and to imports to make up the difference. Now, those sources had been cut off.

The District of Rouen pressed especially hard for its requisitions, but most of the areas levied, if they replied at all, simply argued that they too lacked food.[111] Thus, in the first twenty-nine weeks, only 61 percent of the grain for the District of Rouen arrived, and even that figure was deceptively optimistic. But for the energetic actions of authorities in the Montivilliers district, by far the most productive and militant of the seven Seine-Inférieure districts, which sent 85 percent of its levies to Rouen, the *chef-lieu* would have perished.[112] The tallies from the Caudebec district showed it to be 48 percent in arrears, and 57 percent of the Cany district's levy has late.[113] The little grain that arrived was milled and distributed directly to Rouen's

[110] See the discussion in "Etat des subistances existantes, au 25 floréal an 2 (14 May 1794) . . . ," ADSM L 390. Each of these figures was disputed throughout the Revolution. The department's claim that the Old Regime allowed 7.2 quintals seems a gross exaggeration, and most estimates fell closer to 4 or 5 quintals per person, depending on how women, the elderly, and children were included.

[111] See the many requests sent by the district of Rouen to communes, requesting census information in ADSM L 378; Kaplow, *Elbeuf*, p. 136.

[112] The moderate political leanings of Montivillier's city officials would lead one to expect the requisitions there to have been carried out with less success, although many of the district's smaller towns and Le Havre had more radical tendencies. Danièle Pingué, "Les représentants en mission: les principaux représentants en mission en Haute-Normandie en l'an II et en l'an III (1793–1795)," in *A travers la Haute-Normandie en révolution*, pp. 375–92; Cobb, *The People's Armies*, p. 695.

[113] 27 October 1793, and [10 and 16 Brumaire Year II], ADSM L 1387.

Table 6.3 *Result of the grain requisition of 10 Brumaire Year II for the district of Rouen, as of Prairial Year II*

District	Owed	Supplied	Arrears
Caudebec	54,230	27,931	26,298
Cany	107,300	46,054	61,245
Dieppe	32,770	19,376	13,393
Montivilliers	110,200	93,381	16,818
Total	304,500	186,744	117,755

Source: ADSM L 390; AM Rouen F3 4.

bakers.[114] The Elbeuf hall also received only a fraction of the sacks it needed, and those numbers fell steadily throughout Year II.[115] That city's appeals to the district of Rouen, in which it lay, and to the department yielded only meager distributions.[116] By 18 Prairial Year II (6 June 1794), the Rouen district's requisition totals were further behind (Table 6.3).[117] The decree of 25 Brumaire Year II (16 November 1793), which lifted limits on how much could be requisitioned, yielded no additional help. There was further confusion over conflicting levies. Districts and communes other than Rouen were sending out their own orders, leading to intense disputes when towns were cross-requisitioned.[118]

The laws allowed more decisive measures to provision the cities of the Seine-Inférieure, but they seem to have been of little help. To execute them more forcefully, the Convention and the national Subsistence Commission sent representatives and agents to the department to assess the level of hardship there and to impose greater order.[119] For instance, to shore up the requisitions, the instructions of 18 Brumaire Year II had told districts to follow traditional provisioning networks. Authorities thus had the right to cross political boundaries in search of supplies. The laws also permitted districts to send agents and troops to farms to confiscate grain, even completing the threshing and sorting if necessary, all at the owner's expense. Few used these

[114] AM Rouen I2 4e.
[115] Kaplow, *Elbeuf*, p. 134.
[116] Kaplow, *Elbeuf*, p. 135.
[117] ADSM L 390; AM F3 4.
[118] See the battle between Buchy (in the District of Gournay) and Cailly (in the District of Rouen) for the grain in Vieux Manoir. The Directoire régénéré of the Department S-I to the District of Gournay, 5 Ventôse Year II (23 February 1794), ADSM L 2382.
[119] *La Commission des subsistances de l'An II*, pp. 48–9.

measures to good effect, however.[120] In early 1794, the districts of Caudebec and Cany, followed by Montiviliers, began sending workers to farms to sieze late requisitions, but these were generally not successful missions. The grain confiscated often went for local needs and was not passed on to Rouen.[121] In Richard Cobb's estimate, these departmental armies proved to be weak and inept, "scraping the bottom of already empty barrels," and were unable to break the back of local resistance.[122] To the south, the district of Evreux (Eure) battled with its provisioning areas on the other side of the Seine in the district of Les Andelys, sending its own army, arresting some farmers, and sowing fear. That brought some grain. The success was shortlived, though, for within days, competing districts had sent their own armies, and confusion prevailed.[123] The city of Mantes, closer to Paris, engaged in its own struggles with the commune of Magny for grain, only to find that most of those supplies were also needed for Paris.[124] And in fact, the Paris Revolutionary Army alone seems to have had sufficient organization and energy to collect the supplies and then to protect them en route.[125]

For additional help, the department authorities turned as always to the government. The national Subsistence Commission came back at them with repeated exhortations to be more effective in carrying out requisitions. If the districts feared the levies would drain them of grain before the harvest, they were to be reassured that the government would step into the breach after local supplies had been exhausted.[126] As the shortages became more severe, individual districts, too, appealed to the national Subsistence Commission, which invariably told the department to divide its supplies further and to make additional levies if needed, such as those for Gournay in Nivôse Year II.[127]

There were more valiant efforts to bring order to provisioning and raise supplies during the spring of 1794. Many of the forms of taxes and leases, such as those for unsold *biens nationaux*, for instance, were increasingly to be paid in grain, thus boosting the Republic's granaries.[128] The powers of

[120] Lemarchand, *La fin du féodalisme*, p. 483.

[121] Lemarchand, *La fin du féodalisme*, p. 483.

[122] Cobb, *The People's Armies*, p. 311.

[123] Cobb, *The People's Armies*, pp. 279–81.

[124] Subtil, "Réglementation municipale," p. 287.

[125] Cobb, *The People's Armies*, pp. 281, 287.

[126] 24 Frimaire Year II (14 December 1793), *La Commission des subsistances de l'An II*, pp. 117–8.

[127] 21 Nivôse Year II (10 January 1793), *La Commission des subsistances de l'An II*, p. 227.

[128] See for instance, Citoyen Charles Duthuit, laboureur, to the Commune of Rouen, 21 Frimaire Year II (11 December 1793) and reply from the Directoire of the District of Rouen, 5 Prairial Year II (24 May 1794), ADSM L 2382.

the representatives on missions to set requisitions, with the supervision of the Subsistence Commission, were increased by the law of 12 Germinal Year II (1 April 1794). But by the early summer of 1794, it was clear that the requisitions and other attempts at provisioning had failed. The Bernay (Eure) Popular Society had reported that the Maximum "is without effect," and many others agreed.[129] The Convention announced that the worst was over, and that the harvest would soon appear. The Parisian Revolutionary Army had been largely successful, and so the heaviest requisitions for the capital had ended in Ventôse (February–March 1794). In the Seine-Inférieure, however, distress continued. Real shortages, aggravated by hoarding and sporadic disturbances, persisted. Worse, the initial evaluations of the harvest augured very poorly. If those estimates proved correct, there was real fear for what lay ahead.

The early years of the Revolution had produced a succession of economic reforms, sincere bows to representative government, and earnest – if less-experienced – administrators. They had also brought the trinity of war, severed supply lines, and inflation. Paris had probably survived the shortages in better fashion than other cities. It had had the Commune's and Subsistence Administration's efforts to thank, and perhaps most importantly, its Revolutionary Army. Some of those results had come at the expense of hunger elsewhere. Rouen, certainly, blamed Parisian incursions in the Eure and the Somme for part of its distress. The 1792 collapse of the grain trade, underway since 1789, became more acute each month. The successive legislatures' insistence that there was enough grain contributed to the final recourse to heavy controls. Whether or not there was enough food is nearly impossible to ascertain. The figures from the Seine-Inférieure indicate cereal shortages of some level, but the depth remains a mystery. The department had every reason to claim it had too little grain: Penury might bring more government supplies. Municipal officials in cereal-producing regions, too, would have found under-reporting a good strategy for limiting the requisitions that Rouen could levy. Certainly, hoarding – easy to suspect, but difficult to prove – and increased consumption by rural folk kept supplies from reaching cities. In short, the difficulties of Year II had not yet been overcome, even though Isoré had declared Paris's provisioning crisis over in the spring of 1793 and had disbanded the Revolutionary Army. In the Seine-Inférieure, concern had grown since the early reports on the harvest had appeared. They were alarming, and then were joined by the news of the Paris coup of 9 Thermidor.

[129] *CAC*, 4: 300.

CHAPTER SEVEN

Re-creating the Market: Thermidor and the Directory

The retreat from the Terror was a halting, tortured process. The men who led the coup of 9 Thermidor had not thought beyond the overthrow of Robespierre; exactly how they would reorganize the nation's political structures seemed of secondary importance. At first, some insisted that there was no need for reform: The Robespierrists, and not the government itself, were guilty of the preceding year's excesses. Within a week, however, that view had been shouted down, and the Convention began the unanticipated and formidable task of reforming the government. Quickly, the delegates stripped the Committee of Public Safety of its powers, and purged other committees of Terrorists. Paris prison doors opened and public executions virtually ceased. The revolutionary committees of the departments were told to eject members who could not read or write, which sapped them of their popular support. The Terror had come to a close, but there were no plans for what lay ahead.

The period following the fall of Robespierre is generally characterized as an era of anarchy, famine, and reprisals. Coup followed coup. The polity was divided between royalists, Terrorists, and the alienated. There were violent attacks against former Jacobins, fears of counterrevolution, and a crippling inflationary spiral. The Directory's creators, who had sought only to dismantle a dictatorship, soon found themselves betraying the principles of the Republic itself. By 1797, they had overturned elections, vitiated the modest religious liberty they had sanctioned, and finally installed the "jack-booted-justice" of military courts throughout the nation.[1] Historians have offered several explanations for these decisions, such as the moral failings of those elected, the tragic circumstances of war and inflation, or the struc-

[1] Howard G. Brown, "Breaking the Back of Brigandage," Meeting of the Society for French Historical Studies, Wilmington, Delaware, March 1994; "Public Opinion, Popular Attitudes, and Public Order in France, 1795–1802," Meeting of the Society for French Historical Studies, Boston, March 1996; and his forthcoming article in the *JMH*.

tural inadequacies in post-Jacobin governments. Whatever the reasons, they inevitably led to what one historian has termed a "loss of the middle way."[2]

The Economic History of the Directory: Constraints and Confusion

The political turmoil was accompanied by the further disintegration of provisioning policies and networks. The men of Thermidor and the Directory lifted and reimposed economic controls several times, but the combination of war, the collapse of assignat, local administrative apathy, poor harvests, and unclear administrative responsibilities undid their every effort. The disasters of 1795 and 1796 marked the nadir of the Revolution's provisioning drama. During the spring and summer of Year II (1794), hail, hard rains, and then a drought plunged yields by as much as a third, especially in the north. The lifting of the Maximum on 4 Nivôse Year III (24 December 1794) sent prices soaring upward. Grain prices in Paris in April of 1795 were probably nine times higher than they had been in 1790. Inflation brought further misery. By the summer of 1795, the assignat was worth only 3 percent of its face value. In "*nonante cinq*," the desolation was unfathomable: The hungry waited barefoot in the snow for a few ounces of wheat or rice. What little grain could be spared to plant that winter and spring suffered from freezing rains and a humid summer. The harvest of Year III (1795) produced less than half the amount harvested in 1793, leaving the people of the Seine-Inférieure with less than seven ounces of bread per day, while prices skyrocketed. In 1796, the hungry struck out into the countryside in search of food. The situation improved briefly with the harvest of Year IV (1796), during which the yield increased almost 7 percent and prices leveled off. The favorable impact of Ramel-Nogaret's stabilization of the franc helped immensely. Yet, drought the next year damaged the crops of Year VI (1798), so that during the winter of Year VII (1799), shortages and speculators pushed prices up once again.[3]

Despite the significance of these years, relatively little has been written on the workings of the economy under the Thermidorian Convention and the Directory. In general, discussions of Revolutionary subsistence issues end with the dismantling of the Jacobin controls, often with the lifting of

[2] Sydenham, *The First French Republic*, p. 107.

[3] Harvest records scattered in AN AF III 607; AN BB18 297; AN F10 242; AN F11 447 and 450; ADSM L 389, 390, 403, 2172, 2374, 2379, 2381, 3225, 4512, 4771, 5084, 6375.

the Maximum.[4] The studies that do include the Directory recount the wretched conditions of Years III and IV.[5] As chilling as these accounts are, however, they provide little insight into the strategies used by the regimes, nor the reasoning that may have characterized them. These were crucial years, if only for the levels of despair and desolation that they brought. Their lessons, however hard-won, would be taken up by their successors in the nineteenth century.

The question of the economic ideologies of the Thermidorians and the Directory is an especially understudied area.[6] The first public lectures on political economy were undertaken under the Directory by Vandermonde, and translations of Adam Smith's *Wealth of Nations* had sparked some interest in free-trade ideas. There were several strands of thought in those years. The first, sustained by the *idéologues*, drew on the ideas of Turgot and Adam Smith, and argued that international competition would stimulate the greatest productivity. The second championed internal free trade as a means to boost production, but supported protective barriers such as tariffs and import prohibitions to encourage French manufacturing and agriculture.[7] The state elites were thinking along these lines, and generally favored free trade. They saw controls as a temporary measure, to be used only with enormous misgivings. The Thermidorian Convention wished to regenerate commerce, although it was not certain how to do so.

Many in these years believed that the honest, normal trade of the past had been squeezed out by the corruption and speculation of the controlled economy and wartime provisioning. "*Vrai commerce*" – as the Directory's

[4] An overview by François Hincker, *La Révolution française et l'économie: Décollage ou catastrophe?* (Paris, 1989) devotes four-fifths of its coverage to the period from 1780 to 1794, for instance. The proceedings of four major French Bicentennial conferences on the economy and the French Revolution reflect a similar emphasis on the period before 1794: *La Révolution française et le monde rural*; *La Révolution française et le développement du capitalisme*; *La pensée économique pendant la Révolution française*; *Etat, finances et économie.* Among the exceptions are Francis Démier, "Les 'économistes de la nation' contre l'économie-monde' du XVIIIe siècle," in *La pensée économique pendant la Révolution française*, pp. 281–304; Georges Dejoint, *La politique économique du Directoire* (Paris, 1951).

[5] Richard Cobb, *Terreur et subsistances, 1793–1795* (Paris, 1964), pp. 221–55, 257–95; Denis Woronoff, *The Thermidorian Regime and the Directory, 1794–1799*, trans. Julian Jackson (Cambridge and Paris, 1984), pp. 9–11.

[6] James Livesey's doctoral disseration makes an important contribution, as will his forthcoming article in *Past and Present*. Gerard James Christopher Livesey, "An agent of enlightenment in the French Revolution: Nicolas Louis François de Neufchâteau, 1750–1828," Doctoral Dissertation (Harvard University, 1994).

[7] Jean-Louis Billoret, "L'affirmation et les politiques du modèle consulaire," in *La pensée économique pendant le Révolution française*, pp. 305–21.

first Minister of the Interior Benezech termed it in Year IV – had been encouraged by neither the policies of the Terror, nor the reimposed controls of the Thermidorian Convention or the early Directory.[8] Benezech had been a *commissaire aux armes et poudres* and had grounds for this conclusion. He determined that commerce had been hurt by too little supervision on the one hand – speculation had been allowed to run rampant – and by too much control, on the other. The state's role, wrote this minister, one of the few who had a genuine impact on policies, was to "trace and smooth the path that trade should follow . . . without interfering with its freedom."[9] After Year V, Benezech's successor, François de Neufchâteau, formulated policies to encourage French manufacturing through international fairs, by maintaining its quality, and by erecting protectionist barriers to competition. Under his watch, local agricultural societies were created to help landowners improve their methods.[10] His goals were not purely economic; he wished to find a new basis for political stability. The politics of representative government had failed and the republic was tragically factionalized. His vision of a commercial republic, one based on property owners (such as those who had bought the *biens nationaux*) and trade, was offered to transcend the political disintegration. Exchange between those producing grain and those manufacturing goods could provide the essential bonds that politics had failed to create – a stunning, secularized version of Boisguilbert's thought a century earlier.[11] By the time François de Neufchâteau had arrived to head the ministry, Finance Minister Ramel-Nogaret's sweeping reforms had brought inflation under control and restored the value of French currency. Thus, François de Neufchâteau had greater freedom to begin evaluating ways to restore French trade. No such luxury had arrived in Fructidor Year II, however.

The desires of Benezech or François de Neufchâteau – or of the men of the Thermidorian Convention such as Boissy d'Anglas – to restore the trade could not yet be implemented. The depreciation of the assignat had eroded any funds for outside purchases. Enemy ships waited off the coast. Credit, never fully developed in the Old Regime, had tightened immeasurably. Thus, every effort by the national government or by local authorities, such as those of Rouen, to buy outside supplies crashed against those barriers. In late August 1794, for instance, Rouen's subsistence commission met almost daily and corresponded extensively with the agents it had sent to Baltic ports. Only a small portion of what was needed arrived, however. In

[8] Dejoint, *Directoire*, p. 8.

[9] Dejoint, *Directoire*, pp. 60–62.

[10] Dejoint, *Directoire*, pp. 65–79.

[11] Livesey's work is extremely illuminating on this point. Livesey, "François de Neufchâteau," 204–259.

the spring of 1795, the Convention authorized an emergency loan of 5 million livres to the district of Rouen. Rouen was to make a forced levy on its population, and repay the Treasury at 5 percent interest. Even though the loan was far too little – the Convention had refused to protect Rouen against inflation – the city now could send merchants to the conquered areas of Belgium or to the Baltic to buy grain. Their hopes were dashed when the leading merchants of the city, including the men from the city's 1789 Subsistence Commission, refused to undertake the mission. Even though they had supervised a complicated operation to buy grain from merchants in Hamburg only two months before, the Rouennais merchants could not be persuaded to change their minds. There was too slight a possibility of finding enough grain in Belgium, and it was well known that too many provisioning agents for Paris and the army had drained the region already. The assignat's tumble would make the endeavors all the more frustrating.[12]

One very cunning plan came from the government-commissioned provisioners working in Belgium. They were feeling the deleterious impact of inflation – no owners would accept the assignat, and prices were staggeringly high. The agents suggested strong measures: close down the Belgian ports, make no government purchases for two or three months, direct all efforts toward securing Baltic grain. They believed such tactics would bring the Belgian grain owners into submission, and they soon would be willing to accept the low prices and assignats that the agents could offer. There is no evidence that the government took this advice. Indeed, attempts to find grain in Belgium continued. Nonetheless, these proposals reflected the overwhelming problems that currency depreciation and shortage presented for government purchases. The authors of this scheme were none other than Gabriel-Julien Ouvrard and his associates, whose names would become synonymous with the genius and scandal of the Directory's finances and military provisioning, and eventually with the Napoleonic endeavors.[13]

Thus, the war and inflation cut France off almost entirely from its traditional sources of emergency grain. It was thrown back onto its own resources, resources that in 1795 and 1796 were utterly inadequate. The first years of the retreat from the Terror, then, were marked by catastrophic shortages, and no alternatives for redressing them. Constrained by the desperate circumstances, and having no recourse to outside supplies, the government had to make do with what the Republic itself grew. Its desperate policy about-faces, its inability to reinvigorate trade and restore order, can

[12] Deliberations of the Subsistence Committee of Rouen, ADSM L 381–382; Prairial-Germinal Year III, ADSM L 385; 26 Germinal Year III (15 April 1795), ADSM L 2279; Prairial Year III, ADSM L 2172; *Mémoire*, Fructidor Year III, AN T 1157; see also the records in ADSM L 2380.

[13] "Ouvrard, Banque," AN T 1157.

be blamed on that crippling fact. There was not enough grain at home, and none could be found abroad.

As news of Robespierre's death spread, the remaining *conventionnels* took stock. Despite their general commitment to free trade, their initial impulse was to leave the Jacobins' economic policies unchanged. Empty market halls across the country bore witness to the breakdown of the system of requisitions and price controls, yet news of impending shortages heightened their concern that free trade might bring further political upheaval and collapse. Flagrant dishonesty and disorganization had spread throughout the ranks of agents who ransacked the country buying grain for Paris and the army. Scarcity, chaos, and corruption – these were the challenges that faced the Convention on the morning after Robespierre's execution.

Political Purges After Thermidor: The Men of Property Return

Provisioning concerns, however, pressing, had to take second place to the task of carrying the Parisian coup to the provinces. Over the next months, the Convention and its supporters in the departments purged local administrations of Jacobins and replaced them with more moderate functionaries. In the Seine-Inférieure, the moderate groups were largely comprised of merchants and property holders. The new departmental president, for example was Grandin, a well-known opponent of Jacobinism and the Republic. Most of the purges proceeded calmly in the Seine-Inférieure, with little of the brutality that broke out in western and southern France. From August 1794 through the following summer, one group after another was subject to "renewals," suppression, and arrests: popular societies, surveillance committees, and revolutionary committees.[14] When some briefly resisted these measures in Rouen, representative-on-mission Duport arrested numerous former municipal administrators, including a past national agent and members of the subsistence commission. The greatest opposition to the purges emerged in the western districts of the department, around the textile towns near Yvetot, Montivilliers, and Cany where Jacobinism had been strongest. Politically, the Thermidorians were breaking whatever remaining authority the Jacobins might have held in the department and eliminating the groups that had enforced the Maximum and requisitions. The men of property who found themselves in office had little interest in upholding the controls of Year II. By the spring of 1795, landlords, merchants, and lawyers – some having even less experience in

[14] The majority of these measures were carried out by the representatives-on-mission, Sautereau and Duport. AN D§I 20.

administrative matters than their predecessors – had become the arbiters of the economy.[15]

Repealing the Maximum

Concerning the economy, the inescapable issue remained what to do about the Jacobin controls, especially the Maximum. Despite their belief in free trade, the representatives were loath to dispense with price ceilings and requisitions as long as there were soldiers who needed food and the harvest estimates were so discouraging. The controls, for all their problems, had held off inflation. During the late summer, a rapid series of orders, laws, and warnings poured out of the Convention and its committees. The representatives had two goals. They needed first to buy time by making as few changes as possible, and second, to pry any remaining supplies out of hiding. Substituting speeches for solutions, they denied that any important changes were in the offing. Robespierre might have fallen, but the Jacobin's policies came to no such precipitous end.

The post-Terror strategies, however, were full of contradictions. On 13 Thermidor (31 July 1794), the Convention reinforced controls: Grain owners were to bring late requisitions to market immediately or face stiff penalties. Simultaneously, however, it undermined those pressures: Prisoners who had made fraudulent declarations during the most recent grain census would be freed, as long as the amount of grain in question was less than one month's needs. The desired outcome was to compel farmers, through threats or offers of amnesty, to deliver their requisitions and to complete their harvests.[16] Then, on 15 Thermidor (2 August 1794), the Convention ordered towns to dissolve the community granaries that had served as distribution points for requisitioned grain since the autumn of 1793. Farmers were no longer to deposit requisitioned grain in specified warehouses for later distribution. Instead, they were to take their wares to public markets and sell them, or to turn them over to the appropriate officials for transfer to requisitioning cities.[17] This abrupt change was troubling. Only the granary system had made it possible to keep requisitions going. Yet now the Convention instructed districts and communes to actually step up the level of supervision and levy more carefully calculated

[15] Poret, the former national agent, n.d., AN BB18 806; 28 Ventôse Year III (18 March 1795), AN D§I 17; 4 Floréal Year III (23 April 1795), ADSM L 2347; ADSM L 5711; 26 Thermidor and 4, 11 and 13 Fructidor Year II (July–August 1794), *Délibérations*, pp. 281, 283–4, 288–9; LeMarchand, *La fin du féodalisme*, pp. 499–502; Cobb, *Terreur et subsistances*, pp. 224–6; Pingué, "Les représentants en mission," pp. 375–92.

[16] 13 Thermidor Year II (31 July 1794), *Délibérations*, p. 212.

[17] Caron, *Le commerce des céréales*, pp. 108–9.

requisitions on the surrounding farmlands.[18] On 28 Thermidor (15 August 1794), the Convention increased the proportion of the rent that tenants on confiscated but still unsold émigré lands were obliged to hand over to the state, in the hope that more grain could be found.[19] Taken together, these instructions presented a bewildering series of orders that increased administrators' burdens, but stripped them of what little structure they had had to collect and redistribute requisitions.

The Convention dithered between refusing to help the departments and its fears of what might happen if it did not. The upheaval of the weeks after Thermidor had encouraged grain producers to hoard their supplies.[20] Now safe from the threat of jail, many farmers found that municipal and communal officials could be persuaded to lighten their requisitions during the late summer of 1794 – after all, if officials let them lie on their declarations, more grain would be left for local consumers.[21] Yet now, as requisitioning fell apart, more requests for aid from the Convention poured in from across the nation. The Conventionnels, who wanted to be left free to deal with the more pressing question of political reform, pushed responsibility for the shortages back onto local officials. Thus, the Convention wrote to Rouen officials that the little grain left in the Republic's warehouses was earmarked for Paris and the military alone. If townspeople feared that there was not enough grain, local authorities were to create the illusion of abundance by stocking the halls with sacks – a vague directive that might imply some form of Old Regime tactic.[22] In mid-August, a grain census revealed that Rouen's municipal granaries held only a two-day supply. Multiple requisitions levied on many of the communes, especially in the district of Rouen, had emptied barns. The departmental authorities pressured district officials to prosecute recalcitrant farmers whom they believed were hoarding supplies. The districts replied that it was useless to try – there was no grain.[23] Working

[18] Caron, *Le commerce des céréales*, pp. 107–9.

[19] There is archival evidence that some farmers turned to non-requisitioned crops, and Bouloiseau's research suggests the same. Report by Mulotin, 15 Frimaire Year IV (6 December 1795), Dieppe Dossier, AN F10 242; Bouloiseau, *Séquestre*, pp. 192–3.

[20] See the correspondence in D§I 20.

[21] In Caudebec-les-Elbeuf, the mayor helped organize the resistance to requisitions. Dumon, Mayor, Caudebec-les-Elbeuf to the District of Rouen, 7 Fructidor Year II (24 August 1794), ADSM L 2382. See also the events in Petite Couronne, ADSM L 2382.

[22] Caron, *Le commerce des céréales*, pp. 122–5; Commerce and Maximum Agency, 8 Frimaire Year III (28 November 1794), AN F12 1547 C. See also the numerous similar replies in the same carton.

[23] AM Rouen F3 10. The district of Pontoise was one of the few areas that complied with requisition orders in the summer of 1794. It proved particularly industrious, making numerous searches and rounding up many small-time hoarders. 18 Messidor Year II (6 July 1794), AN F7 3821.

through representatives Guimberteau and Sautereau, the city petitioned the Convention for a temporary supply of 2,000 quintals.[24]

The prospect of riots in the country's fourth largest city – a city prone to seizing Parisian supplies – finally moved the Convention. Guimberteau quickly developed an emergency plan to supply 64,000 quintals to see the department through the autumn, including requisitioned grain from the Eure and shipments arriving in Le Havre.[25] The amounts were insufficient, however. On 8 Vendémiaire Year III (29 September 1794), the Commerce and Provisioning Commission approved further requisitions for Rouen's district, 15,000 quintals of grain per *décade*, half of what the district claimed it needed, for an indefinite period. They were to be wrung from the districts of Yvetot, Montivilliers, Cany, Neufachâtel, and Dieppe.[26] *Cultivateurs*, however, refused to obey. District administrators, aided by the surveillance committees from Rouen and the surrounding communes, did their best. Donning their official sashes, a team of agents from Rouen worked its way across the territory, trying to force *cultivateurs* to obey the requisition of 8 Vendémiaire, and offering to handle the unpleasant task of the harvest census itself, a proposal that infuriated many communes.[27] Sautereau finally demanded that at least half of the overdue grain arrive by 30 Frimaire (20 December 1794), and the rest by 20 Nivôse (9 January 1795). Much less got there, however (Table 7.1).[28]

Throughout these difficult months, the Convention had begun to consider a more massive overhaul of its provisioning system. In Frimaire (November–December 1794), the Commission of Sixteen undertook an inquiry.[29] A *sous-chef* from the Exterior Commerce Agency, Porquier, presented a secret report that argued forcefully for streamlining and professionalizing the many rival bodies involved in supplying the military and in enforcing the Maximum.[30] He pointed out that the various committees, commissions, bureaus, and agencies had proliferated out of control. The infighting among the bodies – some of it originating in political struggles between the Convention, the committees, ministries, Parisian sections, and

[24] Belhoste to the Department S-I, 2–3 Fructidor Year II (19–20 August 1794), ADSM L 385; 19, 23 Fructidor Year II (5, 9 August 1794), *Délibérations*, pp. 291–2.

[25] Guimberteau demanded 1,500 quintals of grain from the district of Evreux (Eure) and 4,600 quintals from the military warehouses and government shipments arriving in Le Havre on American and Swedish ships. *Délibérations*, pp. 281–2, 290–1.

[26] AN D§I 20; "Observations g^{ales}," ADSM L 2381.

[27] 17 and 18 Brumaire Year III (7 and 8 November 1794), ADSM L 2297.

[28] AN D§I 20.

[29] The reports are in AN C 356. I am grateful to Howard Brown for bringing this valuable carton to my attention.

[30] 29 Frimaire Year III (19 December 1794). "Mémoire sécrèt," Porquier, AN C 356.

Table 7.1 *Final totals from the Requisition of 8 Vendémiaire Year III*

District	Brought	Owed	Percentage brought
Yvetot	10,949.26	20,000	54
Montivilliers	17,188.20	40,000	42
Neufchâtel	4,284.64	4,800	89
Dieppe	11,089.62	16,800	66
Cany	14,262.88	38,400	37
Total	57,774.42	120,000	48

Compiled from tables in AN D§I 20.

the Commune – was serving no purpose.[31] The groups refused to communicate with each other, much less cooperate, and were needlessly duplicating activities. Many of the men called to serve on them had either been "patriots" or *négociants*, and neither had distinguished themselves. The seven *négociants* who headed the Convention's Council of Commerce – which Porquier served – "barely understand that the operations, which include the entire commerce of France, must be undertaken with administrative organization, and are quite different from those of a simple businessman or commercial company . . ." What was needed, Porquier advised, were men with "talents, for work, for service, and long experience." The personnel reports on each of the bodies confirm his impression: Each commission, agency, or bureau listed merchant after merchant, even in the bureaus concerned with recordkeeping.[32] It would be many months before these administrations could be pared down and made to work together. The final reforms waited for Napoleon.

The slow process of abolishing and reforming the bureaus had just begun when the Convention was forced to reconsider the Maximum. Even before the fall of the Jacobins, the Commerce and Provisioning Commission had faced an uphill battle to maintain price controls. Trying to counteract the increasing resistance to the Maximum, the convention decided to hold firm. On 21 Fructidor Year II (7 September 1794), it extended price controls through the end of the next summer. Only a few weeks later, however, the

[31] On the critical impact of such battles in the war ministries and committees, and especially of their role in radicalizing the Revolution, see Brown, *War*.

[32] See, in particular, the reports on the Bureaux d'Enrégistrement de Sécretariat, des Rapports, et de la Tenue des Livres. AN C 356.

Convention realized the controls were falling apart.[33] The legislature reasoned that grain owners would be more willing to part with their supplies if offered prices reflecting the costs of production and local practices. On 19 Brumaire Year III (9 November 1794), the first major policy change emerged from the Convention and its Commerce Commission. The uniform Maximum was abandoned, and individual districts were given permission to raise their ceilings slightly, basing the new levels on local prices in effect in 1790.[34] The deputies hoped to pacify grain owners, while still keeping grain prices at affordable levels. This change marked a virtual return to the first Maximum of May 1793. Its principal effect was to reopen the prospects for districts to compete with each other by setting their Maximums just high enough to draw grain from neighboring markets.[35]

Nonetheless, while the Law of 19 Brumaire Year III allowed producers to raise grain prices, it was accompanied by compensatory legislation that placed greater pressure on them to supply requisitions. The Convention ordered more thorough inspections and instructed officials to keep more detailed market records and maintain a regular correspondence with the appropriate commissions.[36] One of the most striking features of the Law of 19 Brumaire was its extension to farmers who grew only enough grain for their households. In effect, this levied requisitions on many poorer producers who had been excused from previous orders.[37] In other words, although farmers could hope to get a better price for their grain, they had fewer legal means of escaping orders from their districts. The Law of 19 Brumaire proved to be an untenable compromise. Prices crept higher as districts competed to offer more advantageous rates to grain owners.

Realizing that the steady stream of revisions was undermining any attempt to maintain controls, the Convention admitted defeat. It abolished the Maximum altogether on 4 Nivôse Year III (24 December 1794).[38] An important clause also reiterated instructions to uphold *libre circulation*, explicitly allowing owners to take grain to any market they pleased, rather

[33] Rapport du 29 Messidor Year II (17 July 1794), AN F7 3821; 8 Thermidor Year II (26 July 1794), AN F12 1547C; 24 Messidor Year II (12 July 1794), ADSM L 384; Godechot, *Institutions*, p. 417.

[34] Each district was to calculate the average price during 1790 and then increase it by two-thirds. Caron, *Le commerce des céréales*, pp. 120–1; Godechot, *Institutions*, p. 418.

[35] "Extrait du procès verbal des séances du District d'Yvetot, Arrêté du 24 Brumaire an 3 [14 November 1794]" and "Arrêté du District de Neufchâtel relatif à la fixation du prix des grains et fourrages . . . 28 Brumaire an 3 [18 November 1794]," in AN F12 1544 (41); Tableau, 1 Frimaire Year III (21 November 1794), ADSM L 2384.

[36] Caron, *Le commerce des céréales*, pp. 126–8.

[37] Further changes included an *arrêté* ordering grain producers to supply their requisitions in full, even if they feared that they would not have enough grain to last the rest of the year. Caron, *Le commerce des céréales*, pp. 123–4.

[38] *Lois*, 7: 444–5; *CAC*, 4: 590–600.

than being constrained to take it to a particular warehouse or market. Moreover, off-market transactions were once again permitted, and any grain owner imprisoned for violating the Maximum was to be freed immediately. As for requisitions, only those issued for the military or for Paris were to continue. Otherwise, the orders were void, unless a district found its markets empty, in which case it could continue to requisition grain for the next month, a provision later extended through 4 Ventôse Year III (22 February 1795).

Requisitions in Year III

Essentially, the Convention was constructing a frail bridge between the economy of controls and that of free trade. The extension of the requisitions for cities created a transitional interval to recreate commercial connections. Cities were to begin making their own arrangements to buy grain, and merchants were to return to their pre-Maximum activities – although the experience in Rouen shows that such efforts were doomed. Price controls were still in effect for overdue requisitions, but any subsequent purchases by national or local administrators were to be paid at the going rate of the nearest market.[39] In Rouen, the news of the Maximum's end was the catalyst for sending a deputation to Paris. The city feared that free trade would expose it "to the horrors of famine."[40] While the group met with little success, it did obtain permission to levy additional requisitions on five of the department's seven districts. On 11 Nivôse (31 December 1794), the district of Rouen was permitted to order 90,000 quintals (15,000 quintals per *décade*) to ensure its survival for two months. The earlier levy of 8 Vendémiaire had failed to bring in even half the cereal ordered. Unfortunately, the requisitions of 11 Nivôse proved equally discouraging.[41] A series of extensions granted farmers until 30 Nivôse (19 January 1795) to send late requisitions, but even those orders eventually lapsed.[42]

The efforts to procure the requisitioned supplies were futile. There was neither sufficient grain nor administrative willpower to combat the owners'

[39] Those instructions led to confrontations between local officials and grain owners over late requisitions and acceptable prices. 23 Nivôse Year III (12 January 1795), ADSM L 2297; Petitions from *cultivateurs* in Bois l'Evêque, Boos and La Rue Pierre, 16 and 23 Nivôse (5 and 12 January 1795) and 28 Pluviôse Year III (16 February 1795), ADSM L 2297; 25 Germinal Year III (14 April 1795), ADSM L 2381.

[40] AN D§I 20.

[41] AN D§I 20.

[42] The district of Rouen was instructed to overlook the 62,000 quintals of overdue grain from the first requisition (a supply of one and a half months). See, especially, the instructions from the Provisioning Commission, 9 Pluviôse Year III (28 January 1795), AN D§I 20.

Table 7.2 *The requisition of 10 Nivôse Year III, totals by 10 Pluviôse Year III*

District	Brought	Owed by 10 Pluviôse	Percentage brought
Yvetot	1,589.46	7,500	21
Montivilliers	1,096.37	15,000	7
Neufchâtel	524.96	1,800	29
Dieppe	689.00	6,300	11
Cany	1,101.74	14,400	8
Total	5,001.69	45,000	11

Source: "Etat des recouvremens . . . de la réquisition [du 11 Nivôse]," [10 Pluviôse Year III] (29 January 1795), AN D§I 20.

refusals. Requests from *cultivateurs* to have their requisitions decreased continued. The department tried to oversee the process, but found it hard to keep tabs on the dozens of municipal cantons under its jurisdiction. It could do little to counteract the local officials' propensity to lighten the burdens on their landowners.[43] At least one district, Cany, formally refused to part with some of the grain it had collected for Rouen's requisitions – as it had refused the year before.[44] In addition, Dieppe's farmers would not sell their grain at the rates specified by the Law of 4 Nivôse.[45] In an extreme effort to uphold requisitions, Rouen tried to make examples of lax administrators by detaining the officers of Fresquienne, but recovered little grain.[46] As prices rose and the value of the assignat plummeted, the farmers' resolve hardened. On 10 Pluviôse (29 January 1795), only 5,002 of the 90,000 quintals requisitioned had been deposited in Rouen's depots (Table 7.2).[47] The shortages were worst in the eastern parts of the department, especially in the district of Neufchâtel. That population relied on requisitions

[43] ADSM L 2172, 2297 and 3098. Georges Lefebvre also noted a rapid increase in the number of requests approved in Year III. *Les Paysans du Nord pendant la Révolution française* (Lille, 1924), p. 661.

[44] 8 Brumaire Year III (29 October 1794), AM Rouen F4 16; Department S-I to Duport, 19 Ventôse Year III (9 March 1795), AN D§I 20.

[45] 7, 18 Pluviôse Year III (26 January, 6 February 1795), AN F[1c] III Seine-Inférieure 8; 21 Nivôse Year III (10 January 1795), ADSM LP 8105.

[46] The municipality was 540 quintals in arrears. [Department to Commune of Fresquienne], 4 Pluviôse Year III (23 January 1795), AN F[1c] III Seine-Inférieure 8.

[47] AN D§I 20.

from across the border in the Somme, which were not forthcoming.[48] Understandably, the district of Neufchâtel's own *cultivateurs* had become increasingly bold in their refusals to supply local markets.[49] Saint-Saëns alone was disposed to exert heavy pressure on its communes. Not only had its officials proved unwilling to hand out reductions, but in Nivôse Year III (December 1794–January 1795), they reported that an inconceivable 94 percent of their requisitions had been received. Even when those amounts fell in the next months, the canton's success was unparalleled, and also – from the documents – inexplicable.[50]

As 4 Ventôse (27 February 1795) and the end of the requisitions approached, arrivals slowed even more markedly. "Invitations, threats," from the department went unheeded. The various laws gave them ample justification for their refusals. Legislation allowed sales from farmers to the local poor to be deducted from requisitions, and so many sought to use that avenue. Other communes sent documents to prove they had supplied grain elsewhere, for Rouen was not the only district imposing requisitions. The list of excuses, some valid, some less so, grew as the winter wore on and inadequate supplies dwindled.

The Convention could not countenance the producers' flagrant disregard for requisitions. Within weeks of the lifting of the Maximum, the deputies realized further measures were needed. The Law of 3 Pluviôse Year III (22 January 1795) gave grain owners ten days to fulfill their overdue quotas or face certain arrest and detention, thereby reinstating imprisonment. Districts were allowed to extend requisitions through 1 Germinal Year III (21 March 1795), if necessary, a provision the districts of the Seine-Inférieure used. Most important, the law concentrated power in the hands of the representatives-on-mission. In the face of a growing popular movement in cities and the stubborn refusal of rural producers, the Thermidorians stripped the districts of the authority to reduce requisitions. Representatives alone could revise requisitions, and they alone could order the arrest or detention of obstinate farmers.[51]

These laws were of little help, however. Rouen sent three agents to districts to press for requisitions. Daily, they became more distressed by the scarcity they discovered. Many of the officials they approached even pleaded for grain from Rouen. The agents returned to Rouen with stacks of *procès-verbaux*, documenting the desperate shortages of the districts that

[48] Not only was the Somme telling its *cultivateurs* to ignore orders from the Seine-Inférieure, but it was continuing to send consumers across the border to buy in Aumale. "Nombre d'individus . . . ," 23 Frimaire Year III (13 December 1795), ADSM L 2172.

[49] 15 Nivôse (4 January 1795) and 30 Ventôse (20 March 1795) Year III, ADSM L 2172.

[50] ADSM L 2172.

[51] Duvergier, *Lois*, 8: 2–3.

Table 7.3 *The requisition of 10 Nivôse Year III, totals by 28 Ventôse Year III*

District	Brought	Owed by 10 Ventôse	Percentage brought
Yvetot	10,030	15,000	68
Montivilliers	7,738	30,000	25
Dieppe	7,912	12,600	63
Cany	15,403	28,800	53
Total	41,085	86,400*	48

* The Neufchâtel district may have been relieved of its requisitions. Representative Casenave attempted to recover those later that spring.
Source: "Requisitions du 11 Nivôse de 90,000 quintals," n.d. [ca. 29 Ventôse Year III (19 March 1795)], AN D§I 20; Department S-I to the Convention, 1 Ventôse Year III (19 February 1795), ADSM L 139; ADSM L 2,297.

were to supply Rouen, and the utter unwillingness of many local officials to help.[52] When the requisitions ended in Ventôse, only one-third of the grain had arrived. A few days later, the district of Dieppe broke with the department, refused to answer letters, and illegally relieved farmers of their requisitions. Rouen continued to press for the overdue amounts, but by the end of the month less than half had been collected (Table 7.3).[53]

Words, Not Deeds: Duport's Mission to the Seine-Inférieure

The department and the districts looked to Deport, the representative-on-mission, for help. He alone had the power to enforce the Law of 3 Pluviôse (22 January 1795). He wrote blistering letters to the districts, ordering them to send the requisitions during that bitter winter. He accused not only producers, but the national agents, of "a great crime," by "[toying] with the lives of their brothers" in Rouen. In comparison with their Jacobin predecessors of Year II, he claimed, the officials of Year III were a worthless lot. The former authorities, he claimed with some exaggeration, had been willing to drag grain away from the *cultivateurs* to ensure the well-being of the people of Rouen. Where was the *cultivateurs*' gratitude, he wondered, now that the revolution of 9 Thermidor had freed them from oppression?

[52] *Commissaire aux subsistances* of the District of Rouen at the Tôtes depot, District of Dieppe to the District of Rouen, 19 Germinal Year III (8 April 1795), ADSM L 2381. See the correspondence from these missions in AN D§I 20.
[53] AN D§I 20; ADSM L 139 and 2297.

The district administrators and their national agents were to bring the overdue shipments within days. Duport would lead the effort himself. He wanted a full listing of every troublesome grain owner, along with an account of every confiscation and prosecution. Moreover, local authorities were to cease their endless finger-pointing. They were all guilty.

His white-hot rhetoric, however, was backed by only the most tepid action. Thoughout the winter, he repeatedly ordered districts to denounce farmers who had not complied, yet did not authorize the confiscation of their grain. The Cany district forwarded names from 28 communes, and 580 grain producers in the district of Montivilliers had been denounced to Duport and sentenced to have their supplies confiscated. In order to proceed with the confiscations, however, the district needed Duport's authorization.[54] Duport, giving no explanation, balked. His moderate nature won out, and he was suddenly reluctant to give those orders. Thus, the grain went uncollected. That technicality served the requisitioned districts well. The districts could halt the confiscations, retain the grain, and still claim to uphold the law. Indeed, the *cultivateurs* appeared absolutely indifferent to the courts' judgments, "view[ing] the sentences . . . as a means of emancipation that exempts them from supplying [requisitions]."[55] Frustrated by Duport's silence, the department on 18 Pluviôse (6 February 1795) instructed each canton to choose two of their most difficult grain owners to denounce to Duport, in hopes of pushing him to proceed with the confiscations.[56]

Finally, in late Ventôse, after repeated pleas from the department, the representative directed the districts to jail the worst offendors. Five substantial farmers from the Epouville, in the district of Yvetot, were held for several weeks in February–March 1795, and thirty-eight in the Montivilliers district.[57] Even with those examples, requisitions continued to slow. By March-April, there was so little grain left that the district deemed it pointless to proceed with the confiscations.[58] Looking for more impressive examples, Duport imprisoned a number of *cultivateurs* who had not sent their requisitions to the market at Caudebec, authorizing the district to release them if they promised to bring their grain to the hall within ten days.[59] For

[54] Department S-I to Duport, 16 Ventôse Year III (6 March 1795), AN D§I 20.

[55] Department S-I to Duport, 8 Ventôse Year III (26 February 1795), AN D§I 20.

[56] 18 Pluviôse Year III (6 February 1795), AN F[1c] III Seine-Inférieure 8.

[57] The Montivilliers farmers were freed quickly, though, as soon as they had furnished some of their overdue cereals. AN D§I 20; [19 Pluviôse Year III] (7 February 1795), AN D§I 17; Municipality of Montivilliers to the District of Montivilliers, 27 Ventôse Year III (17 Ventôse 1795), AN D§I 18; Lemarchand, *La fin du féodalisme*, p. 504.

[58] AN D§I 20; ADSM L 139; Pingué, "Les représentants en mission," pp. 387–8.

[59] Eleven *cultivateurs* accepted his grant of amnesty. Municipality of Caudebec to the District of Yvetot, 24 Germinal Year III (13 April 1795), and the District of Yvetot to the

all Duport's effort, hesitant though it was, the reports did not vary. The department's grain supplies were exhausted and the representative could do nothing to change that.

Fearing the worst for cities and soldiers, the Convention ordered a final emergency requisition on 4 Germinal Year III (24 March 1795).[60] Every district was to gather one-fifth of its surplus grain – the *cinquième* – without exception. The district was to deposit half of those cereals in a granary for local use, while sending the other half directly to the republic's warehouses. These levies were in excess of any late requisitions still hidden away, and took precedence over any other needs communities might claim.[61] The department protested the *cinquième* vigorously.[62] These emergency requisitions were no more successful than the others, however.[63] Thus, all three rounds of requisitions – those of 8 Vendémiaire, 11 Nivôse, and the *cinquième* – had met the same fate.

Riots Resume: The Turbulent Spring of Year III

Violence resulted.[64] Along the Seine, crowds had pillaged barges carrying grain to Paris and Rouen.[65] Roadside attacks terrified bakers and *cultivateurs*.[66] Grain rations in markets had been decreased, as in Fécamp, where the news that purchases had been limited to 2.75 pounds per person "provoked protests and threats against the authorities." Eighteen communes were late with requisitions for that town alone and worse, the farmers "had declared loudly that they would bring no more grain to the city." Local officials were particularly vulnerable, leading to a general feeling that the

Municipality of Caudebec, 21 Germinal Year III (10 April 1795), ADSM L 3098; "Du registre des arrêtés du Directoire du district de Montivilliers," 3 Germinal Year III (23 March 1795) and Michel, National agent of the Montivilliers district to Duport, 4 Germinal Year III (24 March 1795), AN D§I 20.

60 1er Germinal Year III (21 March 1795), ADSM L 1563.

61 20 Floréal Year III (9 May 1795), ADSM L 384; "Extrait des arrêtés du Comité de Salut Public de la Convention nationale, 4 Germinal an 3 [24 March 1795]," ADSM L 2381.

62 Provisioning Commission to Casenave, 24 Floréal Year III (13 May 1795), ADSM L 139.

63 By the middle of Prairial Year III, for instance, the Caudebec council reported in desperation that local farmers had fallen 8,000 quintals behind in their requisitions. Council of Caudebec to the "Représentant du peuple chargé de la partie des subsistances . . . ," 13 Prairial Year III (1 June 1795), ADSM L 3098.

64 See the many disturbances reported in ADSM L 320, including a number in which local officials either refused to protect grain or led the attack.

65 Lemarchand, *La fin du féodalisme*, p. 515.

66 Copy, *procès-verbal*, Officers of Bourgbaudouin, 6 Ventôse Year III (24 February 1795) and Copy, District of Rouen to Duport, 26 Ventôse Year III (16 March 1795), AN D§I 20.

department had succumbed to disorder.[67] As the district officials of Yvetot drafted a plea to Duport for grain, they could hear "an infinite number of people" gathering outside, "demanding bread."[68] In early Germinal, twenty-five to thirty workers confronted municipal officials in the Montivilliers district, "demanding grain with threats." In Boscroger in the Eure, a great number of "citizens *attroupés*" had gathered at *cultivateurs*' homes, seizing whatever grain and money they found. "The invincible slowness of the *cultivateurs*, and even more so, the absolute penury of the department" had destroyed any hopes the officials had of wresting more food from the area. Moreover, many of the communes that had supplied grain to Rouen were now starving; they demanded that the district of Rouen make good on its promises to find grain for them.[69] Indeed, the department pointed out to the Convention, it could be dangerous to press the countryside, and might incite "a general movement." Yet if bread was not found for the 171,000 people of the Rouen district, the department warned it would not be "possible to prevent the disorders that would arise."[70]

The disturbances increased, and worried district administrators reported several women had torn down a flag in Rouen, yelling "*des propos inciviques*," and the city's police and General Security Committee had been completely unwilling to pursue the guilty, leaving the denuciations to the district.[71] In Oissel, famished peasants fell upon a horse that had been dead for twelve days and devoured it.[72] By early Germinal, the people were calling not only for bread, but for the reopening of the churches. "I have only alarming things to report," warned a national agent in the district of Rouen. "We are dying of hunger, and therein lies the trouble and agitation . . . Despair has overtaken our spirits . . . In the midst of all the complaints, everyone demands their church, the religion of their fathers, and it is impossible to contain them . . ."[73]

On 13 Germinal (2 April 1795), Rouen erupted. Groups of hungry

[67] See the many accounts of attacks throughout AN BB 18 297; AN D§I 17; AN F[1c] III Seine-Inférieure 8; AN F11 229; ADSM L 30; L 307; L 320; L 1393; L 2297; L 2347; L 2379; L 2381; LP 6476; LP 6481; LP 6485; AD Eure 238 L 84; AD Somme L1 2446; AM Rouen I5 VI; Cobb, *Terreur et subsistances*, p. 259; G. Leroy, *La Famine à Melun en l'an III* (Melun, 1902).

[68] District of Yvetot to Duport, 29 Ventôse Year III (19 March 1795), AN D§I 20.

[69] ADSM L 320.

[70] Department S-I to the Convention, 1 Ventôse Year III (19 February 1795), ADSM L 139.

[71] 19 Ventôse Year III (9 March 1795); District of Rouen to Duport, 27, 28 Ventôse Year III (17 and 18 March 1795), AN D§I 17.

[72] Commune of Oissel to the District of Rouen, 24 Ventôse Year III (14 March 1795), ADSM L 2381.

[73] Copy, National agent of La Londe to the District of Rouen, 7 Germinal Year III (27 March 1795), AN D§I 17; District of Rouen to Duport, 2 Ventôse Year III (20 February 1795), AN D§I 20.

workers mobbed the municipal hall, breaking into the mayor's offices. Armed with sticks, and calling for bread "*à grands cris*," the crowd worked its way to the department's chambers and forced the administrators to help search one warehouse after another for grain. Alarm bells sounded, but the national guard refused to protect the officials. The temporary commander of the army reported apologetically that he had only ninety-four men, an insufficient number to restore order. During three days of unrest, the people of Rouen tore down liberty trees, shredded tricolor cockades, and even threatened Duport. The administrators, accompanied by Duport, eventually calmed the city with promises of grain from Le Havre. As two ships arrived, Duport recommended that any complaints cease immediately. The city needed no more rioters, he maintained, but instead willing hands to help unload the ships and distribute the grain. The following *décadi*, he orchestrated an elaborate show of republican sentiment, although few attended it.[74]

The Convention, still shaken by the Parisian uprisings of that spring, had little help to offer.[75] Fearing more riots along the Seine, the deputies announced a loan of 5 million livres to the district of Rouen for grain imports, although those efforts ran up against the merchants' unwillingness, the assignat's fall, and the depth of the shortages.[76] No further help was forthcoming from Paris. Rouen would simply have to do what it could with the offer of that inadequate loan.[77] The badly shaken Duport was recalled to the Convention, blamed in part for the riots of Germinal. His successor, Casenave, arrived in the Seine-Inférieure in late April 1795.[78] Casenave, too, would struggle with the limitations that war and commercial disarray had wrought. He had plently of enthusiasm, however. He gave the farmers of the Neufchâtel district five days to bring in their late requisitions.[79] At the same time, he sought without success to exempt the department from the emergency requisition of the *cinquième*, the levy of

[74] AN D§I 17; AM Rouen I5 6.

[75] The daily ration had been cut week by week, so that by mid-May, consumers were offered only 2 ounces apiece with some additional rice. Hungry crowds had attacked the Convention twice while calling for "Bread and the Constitution of 1793." George Rudé and Richard Cobb, "The Last Popular Movement of the Revolution in Paris: The *Journées* of Germinal and Prairial of Year III," reprinted in Jeffrey Kaplow, ed., *New Perspectives on the French Revolution* (New York, 1965), pp. 254–76.

[76] ADSM L 2380.

[77] Deliberations of the Subsistence Committee of Rouen, ADSM L 381–382; Prairial-Germinal Year III, ADSM L 385; 26 Germinal Year III (15 April 1795), ADSM L 2279; Prairial Year III, ADSM L 2172; *Mémoire*, Fructidor Year III, AN T 1157.

[78] Casenave was in the Seine-Inférieure from 9 April to 26 September 1795. Commission des Administrations Civiles, Police et Tribunaux to the Department S-I, 6 Floréal Year III (25 April 1795), ADSM L 139; Pingué, "Les représentants en mission," pp. 386, 388.

[79] *Arrêté*, 5 Floréal Year III (24 April 1795), ADSM L 139.

one-fifth of the remaining grain that had been imposed the preceding month.[80]

In addition, the Convention decreed the Law of 16–17 Prairial Year III (4–5 June, 1795), its response to the nearly universal administrative cowardice and complicity during the food riots of that spring. This law gave elaborate orders for the repression and prosecution of riots, and ordered the twelve richest taxpayers of any commune where pillaging occurred to pay for the grain, unless the guilty could be found and brought to justice.[81] Clearly the Convention wished to force the notables of every town, especially those who were also administrators, to have a personal stake in maintaining order. While every law against *entraves* and other violations of the *libre circulation* of grain had listed fines and severe punishments, this one was different. By forcing the municipal officials to take financial responsibility for a crowd's actions, it had broken the bond between the officials and their *administrés*. And by threatening the more prosperous inhabitants of a community with fines, whether or not they had participated in the episode, the Convention had driven a further wedge between rich and poor. This law was directed at the precise points where the deputies judged there were problems.

With the late summer came the hope that under the Directory and the Constitution of the Year III, a measure of calm and prosperity would bless the country after six years of uncertainty. The harvest was expected to surpass that of 1795, and along with imports from the conquered territories, especially the Batavian Republic and the Rhineland, it might prove sufficient. Any sense of relief soon evaporated, however. The Two-Thirds Decree, which rigged the elections to the new legislature to guarantee a republican majority, undermined the legitimacy of the legislature before it had even convened. The bloody repression of the Parisian uprising against the Convention did not bode well. In the countryside, administrators surveyed the first crops gathered, dismayed to find the wheat was of poor quality and yielded little flour. Between May 1795 and May 1796, the price of grain, driven by inflation and shortage, would increase ten-fold.[82]

[80] Provisioning Commission to Casenave, 24 Floréal Year III (13 May 1795), ADSM L 139.

[81] The authorities were to summon the people to disperse. If they refused, the municipality was to note their names and provide a complete list to the criminal court within twenty-four hours. If the municipality failed to do so, then the pillagers and the authorities were equally responsible for the crimes. From there, the law traced the possibilities for fines and restitution to the grain owners. If the municipal officials had fulfilled their duties, then the entire matter went to the courts, which would levy fines of three times the amount of the damages. The owner would receive twice the sum of his or her losses, and the Republic would receive an amount equal to the stolen grain. *Code criminel et correctionel*, pp. 187–9.

[82] Dubuc, "Le mouvement des prix," p. 431.

The laws passed in the late summer of 1795 encouraged grain owners to hold onto their cereal in the expectation of higher prices in the coming months. In August–September, the Convention had announced that requisitions would cease on 1 Vendémiaire Year IV (22 September 1795). It tried to shore up the last of the orders by authorizing districts to confiscate whatever surplus grain the *cultivateurs* still had. There were by far the most expansive requisitions the Convention had ever demanded, although unlike the Jacobin requisitions, those of the Year IV did not specify which markets the farmers had to supply.[83] The outgoing Convention expected that the national granaries, supplied by the many forms of taxation in kind levied on land owners and the tenants of *biens nationaux*, would offer additional resources that autumn.[84] Owners decided instead to hold onto their grain until after the requisitions had lapsed.

Local Officials and the Law of 7 Vendémiaire Year IV

It took the deputies of the Convention only six days to determine that they had made a grievous error. To confront the worsening situation, the legislature immediately voted the strict Law of 7 Vendémiaire Year IV (29 September 1795), reestablishing requisitions and restricting all sales to public market places. A month later, the delegates even discussed imposing another Maximum, although they eventually rejected that option. The Law of 7 Vendémiaire remained in effect until 21 Prairial Year V (17 June 1797), when the Directory reestablished a largely free market and lifted all but the most critical levies for the army and Paris. Having acquired considerable experience in these matters by now, the deputies crafted this law to overcome the difficulties, which it was all too apparent, had hobbled the preceding year's system. It confined sales to marketplaces and, sigificantly, like the Jacobin requisitions, instructed the cantonal officials to specify exactly how much grain each farmer was to bring, to which halls, and on what days.[85] There were also limits on the amount an individual could buy at a market, and burdensome administrative procedures for recording every transaction and shipment. In the same law, the Convention reiterated its support of the *libre circulation* of grain, a challenge to both over-zealous local officials and would-be pillagers.[86]

The law's shrewdest provision was the way it brought cantonal officials under the government's control. The Thermidorian Convention knew only

[83] Caron, *Le commerce des céréales*, pp. 151, 153–5.

[84] MI to the Department S-I, 7 Frimaire Year IV (28 November 1795), ADSM L 300.

[85] Municipal and cantonal are being used interchangeably, as in contemporary practice.

[86] Duvergier, *Lois*, 8: 296–9.

too well that cantonal officials had cooperated with landlords to lighten the load of requisitions in Year III. Astutely, the Convention placed the ultimate responsibility for the success or failure of the requisitions firmly on those officials' shoulders. As with the Jacobin measures, there were provisions for sending threshers and soldiers to bring grain to market if the owner refused. The officials had powerful incentives to see that grain reached the market: Should the system break down, for whatever reason, the law allowed municipal authorities to be fined a minimum of one-half the value of the grain that they should have requisitioned, a staggering amount.[87] And in case the official thought he could lighten his duties by authorizing reductions as he had before the Law of 3 Pluviôse, the Convention kept that option closed. Instead of deducting the grain from the overall amount a commune owed, after 7 Vendémiaire any reduction would have to be reassigned to other farmers. Thus, permitting reductions would only increase the administrative workload and make the administrator even more unpopular because the amounts would have to be reassigned to other unwilling cultivateurs. And in any case, because so many of the local officials were grain producers themselves, part of the reassigned requisition would probably come out of their harvests. The past year had shown the deputies that the support – whether willing or not – of municipal officials was the only way to ensure the completion of requisitions. The Law of 7 Vendémiaire was designed to guarantee that compliance.

Despite the calculated attempts to bring municipal officials into line, the Law of 7 Vendémiaire had only limited success. Throughout Year IV (1795–1796), the department and the municipal cantons relived many of the more frustrating moments of the preceding year. As soon as the law became known, *cultivateurs* launched the all-too-familiar campaign to have their requisitions reduced by the department. The department scrutinized the requests, frequently finding deliberate miscalculations and fraudulent declarations. Some of these had their origins in the inaccurate harvest reports that the cantons returned, necessitating more work, as the departmental officials searched their files for additional information.[88] At least on the surface, the department tried to combat the *cultivateurs*' attempts to evade requisitions.

[87] I have found only one case of communal agents being arrested for not completing requisitions, in the *arrondissement* of Saint Valéry, 2 Prairial Year IV (21 May 1796), ADSM L 309.

[88] The department caught the municipality of Yvetot in a lie when it checked its figures for Year IV against the ones it had submitted in Year II. As the reduction petitions poured in, the department placed greater pressure on the municipal officials to verify figures. Following the department's lead, some cantonal officials investigated the requests more carefully than before. See the many petitions and responses in ADSM L 2374, 2382, 3225, 3643, 4381, 4771, and 5084.

Continued vigilance by the Department of the Seine-Inférieure, however, did not prevent shortages. The requisitions of Year IV met with the same fate as those of the preceding year, even though the Law of 7 Vendémiaire had been carefully written to prevent hoarding. In nearly every canton, *cultivateurs* evaded the orders for requisitions.[89] The worst news came from the areas that Rouen depended upon for supplies. The cantons north of that city, in the more fertile areas of the Pays de Caux, reported the near exhaustion of their harvests. Closer to Rouen, figures from Duclair were no more satisfactory. To the east, officials in Darnétal complained that many of its communes were late with their requisitions, decreasing Rouen's stocks futher. Even when the department again threatened to send armed guards and threshers to collect the grain at the end of Ventôse Year IV (mid-March 1796), few communes complied.[90] In only one area, the eastern cantons, could the department report any success, although it was not due to the work of the cantonal officials. By then, the Minister of the Interior had agreed to intercede with the department of the Somme, and finally shipments crossed the border with grain for Saint Valéry, Aumale, and Blangy.[91]

Casenave and the Collapse of the Markets

There were several reasons for the failure of the Law of 7 Vendémiaire, and not all of them could be blamed on municipal officials and resistant owners. Indeed, it was the arrival of the department's next representative-on-mission, Casenave, that hastened the law's demise. The energetic representative, enraged by the failure of the Law of 7 Vendémiaire, wasted no time. Unfortunately, on 18 Vendémiaire Year 4 (9 October 1796), he essentially rendered the requisition orders unworkable. He directed the department to confiscate any grain stopped illegally by overzealous municipal or communal officials, and to imprison anyone interfering with grain shipments. Owners who had not complied with the requisitions faced jail as did any

[89] See the problems in Saint-Saëns, Blangy, Neufchâtel, and Duclair. 22 Thermidor Year IV (9 August 1796), AN F11 743; Jean Marc Antoine Gallye, 10 Ventôse Year IV, ADSM L 4771; Department S-I to the Municipality of Blangy, 25 Ventôse and 18 Floréal Year IV (15 March and 7 May 1796), ADSM L 2734; Municipality of Duclair to the Department S-I, ADSM L 3225; ADSM L 3643.

[90] AN F11 209; *Arrêté* of the Department S-I, 17 Ventôse Year IV (7 March 1796) and attached tables, ADSM L 5084; ADSM L 309, 2297, 2379, 2390, 2734, 3011, 3012, 3225 and 4512; AM Le Havre PR F^4 166.

[91] There were further disputes over requisitions from the Somme. "Extrait du registre, administration municipale du canton de Gamaches," 8 Ventôse Year IV (27 February 1796), ADSM L 2734; ADSM L 309.

farmer refusing to supply people who lived too far away from a market, according to the system outlined in the Law of 4 Thermidor Year III (2 July 1795). Finally, he authorized the districts to "use any means in their power" to draw grain from the unwilling countryside. In ways he could not have anticipated, however, those orders quickly spurred illegal arrests and a flourishing black market.

At first glance, Casenave's *arrêté* seemed to strengthen the Law of 7 Vendémiaire. The articles ordering the confiscation of any grain shipments stopped illegally were given a broader interpretation than he had ever intended, however. He had drafted those articles in order to discourage municipal officials from holding up grain shipments for Rouen when there were questions about their papers. In the past, such questions had served as a pretext for local authorities to seize such sacks.[92] Now, following what they claimed to be his instructions, the same meddlesome administrators halted any shipment that they suspected had been bought outside a public market. The department's day laborers also found the decree of use. In Caudebec-les-Elbeuf, fieldworkers watched as wagons left their town. They determined that Casenave's *arrêté* instructed "all good citizens to stop all hoarders who buy outside the public markets, and to lead them to the municipal authorities." So, they "stopped that *affameur*, Pierre Julien," and took him to the nearest municipality, where he was arrested while his grain was returned to Caudebec-les-Elbeuf. The Ministry of Justice ultimately ruled against such an interpretation, but once the *arrêté* had been extended to cover such situations, local officials were quick to exploit it. Intended to help provision Rouen, the orders instead gave municipal officials ways to hold back grain.[93]

A second problem with Casenave's *arrêté* appeared in the way off-market transactions were to be handled. The Law of 7 Vendémiaire Year IV had been clear: All sales were confined to public halls. There was an important exception, however, one that had been incorporated into most of the Revolutionary legislation. If rural consumers lived too far from a market, they could receive certificates from their cantons allowing them to purchase set amounts of grain directly from local farmers. The Law of 4 Thermidor Year III (22 July 1795) had maintained that system, and Casenave reinforced it by threatening to jail any *cultivateur* who refused to honor the certificates.[94] Long before Casenave had issued his unfortunate orders, the certificate system had encouraged a flourishing black market. The frequent

[92] See the examples from the department of the Eure, one of Rouen's traditional provisioning zones, in AN BB18 297.

[93] See the many complications that surfaced in the Ministry of Justice, AN BB18 806. An unofficial copy of the *arrêté* is included in the carton.

[94] Caron, *Le commerce des céréales*, pp. 152–3.

disputes between *cultivateurs* and the rural consumers who appeared at their doors indicate widespread expectations, at least among owners, that price levels would be much higher than in public markets.[95] Casenave broadened the application of the Law of 4 Thermidor by allowing a wider range of consumers to receive the certificates. Instead of restricting them to the poor of outlying villages, he allowed any consumer who had not found enough grain at that day's market to receive one. The measure quickly backfired. Essentially, Casenave had given grain owners carte blanche to operate unsupervised from their barn doors. The result was a rapid rise in the price of this off-market grain and the continuing desertion of public markets. Less than six weeks after his *arrêté* had been signed, a crowd attacked his lodgings, crying for grain, and the cavalry had to be called.[96]

The Convention had not anticipated the problems a strong-minded representative-on-mission might create in his efforts to uphold requisitions. Before the year was out, however, the system was falling apart. "The Law of 7 Vendémiaire is null and void," one lawyer complained, and it was all Casenave's doing.[97] Even the department concurred, directing the canton of Préaux to ignore the articles of the Law of 7 Vendémiaire confining sales to public markets. The department's halls were deserted, it informed Préaux in Nivôse, and the cantons were to send their people directly to the farms if they wanted to find any grain at all.[98] Clearly, many problems of Year IV appeared to parallel the crises of Year III. A severe grain shortage had been exacerbated by rapid changes in policy, political instability, and the ability of grain owners to hold onto their supplies. Nonetheless, Casenave had undermined the very law he wished to enforce.[99]

[95] Ironically, the system of certificates had originally been designed to prevent hoarding. Only farmers who could prove that they had no surplus grain could escape the demands of buyers armed with certificates. Despite the secrecy attending such transactions, some of the quarrels about prices found their way into court. Saint Romain, 16 Ventôse Year IV (6 March 1796), ADSM L 4715; Tableau, Gournay (Seine-Inférieure), 14 Ventôse Year III (4 March 1796), AN F12 1547B.

[96] Department to the Administration of Gournay attached to the petition of François Beauvais et al., 11 Ventôse Year IV (1 March 1796), ADSM L 3790; Department S-I to the Municipality of Préaux, 7 Nivôse Year IV (26 February 1796), ADSM L 4512; 16 Ventôse Year IV (6 March 1796), ADSM L 4715; *Commissaire provisoire* of Cailly to the *commissaire* of the Seine Inférieure, 28 Pluviôse Year IV (17 February 1796), ADSM L 309; ADSM L 298.

[97] 29 Frimaire Year IV (20 December 1795), Neufchâtel, AN BB18 806.

[98] 7 Nivôse Year IV (28 December 1795), ADSM L 4512.

[99] In the Eure, representative Duval had promulgated a similar *arrêté* on 29 Vendémiaire Year IV (21 October 1795). Like that of Casenave, it resulted in a flurry of arrests in which local officials used the lack of correct documents as a pretext for stopping grain shipments. 12 Nivôse Year IV (2 January 1796), Beaumont-le-Roger, AN BB18 297.

The War of the Cantons

Something more was terribly wrong, however. The empty markets of Year IV were not simply the result of a short harvest and a zealous representative. There were other problems buried in the text of the Constitution of Year III – namely, the dissolution of the districts. The process by which these powerful administrative bodies were weakened had started during the spring of Year III (1794–5). On 28 Germinal Year III (17 April 1795), the Convention placed the districts once again under the supervision of their departmental authorities. On 3 Messidor Year III (21 June 1795), the departments regained the power to overrule their districts' legislation, which increased the tensions between the two bodies. With the formal approval of the Constitution of Year III in August 1795, the districts disappeared, and only the departments, municipalities (cities with over 5,000 people), and communes remained. Many of those communes, often small hamlets of a handful of inhabitants, were then grouped into municipal cantons.[100]

The elimination of the districts significantly undermined the requisitions. District officials had directed much of the Terror on the local level. The municipal cantons, much smaller, generally led by landowners, were no substitute for them. Cantonal boundaries, no longer merely marking electoral districts, became barriers to the provisioning process. The Law of 7 Vendémiaire had not anticipated these problems. Article 10 of the law stipulated that municipal administrators were to levy requisitions within the confines of their "*arrondissement*," meaning the areas from which they traditionally drew supplies. In the Seine-Inférieure, the cantons' *arrondissements*, or provisioning zones, cut wide paths across the department, extending many miles into the hinterland. Rouen's provisioning *arrondissement*, for instance, included the fields of the Pays de Caux, almost two days' journey from the city. The law's provisions seemed clear enough, then. The municipal cantons had the right to supply their markets from their customary regions.

Those rights were illusory. Because the Constitution placed the cantons on a strictly equal footing, there was no way for one canton to force another to fulfill its orders. Instead, municipal officials were barred from crossing cantonal boundaries. When requisitions failed to arrive from other cantons, an official could only search farms within his own jurisdiction. If the grain in question came from a neighboring canton, as was very often the case, he

[100] While the municipal cantons had existed since 1790, they were without any real authority, being solely electoral divisions. Godechot, *Institutions*, pp. 106–8, 321–2; Isser Woloch, "The State and the Villages in Revolutionary France," in *Reshaping France*, pp. 221–42.

could only write to that canton's officials, hoping they would proceed with searches and then permit the convoys to depart. If the authorities in the neighboring canton refused, the official could only denounce them to the courts and hope that appropriate action would ensue.

The problems posed by the Law of 7 Vendémaire and the Constitution of Year III quickly became apparent (Figure 7.1). When the canton of Duclair ascertained that the canton of Fréville was 260 quintals in arrears, it sent soldiers to confiscate the grain, and received an immediate reprimand from the department. "Surely you cannot be ignorant of the fact that a municipal authority has no right to take any actions in communes outside its jurisdiction," the department warned Duclair. In the eastern zone of the department, Neufchâtel and Gournay considered themselves victims of the law's strictures and reluctantly ceased attempts to use soldiers to bring in overdue requisitions.[101] The greatest conflicts arose in the Pays de Caux, in the former districts of Yvetot, Cany, and Dieppe. This area of rich wheatfields and poor day laborers was one of Rouen's most precious supply zones. There was intense competition among the municipalities of the area, which needed the grain for their own impoverished citizenry. Yvetot lodged incessant complaints with the cantons to which it had sent its requisition orders. It received neither replies nor grain. In Pluviôse, when Yvetot sent soldiers into the canton of Fauville to confiscate grain, the department fired off immediate orders demanding that Yvetot desist. Incredulous, Yvetot accused the department of reducing it to a level of impotence and starvation that it had not known even in Year III. "You have opened us to the bitter reproaches of the people who gave their time only because we assured them [of your support]," the canton protested. Instead, Yvetot would have to face its hungry citizens with no hope of grain from the canton to its east. A few miles to the north, the canton of Cany tried to wrest grain from uncooperative *cultivateurs* in the neighboring commune of Sassetot, only to find its actions, too, were overruled by the department. The department intended to enforce the laws strictly, even to the detriment of its citizens.

If the cantons could not cross borders to confiscate late requisitions, what was left for them to do? Here the department suggested letters – "fraternal petitions" – to uncooperative cantons. After all, the Law of 7 Vendémiaire had placed complete responsibility on the municipal authorities, and they were liable for arrest and stiff fines if they did not comply. The department

[101] Municipality of Yvetot to the Department S-I, n.d., Pluviôse Year IV, ADSM L 5084; 23 Nivôse (13 January 1796) and 6 Ventôse Year IV (25 February 1796), ADSM L 3009; Department S-I to the Municipality of Duclair, 27 Pluviôse (16 February 1796), 14 Ventôse (4 March 1796) and 22 Germinal Year IV (11 April 1796), ADSM L 3225; Department S-I to the Canton of Gournay, 11 Ventôse (1 March 1796), [15 Ventôse] Year IV (5 March 1796), ADSM L 3790; Department S-I to the Canton of Neufchâtel, 19 Ventôse Year IV (9 March 1796), ADSM L 4318; Ventôse Year IV, ADSM L 309.

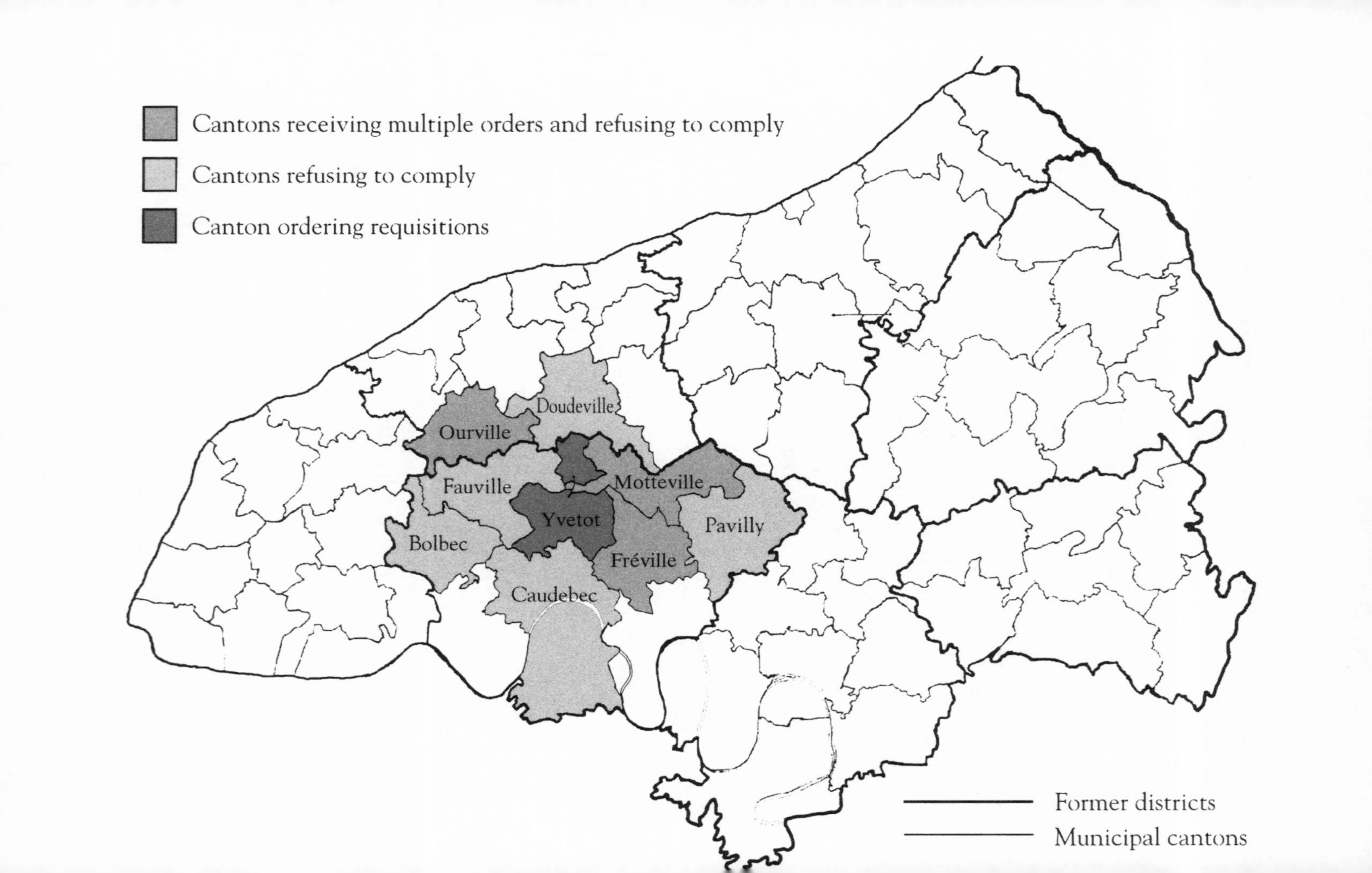
Cantons receiving multiple orders and refusing to comply
Cantons refusing to comply
Canton ordering requisitions
Doudeville
Ourville
Fauville
Motteville
Yvetot
Pavilly
Bolbec
Fréville
Caudebec
Former districts
Municipal cantons

Figure 7.1 Conflicts between municipal cantons over supplies. This map shows how provisioning zones cut across several municipal cantons. Yvetot, the example shown here, traditionally drew grain from at least eight surrounding cantons. In Year IV, it was unable to force the neighboring cantons to comply with its requisition orders. Yvetot was only one of many cantons unable to lay claim to its regular supply zones in the Department of the Seine-Inférieure after the Constitution of Year III dissolved the districts.

told Yvetot and other concerned cities that their only hope was to notify the neighboring cantons that their communes were late with requisitions. From that point on, any responsibility rested with the negligent cantons. The cantons found this advice insufficient at best. "The task you have assigned us is impractical and futile," retorted the officials in Yvetot. "[The powers of] justice and the law must triumph over the obstacles erected by selfishness and evil," they declared, announcing that a delegation from Yvetot was making its way to Rouen to persuade the department to rescind its orders. Saint Romain's officers also pleaded with the department to allow it to take stricter measures against its uncooperative neighbors. "We need a law authorizing us" to "seek out known *détenteurs*," they insisted in the spring of 1796, but no exceptions were forthcoming. Moreover, the department had tired of this endless bickering between cantons. Gournay, cut off from its traditional *arrondissement*, was instructed to put greater pressure on its own *cultivateurs* before complaining about the supposed crimes of neighboring cantons, while Duclair was to follow the same route.[102] Instead of denouncing each other, they should first try to extract whatever grain they could from their own *cultivateurs*, the department advised.

The cantons were furious with the department's view of its very limited responsibility. In the past, such inactivity would have been criminal. The department had been the ultimate arbiter in such disputes, and had often seconded the districts' decisions to arrest and imprison uncooperative grain owners and the authorities that protected them. In Year IV, however, the department claimed that it had only limited means to push local officials to carry out requisitions. It could draft the same letters as the cantons could, and urge the municipal officials to follow up on complaints. But even the department could not send soldiers into the countryside. Only municipal officials had the right to use force against their own farmers. Responding to persistent complaints from Yvetot, the department pressured municipal officials from the surrounding cantons. It guaranteed no grain convoys,

[102] Municipality of Yvetot to the Department S-I, 24 Nivôse Year IV (14 January 1796) and Pluviôse Year IV, ADSM L 5084; [Municipality of Saint Romain] to the Department S-I [spring Year IV], ADSM L 4715; Petitions of François Beauvais and others, 11 Ventôse Year IV (1 March 1796), ADSM L 3790.

however, and cautioned the authorities in Yvetot that it doubted the threats would work. Sending copies of its *arrêtés* of 8 and 17 Ventôse Year IV (27 February and 7 March 1796) ordering communal agents to see that grain was shipped to Yvetot, the department could only "hope that this severe measure might encourage these agents to carry out more satisfactorily the duties the law imposes upon them . . ." The department expressed only vague regret that "we have no help at all to offer you."[103]

The one solution that the Law of 7 Vendémiaire proposed lay with the courts, and through that channel the department sought to supply its markets. The Law of 7 Vendémiaire and the Constitution of the Year III had established a more orderly, supposedly less-politicized procedure for handling administrators who did not fulfill their duties. If the department suspected a municipal official was failing to do his job, it was to report the entire matter to the correctional police, letting them gather the evidence and pursue prosecution. Even though the department often had access to complete figures such as the damning reports from Yvetot, it could do little to combat the farmers' resistance after 7 Vendémiaire, and the new guidelines for prosecuting uncooperative municipal officials became a source of constant frustration. The department pushed the municipal authorities as hard as possible, ordering them to "to let the full force of the law fall on the *cultivateurs*," but the municipalities ignored it. Throughout the year, the department turned in numerous reports, but discovered the courts were reluctant to deal harshly with the officials. In Prairial Year IV (May–June 1796), the department finally arrested the municipal agents of most of the cantons in the *arrondissement* of Saint Valéry, but was uncertain of success. The administrative apathy characterizing the Directory was taking its toll. Regardless of the department's exasperation, there was nothing it could do. It sent complaints to Paris, but could not persuade the Directory to give it any additional powers. Instead, an unsympathetic Minister of the Interior attacked the department for allowing its cantons to fail so miserably, but offered no further advice.[104]

[103] The department's assertions that it wished only to follow the law, whether honest or not, are reminiscent of the strategies of other authorities under the First Directory. Department S-I, 17 Ventôse Year IV (7 March 1796); Department S-I to the Municipality of Yvetot, 23 Ventôse Year IV (13 March 1796); 24 Pluviôse Year IV (13 February 1796) ADSM L 5084; and Colin Lucas, "The First Directory and the Rule of Law," *FHS* 10 (Fall 1977): 231–260.

[104] Court records do not reveal many attempts to prosecute either grain owners or administrators. One *directeur du jury* of the criminal court explained that until the *juges de paix* received copies of the criminal code, they would be unable to properly fill out the indictments, and thus they would have to be nullified. Directeur du jury d'accusation de Dieppe to the Accusateur public près le tribunal criminel du departement, 23 Frimaire Year IV (14 December 1795), ADSM L 298; L 309; L 5084; L 2379.

Municipal administrators had their own response to the changes wrought by the Law of 7 Vendémiaire: No one wanted to hold a government post – even more so now that officials were not being paid their salaries. "You will not have any trouble understanding," the authorities at Cany explained to their departmental superiors, "that they are sorry to have accepted a post that makes them vulnerable to being summoned into court for an error that could only have been committed by the former municipal officials." Such widespread concerns weakened any attempt to reinforce the powers of municipal officials against local farmers in the spring of 1796.[105] The effect of the new political structures and legal system, then, was to further immobilize the provisioning networks in the Seine-Inférieure.

Rural Attacks in Year IV

The breakdown of the provisioning networks brought catastrophic consequences. Mortality rates doubled and almost tripled, leaving 5,700 corpses in Rouen alone in Year IV (1795–1796).[106] In the countryside, crowds of hungry day laborers roamed from barn to barn searching for grain. In the cities, outstretched hands begged for charity that was no longer forthcoming. The unmistakable fact of the spring of Year IV (1796) was that there was no grain in the markets. The deficiencies of the harvest of Year III (1795), compounded by those of the Constitution and the disintegration of local authority, had emptied public markets. Attacks on farms had long been one of the weapons of the weak in Old Regime and Revolutionary France. Rural attacks began in earnest in the late summer of Year III, increasing sharply in the next months, and in Brumaire Year IV, the presence of 150 men from a *colonne mobile* proved worthless. In Frimaire, a *cultivateur* in La Chapelle was shot dead in front of his doorway. Roumare and Anglesquéville, too, reported "murders." Ten men were seen lying in wait for travelers and grain wagons along a wooded road near La Valette. In Belleville, a band of thirty men raided the richest farms of the area, and others were seen marching two by two, "as an organized company," beside grain wagons they had seized. A gang near Gournay repeatedly attacked the farms of owners known to have evaded requisitions. An army of fifty to sixty people from Fécamp was assaulting area farms, undeterred by the bloody battles that erupted when local people tried to defend supplies. One

[105] This had been a problem as early as Year II, but it worsened in Year IV. Canton of Cany to the Department S-I, 6 Ventôse Year IV (25 February 1796), ADSM L 3009; Municipality of Saint Romain to the Department S-I [spring Year IV], ADSM L 4715; Woloch, "The State and the Villages."

[106] Cobb, *Terreur et subsistances*, pp. 307–42.

report of places experiencing "seditious *mouvements*" listed twenty-five cantons. As in the Year III, municipal and communal officials were frequent targets of these raids. Casenave himself pleaded with the government for food and troops, some for his own protection, but to little avail.[107]

The department's *arrêté* of 1 Ventôse Year IV (20 February 1796) sounded the alarm. It urged every able-bodied citizen to join the National Guard in suppressing the violence. A few days later, it reported that many municipal administrators "were terrified" of the crowds' "audacity," and appealed for troops. Minister of Police Cochon repeatedly informed the department that unrest was a local problem to be handled by the local National Guard and existing laws. His ministry began a new campaign to have the *gard champêtre* – essentially rural constables – formed in rural areas. The law of 4 Pluviôse Year IV (24 January 1796) – one of many prohibiting begging and *attroupements* – was to give special force to the department's words, but the law was of no use. The *commissaires du pouvoir exécutif* were completely inert, and the National Guard, along with the gendarmerie and the never-formed *garde champêtre*, had long since failed to answer such calls. In fact, more than one uniform of the National Guard had been spotted among the brigands. And if the guilty were caught, the courts seemed also unwilling to hand out severe sentences if hunger were the cause.[108]

The police bulletins of the late Directory reveal widespread lawlessness until the beginning of the next century. Food riots merged with brigandage, theft, and other disorders and crimes, and continued to disrupt the area. Attacks on farms, in particular, were incessant until the early Consulat. Despite the end of inflation in Year V (1796–1797), accompanied by the better harvest of 1796, unrest continued, and the forces of order seem unable to repress it. Market networks were still fragile, and the intermittent hunger and desperation of many people in the Seine-Inférieure could not be calmed by the few troops available. Religious unrest and political grievances, especially royalism, while not a sharp as elsewhere, were nonetheless present in the Seine-Inférieure.[109] Overall, authorities, especially in outlying areas, were ineffective, and the attempts to create a *garde champêtre* failed repeatedly. A report from the nearby department of the Orne

[107] AN BB18 806; AN F7 7092; AN F11 743; ADSM L 298, 300; LP 6492, 6494, 6501, 6504, 6505, 8072; AD Calvados L 10224.

[108] ADSM L 298, 300; LP 6492, 6494; Octave Festy, *Les délits ruraux et leur répression sous la Révolution et le Consulat* (Paris, 1956). The court of appeals records some very heavy sentences, but they were generally for cases of outright theft of non-food items or for crimes that were extremely violent and involved injury to servants, women, and children. ADSM LP* 6652.

[109] Department of the S-I to the Minister of Police, 29 Floréal Year V (18 May 1797), ADSM L 309; AN F^{1c} III Seine-Inférieure 1; ADSM L 342, 343.

alleged that the men chosen for the guard "were as likely to take part in crime as to repress it."[110] Even after the promulgation of the Law of 21 Prairial Year V (9 June 1797), which once again freed the grain trade, the unrest continued. By then it had joined with hostility over religious and political issues, and the collapse of local government could not contain it. The shortages of Year VII (1798–1799) triggered more disturbances and the government's attempts at repression again failed.[111] When the 100 members of National Guard of Neufchâtel were mustered, for instance, only 62 appeared, and of those, only 16 had firearms.[112] In Year VIII (1799–1800), many departments decided they could not afford the *garde champêtre*, and cut off funds altogether. As late as Year IX (1800–1801), when the first Napoleonic regime decided that the proper personnel for that guard should be former soldiers, most communes ignored the law.[113]

The war and the disintegration of the assignat had devastated provisioning networks. Given that no emergency supplies could be found elsewhere, the uneasy Republic was forced to rely on its own insufficient harvests. Unfortunately, its political organization was in shambles. The legacy of the Thermidorian Convention and the Directory, then, is one of utter frustration and failure. It was a paralyzing interim, characterized by greater famine than than anyone could remember, and by desperate and extreme efforts to ensure urban food supplies. These failed, and when added to religious and political strife, they brought nearly universal lawlessness by the late Directory. The disorder of those years yielded a deep desire for more reliable results. Solutions, when they were found, came only under Napoleon in the early nineteenth century.

[110] Festy, *Délits ruraux*, p. 120.
[111] AN F^{1c} III Seine-Inférieure 8.
[112] ADSM L 299.
[113] Festy, *Délits ruraux*, pp. 158–93.

PART THREE

The State Learns, 1800–1860

CHAPTER EIGHT

The Last Maximum: 1812

"Je ne veux pas qu'un tribun du peuple puisse me demander du pain; formez les approvisionnemens."[1]

Veiller à la subsistance du peuple est le premier soin du gouvernement . . .[2]

During the first week of May 1812, Napoleon surveyed the dismal reports piled on his desk. Parisian warehouses were empty, flour prices neared the almost unprecedented level of 150 francs per sack, and the map of food riots traced the paths bringing grain from the fields to the capital: Bernay and Les Andelys in the Eure; Rambouillet, Etampes, and Versailles in the Seine-et-Oise; Dreux in the Eure-et-Loir; Bellême in the Orne.[3] In March, a very bloody melée in Caen required 4,000 troops to calm it, and only ended with a wave of arrests and harsh sentences.[4] As many as 2,000 "walking skeletons" roamed from farm to farm in search of food near Dieppe in the Seine-Inférieure.[5] In the neighboring Somme, 50,000 beggars demanding handouts had been counted.[6] The Prefect of the Eure

[1] Napoleon, cited in [G-J Ouvrard], *Mémoires de G-J Ouvrard sur sa vie et ses diverses opérations financières*, 3 vols. (Paris, 1826–1827), 1: 70.

[2] Subprefect of Le Havre to the Prefect S-I, 25 November 1811, AN F7 3639.

[3] AN F7 3619, 3926, 3634, 3640.

[4] J. Vidalenc, "La crise des subsistances et les troubles de 1812 dans le Calvados," in Ministère de l'Education nationale. Comité des travaux historiques et scientifiques, *Actes du 84e congrès national des sociétés savantes. Dijon, 1959. Section d'histoire moderne et contemporaine* (Paris, 1960), pp. 321–64.

[5] Subprefect of Dieppe to Prefect Girardin, 12 August 1812, ADSM 6M 1193.

[6] Somme dossier, AN F7 3619.

had included a moving piece of evidence in his monthly report: a sample of miserable bread, "only crudely fashioned wheat husks," the most his department's people could afford.[7] In Paris, rumors had flown for months that the city was nearing starvation and emergency measures were imminent. Dreading what might follow, Napoleon determined that only direct intervention would stave off greater disorder. By the morning of May 4, he had drawn up his master plan, and by the end of the week, one might have concluded that the Jacobins had returned to the Convention. Two Imperial decrees, the first on May 4 and a second on May 8, were issued from his chambers. The former once again restricted all sales of grain to public marketplaces, and the latter imposed price maximums on all cereals. Leaving the details to his stunned ministers, His Imperial Highness rushed off the next dawn to join his troops and lead them to Russia.

His measures were the culmination of months of frustration over grain shortages and followed nearly a decade of experiments to solidify urban supplies. The abrupt return to regulation proved to be a spectacular failure, convincing administrators once and for all of the utter futility of extreme measures.[8] That experience – more than any other, more even than Jacobin Maximums – seemed to confirm the value of the less dramatic, but infinitely more effective solutions to be found in exerting more subtle pressures on the grain trade. These conclusions provided the first tentative resolutions to the sharp twists and turns of subsistence policies since 1789. The difficult Revolutionary years had yielded an abstract free-trade ideology and the fragmented framework of an experienced supply bureaucracy, honed by the needs of a hungry army. The administrators of the Napoleonic era built on these, consciously testing strategies and ultimately creating firmer policies, guiding the food trades to benefit the growing cities of a still precarious regime.

[7] Prefect of the Eure to Réal, 28 February 1812, AN F7 3626.

[8] Because the secondary literature is still scant, memoirs and correspondence collections, even if polemical and self-serving, provide the most informative sources for Napoleonic administration. Etienne-Denis Pasquier, *A History of My Time: Memoires of Chancellor Pasquier*, 3 vols., Duc d'Audiffret Pasquier, ed., trans. Charles E. Roche (London, 1893–1894); [Etienne-Denis] Pasquier, *The Memoires of Chancellor Pasquier, 1767–1815*, ed. Robert Lacour-Gayet, trans. Douglas Garman (Rutherford, Madison, and Teaneck, N.J., 1968), pp. 136–7; [Jacques-Régis de] Cambacérès, *Lettres inédites à Napoléon*, Jean Tulard, ed. (Paris, 1973), 1: 150–1; [Agathon-Jean-François] Fain, *Mémoires du Baron Fain, Premier Secrétaire du cabinet de l'Empéreur*, P[aul] Fain, ed., 3rd ed. (Paris, 1909), pp. 119–21, 147–57; Napoléon, *Correspondance officielle* (Paris, 1970); *Napoléon sténographié au Conseil d'Etat*, Alfred Marquiset, ed., (Paris, 1913); Michel Roussier, *Le Conseil général de la Seine sous le Consulat* (Paris, 1960); J[acques] Peuchet, *Mémoires tirés des Archives de la Police de Paris, depuis Louis XIV jusqu'à nos jours*, 5 vols. (Paris, 1838), 1: 14–15.

The Napoleonic Legend

The development of the Napoleonic state, and in particular, its economic contours, has been woefully neglected by historians.[9] Even the 1812 Maximums have gone virtually unnoticed. This is a startling oversight, given the importance of early nineteenth-century policies in ending the Revolution. Nonetheless, the problems of reestablishing order in a country engaged in a protracted civil war, one in which elections and coups had followed one after another, has drawn little historical attention. The heated rhetoric that destroyed the monarchy and then the Republic has seemed of greater importance than the forces that ushered in the cold silence of the Napoleonic era.[10] The image we have of the Napoleonic regime rests largely on its military victories and defeats. Napoleon's skill has dazzled enthusiasts for almost two centuries, creating a hagiography without rival, yet leaving a desperately impoverished historiography. The Empire consolidated some of the successes of the Old Regime and the Revolution, paying lip service to democratic ideals while drawing together an elite based on wealth, service, and dedication to Napoleon. Administratively, Napoleon advanced the work of the Old Regime and the Revolution, creating a departmental structure of prefects, men more dedicated and efficient than any intendant or representative-on-mission. In all of this, he consciously sidestepped politics, putting aside ideology in the interest of results.

From the first years of the Consulate, Bonaparte deemed the free-trade theories of the *idéologues*, as the most prominent set of liberals was known, too "speculative" and "too abstract." They were not flexible enough to meet the ever-changing needs of the state, and he rebuffed their ideas. Protectionism, also, was too rigid in his eyes. He favored instead a third

[9] The classic account by Georges Lefebvre focuses on the Continental blockade and makes passing reference to the Emperor's distrust of merchants and his somewhat mercantilist outlook. Georges Lefebvre, *Napoleon*, 2 vols., vol. 1, *From 18 Brumaire to Tilsit, 1799–1807*, trans. Henry F. Stockhold (New York, 1969), 1: 164–82. See also the research agenda outlined by Guy Thuillier and Jean Tulard, "Conclusion," in *Administration et contrôle de l'économie*, pp. 161–7. An important recent synthesis on the period presents an effective overview of research on agricultural and industrial activity, but does not look closely at the workings of Napoleonic political economy. Martyn Lyons, *Napoleon Bonaparte and the Legacy of the French Revolution* (London, 1994). See also Stuart Woolf, "Contribution à l'histoire des origines de la statistique: France, 1789–1815," in *La statistique en France à l'époque napoléonienne* (Paris and Brussels, 1891), pp. 45–126.

[10] Louis Bergeron, in his penetrating overview, has described the regime aptly as one that relied more on appearances and symbols than on substance. "Napoléon ou l'état post-révolutionnaire," in *The French Revolution and the Creation of Modern Political Culture*, vol. 2, *The Political Culture of the French Revolution*, Colin Lucas, ed., (Oxford, 1988), pp. 437–41.

line of economic thought, recently termed "rational administration."[11] Here he found the economic views that would allow him to bend policy to meet the perceived needs of his Empire. The grain trade was not, for instance, a subject for ideology and immutable laws, but instead should "vary at each instant. That is why the government governs," he explained.[12] In the legislatures and in his ministries, he steadily confronted and removed the opposition, including the *idéologues* and other liberals. The editor of J.B. Say's *Traité d'économie politique*, which had gained the author a large following, for example, was denied permission to publish a second edition, and Say himself was ushered out of the legislature in 1802. Minister of the Interior Chaptal – chemist, theorist, industrialist, and liberal – resigned his post when Napoleon claimed the title of Emperor in 1804. By then, public, ministerial, and legislative discussion had been well-nigh extinguished. Economic theory was relegated to the domain of intellectuals and barred from that of practice.[13]

Recently, a few historians have focused on the underpinnings of the Napoleonic state, criticizing the dictatorship's cynicism and brutality.[14] The Emperor scorned the legislative and judicial bodies and offered military glory in the place of elections. The images we now have of the Empire are of young men who willingly chopped off toes to avoid conscription and a march to the bitter cold of the Russian steppes; of police informants hiding in dark cafes; of women stripped of the few rights the Revolution had given them, once again subject to the rule of fathers, husbands, and brothers; of foreign countries suffering harshly under their supposed "emancipation" by French armies. The regime failed to win allegiance, except through the intimidation and passivity of a citizenry exhausted by almost a decade of political upheaval and war. Police *mouches*, military deserters, and wooden administrative reports, as well as intrepid battle plans, have become emblematic of the Empire.

[11] Billoret, "L'affirmation et les politiques du modèle consulaire."

[12] Cited in *Napoléon sténographié au Conseil d'Etat*, pp. 60–1.

[13] There has been little written on economic thought before the mid-century. Several articles are worth noting – in particular, Billoret, "L'affirmation et les politiques du modèle consulaire;" D. Dammane, "L'économie politique sous le consulat et l'empire. Misère de l'écomomie, sciences des richesses," *Economies et sociétés* 20 (October 1986): 49–62; Facarello and Steiner, "Prélude: une génération perdue?" in *La pensée économique pendant le Révolution française*, pp. 9–56; Francis Démier, "Avant gardes économiques et diffusion de l'économie politique en France de 1815 à 1914," p. 142; and Denis Woronoff, "Economie de guerre et intervention de l'état," in *Etat, finances et économie*, p. 293. An illuminating analysis of *idéologue* thought is Cheryl B. Welch, *Liberty and Utility: The French Idéologues and the Transformation of Liberalism* (New York, 1984).

[14] See, especially, Sutherland, *France*; Alan Forrest, *Conscripts and Deserters: The Army and French Society during the Revolution and Empire* (Oxford and New York, 1989).

The Empire's Administrators

While the regime's betrayals are clear, its administrators' hesitant efforts to reconstruct the derelict state the Directory had left them are less so. The officials of the Napoleonic regime inherited the wretched failings of the Directory. They continued the work of depoliticizing the country, consciously substituting administrative and military solutions for political debate and factionalism.[15] This was especially the case for provisioning policies. Ideology held little appeal for the early nineteenth-century bureaucrats who had to supply cities and armies. They could agree or disagree with free-trade theories or Jacobin controls, but were intent on finding policies that would work. Experience, not theory, guided them.

Some of that experience came from the military bureaucracy. Nearly two decades of conflict yielded legions of *commissaires*, *commissionnaires*, and *fournisseurs*. The war probably absorbed almost 60 percent of the nation's expenditures in the mid-1790s, and the administrative energies that it took to feed, clothe, and force an army to march were considerable.[16] As the troops passed each frontier, new men were needed to supervise the conquered territories. The war ran on credit and food. Not surprisingly, the areas of provisioning, banking, and taxation are those with the greatest continuity from at least the Directory, and in some cases, the Old Regime or early Revolution. Accountants, suppliers, and tax collectors found safety behind their ledgers. They ranged from ambitious and scheming to tedious and exacting, but had learned how to get food to armies and sometimes to Paris. Among these cadres were a number of future administrators and provisioniers – including Ouvrard, Paulée, and Vanlerberghe – who would direct the supply apparatus for French cities.[17] Within the Seine's departmental council were others, such as Jacques Constantin Perier and the Swiss banker Bidermann, who had held posts in both the capital's and the

[15] The Napoleonic regime used professional rather than political criteria in selecting generals for the army, in contrast to the confused and contradictory policies of the Directory. Howard G. Brown, "Politics, Professionalism, and the Fate of the Army Generals after Thermidor," *FHS*, 19 (Spring 1995): 133–52.

[16] Jean-François Belhoste, "Le financement de la guerre de 1792 à l'an IV," in *Etat, finances et économie*, pp. 317–45; Brown, *War*, pp. 98–123.

[17] Louis Bergeron, *Banquiers, négociants et manufacturiers parisiens du Directoire à l'Empire*, *Civilisations et Sociétés* 51 (Paris, 1978), pp. 25–6, 132–3, 147–9, 150–1, 343–4; Michel Bruguière, "Finance et noblesse: L'entrée des financiers dans la noblesse de l'Empire," *AHRF* 199 (January–March 1907): 161–170; Michel Bruguière, *Gestionnaires et profiteurs de la Révolution: L'administration des finances françaises de Louis XVI à Bonaparte* (Paris, 1986). See the essays by Woronoff, Belhoste, Alain Plessis, Pierre-François Pinaud, and Georges Naquet in *Etat, finances et économie*; and Jean-Charles Asselain, "Continuités, traumatismes, mutations," *Revue économique* 4 (November 1989): 1137–88.

Republic's provisioning committees and who could muster many years of experience for the new regime.[18]

These men sought to guide and shape the economy, deflecting attention from their operations and finding in clandestine means a greater ability to shore up the regime. There was much that was reminiscent of the eighteenth century in their methods. The Napoleonic regime's administrators returned to the pre-Revolutionary goal of smoothing supply routes, forging commercial ties, and shaping the grain trade in northern France. Their efforts to craft such solutions developed one step at a time, as they reestablished baking guilds across the country, secretly bought and sold supplies, and built up government-run warehouses. Their greatest efforts were exerted in 1812, when they struggled to overcome severe shortages and then to navigate one last, failed attempt to impose maximums and requisitions. The lessons of those years were that secrecy held more promise than publicity, and long-range planning was more successful than desperate, abrupt strokes carried out at the last minute.

The origins of the nineteenth-century provisioning strategies lay in the government's response to four rounds of shortages: two mild, in 1801–2 and 1805–6, and two severe, in 1811–12 and 1816–17. Each one brought about conscious attempts by bureaucrats to bring their experience to bear on the crises. Bonaparte gave his administrators little choice, having but one rule: Bread prices in Paris must not rise. "Napoleon was afraid of the people," the once Minister of the Interior Chaptal reminisced years later. "He believed that wheat should be at a very low price, because riots are almost always the result of high prices or a shortage of bread."[19] But low bread prices and urban calm, however desirable, depended on much more than the commands of emperors and kings. It took significant administrative skill and much experimentation during the numerous *disettes* to forge policies holding any promise.

Prospects and Limitations

These efforts must be set within the limits imposed by a French economy hard-hit by war, requisitions, and intermittently poor harvests. The data on agricultural growth before the mid-nineteenth century are barely sufficient to hazard any sure conclusions. Any post-Revolutionary agricultural recovery was slow in coming to much of France, and viticulture alone seems to

[18] Roussier, *Le Conseil général de la Seine*, p. 10.

[19] Chaptal added that Napeoleon's "fear [of insurrections] led him always to take unsound measures." Chaptal, pp. 275, 291; Chaptal, *De l'industrie française*, Louis Bergeron, ed. (Paris, 1993).

have improved during the era. Generous estimates suggest that harvests were uneven and only returned to the level of Old Regime production at the end of the Empire. According to Le Goff and Sutherland, the Revolutionary period overall represented a "lost generation for economic growth and development," and the Napoleonic era one of only slow renewal, perhaps "stagnation." The Paris basin and much of northern France may have weathered the period better than elsewhere. Since the mid-eighteenth century, a slow shift from fallow to artificial prairies had been underway. Close to the capital, an ample supply of urban night soil and the demand for foodstuffs created areas of flourishing and varied agriculture by the mid-nineteenth century. North and west of the capital, however, where the soil was less fertile, there were fewer breakthroughs. The agricultural transformation of the Seine-Inférieure was uneven. The department remained heavily dependent on grain from the Somme to the east or the Vexin to the south, or on whatever imports it could secure. In the Eure, where agricultural increases were greater, tensions over conflicting demands from Rouen, Paris, and the Eure's own consumers only heightened.[20]

Provisioning problems in the Napoloenic era lay within the domain of several influential bodies, some with overlapping jurisdictions. The most significant was the standing Wednesday roundtable that Bonaparte conducted with his ministers, and a larger Council of State, where committee (*section*) presidents presented their briefs. In the period before August 1811, subsistence policies were handled by a succession of ad hoc councils with experts appointed from the Ministries of the Interior, Police, and War, aided by representatives of Parisian government and departmental administrations. Like other areas of Napoleonic government, the subsistence bureaucracy suffered from unclear jurisdictional boundaries and frequent reforms. The heavily staffed subsistence bureau, for instance, long part of the Ministry of the Interior, was moved in 1811 to the newly created Ministry

[20] Harvest information of varying degrees of accuracy is contained in AN F[1c] III Seine-Inférieure 8; AN F[1c] V Seine-Inférieure 1 and 2; ADSM 6M 1070; AN F7 3639. See also Le Goff and Sutherland, "The Revolution and the Rural Economy;" Paul Butel, "Crise et mutation de l'activité économique à Bordeaux sous le Consulat et l'Empire," *AHRF* 199 (Jan-March 1970): 110–34; and "Revolutions and the Urban Economy: Maritime Cities and Continental Cities," Forrest and Jones, eds. *Reshaping France*, pp. 37–51; and Woolf, "Statistique," pp. 51–2; Etienne Dejean, *Un préfet du Consulat: Jacques-Claude Beugnot* (Paris, 1907), pp. 288–91; Hugh D. Clout, "Agricultural Change in the Eighteenth and Nineteenth Centuries," in Hugh D. Clout, ed., *Themes in the Historical Geography of France* (London, 1977), pp. 407–446; Dominique Margairaz, *Foires et marchés dans la France préindustrielle*, *Recherches d'Histoire et de Sciences Sociales/Studies in History and the Social Sciences* 33 (Paris, 1988), pp. 84–92; Grantham, "Agricultural Supply;" Vardi, *Land and Loom*. For Chaptal's goals in the survey of Year IX, see Marie-Noëlle Bourguet, *Déchiffrer la France: La statistique départementale à l'époque napoléonienne* (Paris, 1988), pp. 70–1.

of Manufacturing and Commerce. The agricultural bureau, intimately tied to *subsistances*, however, remained part of the Interior. The two were not rejoined until the Second Restoration, after several years of confusion and inadequate communication.[21]

Vanlerberghe's Entrance: The Shortages of Year X

The administrators' struggles began with the harvest shortfalls and industrial contractions of Year X (1801–2), a "*banale panique*," a run-of-the mill scare. While it cannot be compared with the catastrophes of 1792 or 1795, this shortage nevertheless brought enduring shifts in provisioning policies. The origins of the scarcity remain unclear, and are generally attributed to a poor harvest, military purchasing, and exports.[22] Intent on keeping Parisian bread prices to 65 or 70 centimes for four-pound white loaves, or 50 to 60 centimes for poorer-quality working-class bread, Bonaparte set aside 100,000 francs per month.[23] The funds were used for ration cards to help the capital's poorest inhabitants buy subsidized low-quality bread. While the plan did not work as smoothly as hoped, the city remained calm.[24]

The second, more ambitious, part of the plan created infinitely greater frustration in the corridors of the Ministry of the Interior but nonetheless created a foundation for future plans. Bonaparte ordered the covert importation of thousands of sacks of grain under the auspices of a *régie*, a private, government-supervised agency, led by an influential association of merchants and financiers known as the Cinq Négociants.[25] The grain was to

[21] Godechot, *Institutions*, pp. 656–8; Roussier, *Le Conseil général de la Seine*.

[22] Le Goff and Sutherland, "The Revolution and the Rural Economy," pp. 57–9. The shifts in export laws, problems with military purchasing agents, and the resulting anxieties can be followed in AN F11 322, 338, 402, 403; ADSM 6M 1204; AD Finistère 8M 49 and 52.

[23] Even after the implementation of the metric system during the Revolution, Old Regime weight and volume measures and currency were still in wide use. The Imperial correspondence generally discussed bread prices in sous and pounds. A sous represented 5 centimes in the new currency and the pound was considered a half kilogram, even though it was actually only .45 kilograms. The Old Regime 4-pound loaf, then, was treated as the equivalent of the nineteenth-century 2-kilogram loaf. The difficulty in finding genuine equivalencies and in establishing new terms and practices persisted into the mid-nineteenth century.

[24] Designated bakers sold loaves for 70 centimes to buyers deemed sufficiently poor to merit ration cards. Many indigent Parisians added a few centimes of their own money, and purchased more expensive white loaves, which threw off the council's calculations. AN F11* 3049.

[25] Louis Bergeron and Michel Bruguière have explored these men's "brutal and spectacular social ascent." Since the Directory, they had been enmeshed in almost every area of war

come primarily from conquered territories, or through merchants who were prepared to accept the now-stabilized French currency or credit. Estimating that the government needed to supplement French diets by at least 100,000 metric quintals[26] per month, Bonaparte provided a monthly budget of 1.5 million francs to underwrite the *régie*'s expenses. The *négociants* were to bring in 45,000–50,000 metric quintals of grain each month to Paris. The supplies would be sent to market under the merchants' names at prices below the going rate.[27] The entire operation was to remain a tightly held secret. In fact, the government even disavowed any knowledge of purchases when bargemen inquired about the presence of unfamiliar ships and unidentified shipments in French ports.[28] It was hoped such schemes would lower prices temporarily. While there was no indication that either the council overseeing the operation or Napoleon recognized these as partially successful Old Regime methods, the plan strongly resembled the simulated sales of the Old Regime.

Secret or not, the 1801–02 *régie* arrangements disintegrated rapidly, due in no small part to the *négociants*' immediate realization that they had everything to gain from the most questionable methods. They subcontracted the orders between several companies and sent their agents scurrying across the continent. The appearance of French agents immediately raised prices in markets and ports across northern Europe. The monthly accounts the *négociants* presented were rarely filled out, and the ones that did arrive never tallied. The amounts of grain the *régie* imported fell far below what the government had paid it to bring. Further losses for the government

and state finances. Ouvrard, pp. 58–59. Bergeron, *Banquiers*, pp. 25–6, 132–3, 147–9, 150–1, 343–4; Bruguière, "Finance et noblesse;" and *Gestionnaires et profiteurs*; Romuald Szramkiewicz, *Les Régents et Censeurs de la Banque de France*, *Hautes études médiévales et modernes* (Geneva, 1974).

[26] A metric quintal weighed 100 kilograms. The most common volume terms employed for grain were the metric quintal and the hectoliter, about 70–80 kilograms. A sack of flour, the standard measure, weighed approximately 325 pounds, or 159 kilograms, although in some towns sacks weighed as much as 200 kilograms. There were problems in determining workable standardized measures for both flour and grain throughout the nineteenth-century. These difficulties generated a great deal of tension over the ways in which their prices could be compared and over the corresponding appropriate bread price rates.

[27] Additional supplies were to travel up the Loire and overland to the capital. The Cinq Négociants were to receive a commission of 2 percent along with funds to cover insurance and shipping. The merchants could find additional profits in the sale of the grain at Parisian markets "by the usual channels," claiming any revenue above what they had paid for the imports. "Arrêt," 9 Frimaire Year X (30 November 1801), AN F11* 3049; "Extrait des Registres des Délibérations des Conseils de la République," Dossier "Police des Grains," 9 Frimaire Year X (30 November 1801), AN F11 209.

[28] Prefect of Police to MI, 24 Vendémiaire Year X (16 October 1801), and Traité with Carié, Year X, AN F11 208; Le Blond, Commissionnaire de roulage, Le Havre, 5 Floréal Year X (25 April 1802); MI to Le Blond, 14 Floréal Year X (4 May 1802), AN F11 402.

resulted from the careless, and probably criminal, mishandling of shipments once they arrived. Ship crews and dockworkers alike colluded with the agents to fleece the government. The First Consul's administrative council struggled to verify the amounts, while the *négociants* fended off each accusation of incompetence and fraud, losing as much as 20 francs per quintal and unashamedly presenting exaggerated claims for commissions and transport costs. The scheme to use the *négociants* to bring grain to Paris had backfired. Instead of lowering prices by increasing supplies, the merchants had publicized their activities widely, conspired to raise prices, and appropriated the profits. According to the Council, there was hardly one merchant who did not see the venture as a "money-making affair." Disgusted by "the truly reprehensible neglect" that characterized the *négociants*' efforts, the council members abandoned the plan in Floréal Year X (1802).[29]

Unfortunately, the crisis did not go away. Throughout the spring of 1802, what little grain the *régie* had imported was threatened by hungry townspeople along its route. Prefectoral reports from along the Seine, especially near Rouen, were filled with accounts of pillaging and attacks at farms, many directed at mayors. The officials demanded as many as 600 troops to help keep calm in the markets.[30] The *négociants*' failure to bring enough grain to Paris and the government's inability to ensure its safe arrival brought about a complete overhaul of the provisioning procedures. The man charged with this operation was Joseph Ignace Vanlerberghe, a cunning naval munitioner from Douai.[31] Since Pluviôse Year III, when Vanlerberghe and his associates had bought Belgian grain for the government, he had supplied all manner of goods to the French military.[32] Later, in 1800, he and the notorious Gabriel-Julien Ouvrard had been induced to help lend the government approximately 12 million francs.[33] The two men and their partners had been among the prime beneficiaries of the Directors' decision to decrease requisitions for the military and to turn to private operations and state supervised agencies for supplies. In Year X, the First Consul

[29] See the lengthy correspondence in AN F11 402 and 3049; and ADSM 6M 1189 and 1190.

[30] AN F11 402; ADSM 6M 1189.

[31] Vanlerberghe probably had played a role in the original Frimaire Year X contract with the *régie*.

[32] General Morgan to Representant Lefebvre, 13 Fructidor Year III (30 August 1795), AN T 1157.

[33] Ouvrard, one of the most powerful, mercenary financiers of the era, had made his early fortune in paper and colonial goods. He moved on to loans to the Directory and then to Bonaparte. He was involved in the speculation on *biens nationaux* under the Directory, and was an official supplier for the Spanish fleet. Ouvrard to MI, 27 Ventôse Year IX (18 March 1801), AN F11 338; scattered records in AN F11 208, 1594 and AN F11* 3049; Cambacérès, 1: 45, 310; Bergeron, *Banquiers*, pp. 28, 153–7, 160–1, 291, 344–5; Tulard, *Révolutions*, p. 179; Bruguière, *Gestionnaires et profiteurs*, pp. 83–7, 131.

resurrected Ouvrard, who had been spectacularly disgraced in a run of criminal bankruptcies, and with him, Vanlerberghe. Bonaparte wanted them to save the *régie* operation. Of the two, Vanlerberghe was most intimately involved in Parisian and military provisioning, although his plans were frequently mixed with Ouvrard's more perilous speculations.

The two men devised a new system that left the burden of risk to the government. Under this plan, the government provided funds and credits for the grain. Ouvrard and Vanlerberghe then purchased the cereal, deposited the sacks in government warehouses, and collected a 2 percent commission. It would be the government's responsibility to sell the grain when and how it deemed best. Once the contract of 15 Floréal Year X (May 5, 1802) was drafted, the two *munitionnaires* went to work. "I invite you," wrote Chaptal to Vanlerberghe, "once more to give orders that are prompt, extensive and unlimited. A great state should employ great means." Vanlerberghe had the connections and expertise to employ those means. He was to bring 300,000 metric quintals of wheat to Paris within two months, and an additional 100,000 metric quintals of rye. These amounts would go a long way toward covering the deficit left by the *régie*. Within weeks, ships from Danzig, Liverpool, and Hamburg lined ports along the Channel and the Seine. Vanlerberghe's efficiency overwhelmed the government and showed the Ministry's calculations to have been seriously flawed. His ships waited for weeks in ports because not enough barges could be found to unload the grain and take it to warehouses. To his great irritation, the government required a long list of formalities. As many as twenty-six ships were stalled in Rouen's port alone, and new summonses to appear before the local tribunal to answer charges of heat-spoiled grain arrived daily. "If this continues," Vanlerberghe fumed, "I will be the only person seen in court."[34]

He was not alone, however. Many other merchants had sensed the possibility of gains, and according to Seine-Inférieure Prefect Beugnot, had overcome "the repugnance that [earlier] had caused them to withdraw from all speculation on *subsistances*." The entire situation was exacerbated by Ministry of the Interior's earlier refusal to help provide grain for Rouen. If the city needed grain, its merchants were to do what the Ministry claimed Paris merchants had done: buy it. Within weeks, almost 700,000 metric quintals, destined for Paris, Rouen, and other towns in between, jammed the Seine-Inférieure ports, and other merchants joined Vanlerberghe in

[34] Even Seine-Inférieure Prefect Beugnot criticized the procedures as "contrary to the free circulation of *subsistances*, a principal consecrated by law, by the government and by many years of experience." Vanlerberghe to MI, 14 Messidor Year X (3 July 1802) and Procès verbal, 9 Thermidor Year X (28 July 1802), AN F11 1594. Other records are in AN F11 208 and AN F11* 3049.

inveighing against the "inconceivable slowness" of the port authorities and dock workers.[35] By early July, when the grain was being unloaded, it became clear the government had overestimated the extent of the shortages. Local landowners, who had withheld large stores earlier in the year, suddenly realized the government's grain would compete with their cereal. Their wagons appeared at markets that had been deserted all spring. Then the harvest came in early. Vanlerberghe's operation was complete, but the markets were glutted with the new grain, and he could find no takers for the old. The government had fallen 9 million francs behind in its payments of 13 million to him.[36]

The entire misadventure resulted in thousands of francs of losses. The difficulties in no way altered the Council's belief that Paris needed a steady supply of flour at its disposal, however. Instead, the mishaps of Year X only further convinced the Council that steadier procedures, better information, and more efficient, more secretive agents were the surest means to safeguard cities.[37] Over the next years, the annual contracts were modeled after the emergency agreement of Floréal Year X. In general, the government kept a supply of as much as 250,000 metric quintals of grain and flour, known as the Paris Reserve, although the amounts fluctuated each year. As Ouvrard and Vanlerberghe had counseled, the sale of the sacks was no longer left to the questionable wisdom and accounting practices of *négociants*. After Year X, until the reforms of 1807, the government advanced the money for the purchases and laid claim to the grain as it arrived at the warehouses. It paid Vanlerberghe, as *munitionnaire général*, the expected 2 percent commission and additional storage costs for the warehouses he leased.[38] The Council then directed him to sell the Reserve grain and flour at the market, frequently at levels of 500 or 1,000 sacks per day, using the shipments to lower prices at the Paris hall. The arrangement still left much to be desired: Vanlerberghe's efficiency consistently outpaced the government's expectations, resulting in losses and frustration for the *munitionnaire*. Nonetheless,

[35] The harbor reports from Le Havre and Dieppe show more than 230 ships arriving during the month of Messidor Year X alone. AN F11 402; Gougeon, père, 21 Thermidor Year X (9 August 1802), AN F11 628; Vanlerberghe to MI, 14 Messidor Year X (3 July 1802), and Procès verbal, 9 Thermidor Year X (28 July 1802), AN F11 1594. Other records are in ADSM 6M 1190; and 6M 1204.

[36] "One can see already . . . ," Ouvrard wrote in his memoirs, "what plan the government had adopted with regard to me; I was always called into public affairs by *necessity*, and I was always repaid with bankruptcy." Vanlerberghe to MI, n.d. [ca. Thermidor Year X], F11 628; also, AN F11 1594; Ouvrard, p. 71.

[37] In fact, within a few weeks, the Council had drawn up the contracts for a new round of purchases. Subsistence Bureau, "Rapport au Ministre de l'Intérieure," Germinal Year X (March-April 1802), AN F11 402. See also the discussion of this in the late summer of 1812, as a later Subsistence Council dealt with similar problems, AN F11* 3054.

[38] Among his associates were Maurin, Paulée, Bendecker, and Ouvrard.

Vanlerberghe continued to accept the contracts, and his commissions and his methods met with grudging, if not universal, approval through the shortages of 1806.[39]

A second series of chaotic purchases in 1805–06 led to a further complete reorganization of the Reserve in 1807. This time, it was Vanlerberghe's projects that endangered both the military and Parisian supply lines. The crisis came in 1805, just after a bumper crop and plummeting grain prices had convinced the government once again to permit grain to flow out of the country.[40] In that year, Vanlerberghe and his associates undertook to provision both the home army and the capital, receiving initial advances of at least 60 million francs. The 1804 grain surplus had given rise, though, to one of the most monumental gambles of the era, Ouvrard's plans to ship French grain and Mexican gold to Spain. In early November 1805, just as the Minister of the Interior realized exports had reduced Parisian grain supplies to the dangerously low level of 10,000 sacks of flour, Ouvrard's venture crashed. Vanlerberghe's Reserve advances were mingled with Ouvrard's schemes, "mixed without prudence and in a gigantic manner," and his company owed the government at least 140 million francs, not including interest. Upon his return to Paris in January 1806, Napoleon immediately set in motion a number of reforms in finance and banking. Vanlerberghe and Ouvrard declared bankruptcy, and Ouvrard eventually was imprisoned yet again for bad debts, among other crimes, in 1809.[41]

The crisis of 1805–06 brought about an overall discussion of the Reserve's value and further changes in its 1807 contracts. No one had expected Vanlerberghe's society to fold so suddenly. It had been buying

[39] In Year XI, for example, the government ordered him to mill "as much grain as possible" to be sold at the hall. He produced almost 25,000 sacks of flour, enough for almost three weeks. When the flour arrived, Vanlerberghe found commercial merchants had oversupplied the Paris hall. The greatest number of complaints against his purchases were raised by bakers, who alleged that he bought grain in the Paris *rayon* and that his activities thus had raised their flour prices. Prefect of Police to MI, 9 Frimaire Year 14 (30 November 1805), and 25 February 1806 and "Note pour Son Excellence," n.d. [January–February 1807], AN F11 208; AN F11 1594; AN F11* 3049; Cambacérès, p. 46.

[40] There had been much pressure for such reforms, some of it based on estimates of the price levels needed to make grain production profitable. 22 July 1804, AN F11 403; "Motifs au Projet de Loi sur le Mode et les Conditions de l'Exportation des Grains," Ministry of the Interior, 2 September 1814, AN F11 435–6.

[41] Vanlerberghe had received at least 80 million francs in advances from the Treasury for his military contract alone, much of it alleged to have been channelled "to his own uses." Further funds had gone into his pockets on the sole authorization of the Minister of the Treasury. See the correspondence in AN F11 208, especially the "Rapport à l'Empereur," 5 Pluviôse Year XIII (25 January 1805); Vanlerberghe to "the facteurs at the hall," 11 March 1806; and Prefect of Police to MI, 13 March 1806. See also J.-P. Merino, "L'affaire des piastres et la crise de 1805," in Ministère de l'économie, des finances et du budget, *Histoire économique et financière de la France*, *Etudes et documents*, pp. 121–26.

grain at reasonable prices for over five years, and in the words of the Minister of the Interior, the government contracts "should have been the foundation of one of the most useful fortunes in France." Instead, its failure sent waves of bankruptcies ripping across the continent.[42] The corruption and disorganization, repeated far too often and now with overwhelming consequences, only confirmed Napoleon's deep distrust of merchants. He was convinced that commercial efforts alone would never completely supply the city. Despite Vanlerberghe's mishandling of the Reserve, Napoleon determined that it was essential to maintain a steady food supply for the capital.[43]

The contracts of the spring of 1807 reflected Napoleon's decision to place heavier responsibilities on Vanlerberghe, whose expertise was desperately needed, and then to hold him to them. The most important adjustments concerned the ownership of the cereals: The grain and flour henceforth would belong to Vanlerberghe alone, a complete reversal of the contracts drafted since 1802. Vanlerberghe was to keep an increased supply, 300,000 metric quintals, ready to be milled and shipped to the Paris hall at a moment's notice. He received a hefty commission, but the risks were all his to take, with few for the government. In years when commercial houses alone sufficed to supply the city, Vanlerberghe received instructions to decrease the amounts in the Reserve. Such fluctuations aside, by 1809 the Reserve had grown to about 30,000 sacks of flour and 300,000 metric quintals of grain. Thus, in the decade following the shortages and confusion of Year X (1801–2), the government had established a permanent network of warehouses holding enough grain and flour to meet almost half the city's annual needs.[44]

The warehouses of the Reserve remained an integral part of the Parisian provisioning plans, although they underwent one last series of reorganizations in April 1810. Napoleon, still deeply suspicious of the financiers' operations, became convinced that the private contractors, Vanlerberghe among

[42] Circulaire, Ministry of the Interior, 22 October 1807, AN F11 208.

[43] MI, "Notes on His Majesty's Observations of September 2, 1807," AN F11 208.

[44] The council's earlier contracts may have included some similar provisions, but the spring 1807 agreements had the clearest statement of Vanlerberghe's new status. If, in periods of abundance, he decided to sell it commercially, rather than to the government, he stood to lose a great deal of money, because buyers would be unwilling to pay top price for grain they knew had been stored in government-run warehouses. At the end of each summer, when new supplies became available, he could dispose of the surplus as he saw fit. The 1807 contracts gave Vanlerberghe 7.5 centimes per quintal per month to store the 300,000 quintals of grain (270,000 francs per year). He was responsible for renewing the supply and making sure it was kept free from spoilage. In good years, surpluses went to charity bureaus and prisons. In the spring of 1808, a bumper crop had eased concerns about the capital's welfare, and Vanlerberghe let the amounts in his warehouses fall to only 15,000 sacks (23,850 metric quintals). Correspondence in AN F11 208, 1363 and 3049.

them, were making too much profit, "a form of theft." He cancelled the company's agreement, effectively removing the Reserve from the hands of Vanlerberghe and his associates, the Paulée Company. The Emperor then brought the Reserve permanently under direct state control. A new Reserve chief, Count Maret, was drawn from the bureaus of the military suppliers.[45] The Reserve warehouses, having survived numerous reforms, were now firmly part of the government's routine plans to provision Paris, and were poised to save the city in the shortage of 1812.

The Paris Reserve and the Crisis of 1812

The subsistence crisis of 1811–12 came on the heels of an industrial downturn. The hard times had been felt throughout northern France since 1810. The Continental Blockade had made imports difficult or illegal, and in town after town, factories and workshops closed down, lacking capital, raw materials, and machines.[46] Across the Seine-Inférieure, many textile workers lost their jobs. By the spring of 1811, bands of beggars wandered from farm to farm. As many as 800 were seen in a single day in just one village.[47] In the Eure, the slump in the lumber trade halved workers' incomes.[48] There was little hope for improvement. The Seine-Inférieure Prefect's answers to administrative surveys were monotonously pessimistic: There were no new industries, no new techniques, no "*nouveautés*" in commerce, only more closed shops and greater "*misère*."[49]

The manufacturing failures of 1810 were not alone in creating the hardships. As in Year X, fluctuations in both crops and exports had contributed to the crisis. The harvest of 1808 had been one of the best during the Empire, and the government had allowed grain to be exported. In February 1809, the Emperor and his ministers, especially Collin de Sussy, the head of Customs, reasoned that if landowners could export their grain for higher profits, and if the government could continue to collect an export duty, the Treasury's overall receipts would fund the looming campaign against Austria.[50] Exports in 1810 reached their highest levels since 1805. Within

[45] Maret, the brother of the Minister of Foreign Affairs, the Duc de Bassano, had signed the disastrous *régie* contract of Year X. "Extrait des Registres des Délibérations des Conseils de la République," 9 Frimaire Year X (30 November 1801), AN F11 209.

[46] Lefebvre, *Napoleon*, 2: 121–4, 129, 142–3; Tulard, *Révolutions*, p. 251. See also Jean Bouvier, "A propos de la crise dite 'de 1805'. Les crises économiques sous l'Empire," *AHRF* 199 (Jan–March 1907): 100–109.

[47] Prefect S-I to MI, "Comptes Trimestriels," April 1811, AN F^{1c} III Seine-Inférieure, 8.

[48] Prefect of the Eure to Réal, 28 February 1812, AN F7 3626.

[49] "Comptes Trimestriels," AN F^{1c} III Seine-Inférieure, 8.

[50] The final February 1810 version of the policy allowed the export of both agricultural

only a few months, however, the entire situation reversed. The June 1810 realization that the upcoming harvest would be short caused a swift prohibition of the export of any grains or flour and the resumption of emergency provisioning plans.[51]

Such was the grim situation confronting Maret when he finally assumed his position as chief of the Reserve in November 1810. He found shipments months in arrears, old supplies spoiled by the summer heat, and depressing reports from the farmlands of the Empire. By the late spring of 1811, the Reserve had shrunk to 115,000 metric quintals of wheat and 11,000 sacks of flour, less than half of what was needed.[52] Anxious, Maret then embarked on a series of emergency purchases. He hurriedly bought up all the grain he could in the regions around Paris, a debatably reckless decision. The government's purchases sent grain prices up and brought the bakers into Prefect of Police Pasquier's office, clamoring for an increase in the *taxe*.[53] Clearly, the Reserve procedures needed to be better coordinated with the capital's daily operations, especially given the dismal expectations for the harvest of 1811.

This impasse, along with overall fears of shortages, brought about the creation of the Subsistence Council in 1811 and the further integration of the Reserve into the state. Napoleon directed the Minister of the Interior to establish a council to oversee an ambitious series of grain purchases, in particular imports from areas along the Rhine. He assembled a group of men who had long experience in provisioning, although they were not necessarily in agreement on many points and frequently fell to arguing about

and industrial goods, the last provision, in particular, the result of heavy lobbying by the prefects. Three-quarters of every shipload exported was to contain agricultural goods, with one-quarter allowed for manufactured goods. Lefebvre, *Napoleon*, 2: 123–4; AN F12 2056; AN F11 592; MI to the Prefect S-I, 7 February 1809, ADSM 6M 1204.

[51] 1, 15, and 22 June 1810, ADSM 6M 1204.

[52] Upon hearing that its contract for the Reserve was about to be annulled, the Paulée company, associated with Vanlerberghe, had stopped all shipments. Maret responded by refusing to accept his post until Montalivet, the Minister of the Interior, could conduct a thorough inventory of the stocks and guarantee that new shipments were on their way. He and Montalivet remained at loggerheads for six months. "Extrait des Registres des Délibérations des Conseils de la République," 9 Frimaire Year X (30 November 1801), AN F11 209; Pasquier, *Memoires*, p. 113–14.

[53] Paris had no mayor, but instead was run by the prefect of the Seine department and the prefect of police, who had the most direct control over the city. The prefect of police oversaw a number of areas of economic life, with responsibilities similar to those of the Old Regime concept of *police*. Initially, Pasquier was unhappy with Maret's appointment; the reorganization removed the Reserve's supplies from Pasquier's direct supervision. He blamed Maret's foolhardy purchases for throwing the system out of kilter and refused to raise bread prices. Pasquier, *My Time*, 1: 530–1; Pasquier, *The Memoires*, pp. 89, 113–5; J[acques] Peuchet, *Mémoires*, 1: 14–15.

matters both substantive and personal. Even in the heat of the crisis, when situations required quick decisions and consensus, their antagonism consumed page after page of the minutes. The most powerful members of the Subsistence Council were Frochot, the Prefect of the Seine department, and Réal, a councillor of state in the Ministry of Police. Paris Prefect of Police Pasquier maintained that free trade was the best long-range solution to his city's provisioning problems. He wished to use high prices to draw grain to his city, even if it risked unrest. His greatest scorn was reserved for the men of the Ministry of the Police, his administrative rivals, and in particular its Minister, Savary, the Duke of Rovigo, who, according to Pasquier, "knew no other law and rule than his blind devotion to Napoleon, and had little thought for anything not connected with military glory."

Réal, the ministry's representative and thus Pasquier's nemesis, firmly upheld Napoleon's ideas, especially his belief that high bread prices threatened public order, and hence the stability of the Empire. At each turn, he countered Pasquier's more *laissez-faire* views with his own belief in a substantial government role to keep urban peace, a conclusion he drew from his long experience on earlier subsistence councils. Louis Nicolas Dubois, who had preceded Pasquier as the first Prefect of Police and had served on many previous provisioning councils, was among the most outspoken in meetings, offering detailed advice about the potential impact of any decisions the group might make. Maret, the head of the Reserve, emerged as the pragmatist of the group, forging compromises between the extremes that Pasquier and Réal often demanded.[54] The Council began with a general idea of what its responsibilities were: to oversee government grain imports and to keep Paris calm.[55] The specific strategies were less clear. The members could agree that extremes were to be avoided, that neither Maximums nor

[54] Réal had served on the council that drafted the 1807 contracts making the Reserve, and Maret had run the warehouses for the past eighteenth months. Although not a royalist himself, Pasquier came from a royalist family, and felt less personal loyalty than other Napoleonic administrators. He approached his responsibilities seriously and reestablished lines of communication with the Seine Prefecture and a professional atmosphere in the Prefecture of Police. (Dubois had gained a reputation for hasty and inaccurate accounts and an unwillingness to submit them to the municipal council.) Under the Restoration, Pasquier would be named Minister of Justice in the ultra-conservative Chambre Introuvable, among other appointments. Thus, he earned his nickname "the inevitable one" for his ability to make his way into administrative and ministerial offices under each regime. 19 September 1807, AN F11* 3049; Pasquier, *Memoires*, pp. 49–50, 80, 111–14; Pasquier, *My Time*, pp. 434–43, 445, 452; Charles de Rémusat, *Mémoires de ma vie*, 5 vols., ed. Charles-Henri Pouthas (Paris, 1958–1967), 1: 247–60; Baron Fain, pp. 122–7; Roussier, *Le Conseil général de la Seine*.

[55] The minutes of weekly meetings of the Council from 1812 to 1815 are in AN F11* 3050–56. Unless otherwise noted, the following account is drawn from those *procès-verbaux*.

unimpeded free trade would solve the shortage. Yet the Revolution and the Old Regime offered too many conflicting examples of failed policies.

Reserve Sales Begin

The strategies these men of 1811–12 chose were linked in fundamental ways, nearly impossible to describe in isolation. Nonetheless, two general strategies emerged. The first of these was to hold Parisian bread prices as low as possible. Not only had the Emperor commanded it, but their own fears of unrest in the capital confirmed it. There were as many as 50,000 impoverished workers in Paris alone, 20,000 of them without work, bearing the brunt of the run of 1810–11 bankruptcies from the uncontrolled speculation on colonial goods. Concerned about their welfare and the possibility of riots, Napoleon balked at raising prices, a policy that nonetheless severely handicapped the Council's flexibility during the next months. The second strategy consisted of distributing the grain of the Reserve, a process already somewhat refined, but not yet proven by a shortage as serious as the one the Council now faced.

Napoleon's charge to the Council in August 1811 was clear: Keep Parisian bread prices down. If the Council could not juggle bread prices, then it would have to choose some combination of grain imports and subsidies to either bakers or consumers. Any of those alternatives would strain a Treasury already burdened by the demands of the Emperor's armies. The members did not even have a chance to get their bearings before the shortages began to take their toll. Flour prices rose consistently at the Paris hall, nearing 70 francs per sack by early October 1811, almost 30 francs above the price in a good year. Each of the city's flour supplies was starting to run low, and bakers were reportedly halting their grain purchases.[56] For the next four months, the Council scrutinized bread prices, bitterly disputing each point of view, seeking formulas to keep bread prices low without sending bakers into bankruptcy.[57]

The Emperor himself weighed in. He decided that there was no true grain shortage, only a temporary flour deficit. Accordingly, he refused to increase bread prices. Instead, he insisted that a 5-franc bonus be paid to Paris-area bakers for every sack of flour bought after October 20, 1811, a decision

[56] By October 12, the hall, normally stocked with 7,000–8,000 sacks of commercial flour, held only 2,798. The bakers' private stores, usually 40,000 sacks, held only 31,124, and the Saint Elizabeth *dépot*'s weekly statement showed that 111 of the city's 654 bakers had fallen behind in replenishing the supplies kept there. "Mercuriales," AN F7 3619; AN F11* 3051.

[57] Pasquier argued for raising the ceilings, while Réal claimed that it was worth emptying the state's coffers to keep peace in Paris. AN F11* 3051–3052.

that the Council deemed a bookkeeping nightmare.[58] The worst news came in early November, when the first harvest figures arrived, showing France with a deficit of at least one-tenth of its consumption.[59] The Council members pored over local market statistics, police reports, and records from the shortages of 1789, 1795, and Year X, discussing each potential course of action and comparing their own experiences. Pasquier, of course, pressed for higher bread prices and an end to the system of bonuses. He estimated that bakers were losing 10 centimes per loaf and that government subsidies would be difficult to sustain at a rate of 6,735 francs a day.[60] Réal, on the other hand, received daily reports of unrest. He feared that Paris and its suburbs could soon explode.[61] Apprised of the council's frays, Napoleon wrote testily from Basel, imposing a compromise: He would permit raising first-class loaves to 75 centimes or, if absolutely unavoidable, 80 centimes, but no higher.[62]

Napoleon's demands and the development of other plans to provision the capital notwithstanding, the Council continued to wrestle with the problem of bread prices at every session. Over the next months, they haltingly moved the price upward, finally abrogating Napoleon's commands in late January 1812 by setting the ceiling at 85 centimes. In the meantime, the system of bonuses to bakers far surpassed the credits at hand, and the Council fell many weeks behind in paying them.[63] In January came news that the credit networks of the hall were disintegrating. Merchants were refusing to extend their customary credit, leaving the bakers without any means to buy flour. By early March 1812, despite the Council's efforts, bakers had reached a state of exhaustion. The reports from the depots showed that their stores, earlier exceeding 40,000 sacks, had plunged below 17,000. By April 30, they had fallen to 13,000 sacks. Police spies reported

[58] While the Emperor correctly predicted that the bonuses would encourage Parisian-area bakers to buy more flour, the plan offered many possibilities for fraud. The news of the *primes* spread rapidly to the *banlieues*, and within a few days, the mayor of Saint Denis demanded similar bonuses for his bakers. AN F11* 3051.

[59] The Council suspected some underreporting. AN F11* 3051.

[60] By mid-November 1811, the cost of the bonuses had surpassed 50,000 francs, and the Council's credit was overdrawn by almost 8,000 francs. AN F11* 3051; Pasquier, *Memoires*, p. 112.

[61] See the many reports in the police files in AN F7 3619, 3624, 3625, 3626, 3634, and 3639; F. L'Huiller, "Une crise de subsistances dans le Bas-Rhin (1810–1812). Origines, aspects principaux, évolution," *AHRF* 14 (1937): 518–36.

[62] AN F11* 3051.

[63] During November and December 1811, the Prefect of Police "had been assailed constantly by the bakers' claims [that they needed higher bread prices]." Even though he had been equally forceful in "rejecting them as severely as possible," he could no longer condone holding bread at 75 centimes. That forced the penultimate increase to 85 centimes in January 1812. AN F11* 3051 and 3052.

bakers who were selling off personal items, pawning family silver, and showing other signs of distress. That information was accompanied by reports of public discontent. The bakers' desperation had become all too apparent in the winter's avalanche of complaints to the police about short-weighting. If the Prefect of Police did nothing in response, the Parisian populace would accuse the government of siding with the bakers. If the police began prosecutions, however, they could stir up anger against the bakers, leading to violence. Instead, the Council voted, not without many misgivings, to disregard Napolon's orders. It raised the bread price to 90 centimes on March 3, 1812. That would be the last price rise.[64]

Any attempts to hold off bread price rises necessarily relied on a complementary increase in the city's grain supply. Each one of the Council's arguments took place in the larger context of overwhelming concerns about how to supplement supplies. Commercial networks, hindered at least in part by a lack of financial resources resulting from the economic contraction of 1810, had failed. The Council concurred with the Emperor's late summer 1811 sense that more grain would appear only under the government's aegis. The capital needed at least 1,300 sacks of flour – 2,067 metric quintals – per day. Looking over the records of the administrative council that had overseen the ill-fated bankers' *régie* of Year X and the speculations of 1805, Réal insisted "the most essential of all the measures is to keep the supply of the Reserve intact." The best policies, he asserted, ensured both a sufficient grain supply and the appearance of calm in administrative corridors. The other members, while they might disagree with some of the methods he proposed, certainly believed all would be lost if the government panicked in the face of shortages.[65]

During the course of the autumn of 1811, the Council was forced to bolster the Reserve warehouse network on a scale never before imagined. The everpresent Vanlerberghe pressed to begin purchases from abroad. Such overtures were clearly self-interested, but hardly contradicted the Council's sense of its straits. On Napoleon's orders, it signed a contract in late August to import 136,000 metric quintals of grain from the regions near the Lower Rhine, Moselle, and Ruhr rivers held by the French.[66] By the time the pur-

[64] In general, a certain leeway was allowed the bakers, termed the "*tolérance*," because their ovens and leavening could not be counted on to produce uniform loaves. As the loaves cooled and dried, they lost more weight. In 1812, however, the bakers' attempts to cheat customers had exceeded acceptable limits. "For the past week or ten days," Pasquier reported on March 3, "this fraud . . . has become so great, that it has given rise to *murmures* of a very dangerous character . . ." 23 November 1811, AN F11* 3051; AN F11* 3052–3.

[65] 7 September 1811, AN F11* 3051.

[66] The most pressing problem seemed to be shortages in southern France, evidenced by the fact that every day, merchants bought grain and flour in Paris to ship to Lyon, Bordeaux, and Marseilles. 7 September–5 October 1811, AN F11* 3051.

chases were underway, the Council was locked in battles over how to distribute the Reserve's supplies. Réal maintained that the Council should supply bakers directly, thereby forcing commercial grain and flour merchants to lower their prices. That strategy had the added benefit of secrecy, preventing the restive Parisian populace from becoming alarmed and agitated by news that the government too was concerned about the dearth. Pasquier contradicted him. Explaining that the best long-range policy was to reassure merchants that they could sell at the prices they wished, Pasquier demanded bread price increases so that bakers could meet the more expensive commercial rates. Maret, assessing the potential for riots, refused, but found a compromise. He cobbled together the modest mid-October 1811 increase in bread prices and a plan for Reserve flour sales. The flour would be made available directly, but covertly, to the bakers at a price slightly below the going rate at the hall. This was reminiscent of the Old Regime's simulated sales, although – as in Year X (1801–2) – there was no mention of any realization of this by the Council.[67]

Napoleon fired off precise directions from his camp, demanding that the Council henceforth have a steady supply of at least 500,000 metric quintals of grain on hand every autumn. These instructions stunned the Council. It would need thousands of francs to advance to merchants and additional shipments of at least 300,000–400,000 metric quintals arriving throughout the year. In effect, Napoleon wished to double the Reserve.[68] Still not trusting the Council to handle the decisions about the sales and pricing of Reserve flour, Napoleon gave exacting guidelines: 500 sacks a day were to be taken to the Paris hall and sold by the Hall Controller directly to the bakers for 73 francs. The Emperor judged that price to be just below the going rate for commercial flour. As the grain imports arrived, the Council hurried them to the mills around Paris and then turned them over to the Hall Controller for sale. All the while, they hoped that flour merchants and the Parisian population would suspect nothing.[69]

Within weeks, the Controller's sales of Reserve flour were providing at least half of the city's needs. The market report of December 10, 1811 revealed that 984 of the 1,174 sacks sold had come from the

[67] The Council authorized Pasquier to prosecute those who were furthest in arrears to the *dépot*, and to use municipal funds to buy out the most insolvent bakers, making the whole effort appear part of the daily operations of local government. AN F11* 3051; Pasquier, *Memoires*, p. 112.

[68] Such direct intervention in the plans was necessary, in his estimation, because the "ignorance" of his ministers, especially that of Minister of the Interior Montalivet, had rendered the Reserve useless. He claimed that they had allowed the supplies to become dangerously depleted the past summer, and then their haste to buy had raised prices across the Empire. 6–7 November 1811, AN F11* 3051.

[69] November 1811, AN F11* 3051.

Reserve.[70] The skyrocketing Reserve sales were accompanied by the apparent collapse of commercial networks. Frustrated, the merchants held off shipments to Paris.[71] The disarray in the hall meant that the Council had to exert even more caution in juggling the prices and supplies available to the bakers through the Reserve. By the early December 1811 meetings, the city's needs exceeded the 500-sack allotments that Napoleon had allowed. The Council was distressed to find its warehouses held perhaps only 18 days' supply of flour. Even with rationing, daily sales of Reserve flour surpassed 500 sacks, and commercial prices edged above 80 francs, with an unsubstantiated report of 90 francs from Chartres. The Council begged Napoleon for more flexibility, adding that it soon expected to see the positive effects of its efforts. If the Reserve could buy more grain, mill more flour, and set lower prices, there was a real possibility that the shortage and high prices would end. Commercial merchants would be forced to accept lower prices, bring their wares to the Paris hall, albeit grudgingly, and the crisis would abate.[72]

"Third-Party Sales"

The meetings of December 7 and 11, 1811, represented a breaking point, and brought about a fundamental change in Reserve policy. The Council concluded that news of the government's imports had spread too widely. The Reserve sacks, with their large lettering, had made the sales too obvious. Disgruntled merchants, trying to sabotage government sales, were spreading rumors that the imported flour was spoiled. Such allegations allowed them to drive their own prices higher. To counter this, the Council decided to sell some of the Reserve flour through the Hamot and Watin Company. Like Miromesnil and De Crosne forty years before, the Council members wished to convince both buyers and merchants that there were independent sellers willing to lower rates. Starting with a small shipment of this "third-party" flour, the Council placed its price 1 franc above the rate for Reserve flour. Buyers fell for the ruse, and bought sack after sack. By mid-January, the third-party shipments had far outstripped both commercial and Reserve sales, furnishing a large portion of the city's needs[73]

[70] Many of the 194 sacks of commercial grain were destined for other cities, especially in Normandy and the south, meaning that the capital was even more dependent on the Reserve supplies than the reports alone indicated. 16 and 23 November and 7 December 1811, AN F11* 3051.

[71] 16 and 23 November and 7 December 1811, AN F11* 3051.

[72] The Reserve milling operations had shaken up the flour merchants' speculations, including those of "Le Loup" (Le Leu) from Saint Denis. AN F11* 3051–2.

[73] There were no recognizable Reserve sacks at the hall on January 17; instead, independent

(Figure 8.1). The sales could not continue at that pace. Napoleon had permitted the Council to sell 5,000 sacks per week, but each market report showed sales creeping beyond even that threshold. Flour merchants, especially the millers from the Seine-et-Oise, insisted loudly that the government's imports were destroying commercial profits, and they would no longer send grain to Paris.[74] Consequently, throughout the winter, the bakers increasingly bought almost all their supplies from the Reserve, directly or through third-party sales. They purchased as many as 1,300 sacks per day, with staggering losses for the government.[75] The third-party sales were emptying the Reserve's warehouses and the state's meager coffers, and, as the Council knew only too well, no relief was expected.

Having embarked on a policy of intervention, the Council could not turn back. Despite their disagreements, the members knew they could not abandon Paris to commerce, especially in wartime. As the Reserve supplies dwindled, they had no choice but to seek out whatever grain they could still find and to distribute it as best they could. Their nightmares were confirmed in early January 1812: No more grain would arrive from Germany or the Baltic before the summer.[76] To fill the barren warehouses, the Council turned to the Paris *rayon*, the officially circumscribed area around the capital, which had been traditionally off-limits to government purchasing agents. Napoleon concurred. On January 24, the Council imposed a bread price increase and simultaneously ordered agents to buy 100,000 sacks of flour at the outer boundaries of the *rayon*. The selected merchants were to keep the measure secret so as not to create "any unfortunate competition."[77]

Rayon buying, however necessary, sent the Council's careful arrange-

flour merchants sold only 328 sacks for an average price of 81 francs, while Hamot and Watin sold 1,029 unmarked Reserve sacks at 75 francs each. AN F11* 3052.

[74] "Note," 4 January 1812; "Note," 15 February 1812, AN F7 3640.

[75] Maret estimated that the Reserve was losing as much as 31 francs per sack, and was running a deficit of at least 3,120,000 francs on its sales and bread bonuses. Some of the losses would be recovered when all of the imported sacks had arrived and the bakers had repaid the Reserve in full for its flour. In February, Maret proposed limiting the sales from the Reserve to 800 sacks per day. The other Council members disagreed, and decided instead to allow bakers to buy what they needed, and to demand in exchange that the bakers continue to boost their output. 8 February and 3 March 1812, AN F11* 3052; Pasquier, *Memoires*, p. 116.

[76] Of the original order for 136,000 sacks of grain, 84,470 had reached the Reserve warehouses; a further 40,000 sacks from the December order for 100,000 had arrived. AN F11* 3052.

[77] Maret supported *rayon* buying through third parties. Réal and Dubois, anxious about the effect on prices in the Parisian hinterland, objected streuously. The orders were placed with substantial flour merchants and millers in the Paris region, Chambaut (Pontoise), Chappon (Meaux), Truffaut (Pontoise), Vaudeville (Chartres), and Watin. AN F11* 3052.

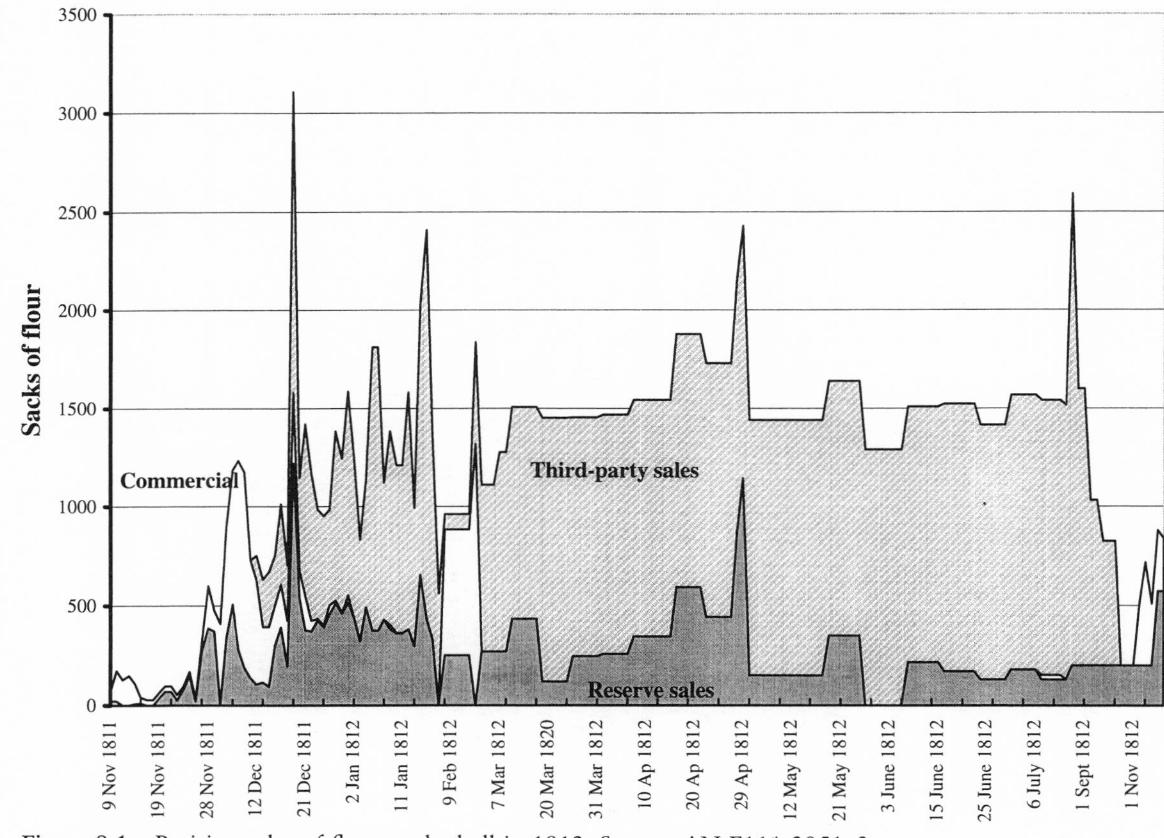

Figure 8.1 Parisian sales of flour at the hall in 1812. Source: AN F11* 3051–3.

ments spinning out of control. First, purchases in those areas triggered an illusory war between the Reserve agents and miller-flour merchants in northern France and the Paris basin, rendering the arrangements chaotic and impossible to supervise. For months, the merchants had been protesting that the subsidized Reserve operations made it impossible to get a fair price in Paris. *Rayon* buying only added credence to their allegations that government buyers were driving up prices in the *rayon* and then taking their wares to Paris where they sold them for a low price. No merchant could compete against such tactics.[78] The Prefect of the Seine-et-Oise suspected, however, that beneath those facades lay merchant schemes to sow fear, raise prices, and make the government pay for its policies with equal parts of money and anxiety. He had noticed that "the speed with which the millers buy [grain in area markets] is in direct contradiction to their words and laments." The pattern he had observed was increasingly prevalent. Throughout northern France, grain disappeared from the markets. Owners appeared at halls with small pouches of sample wheat and disappeared into the depths of inns and cabarets to deal with millers. Any sacks that did arrive at market were sold immediately and milled and the resulting flour then passed to the hands of the Reserve's agents. This invisible trade spread throughout each department that Maret's intermediaries reached. The merchants and millers continued to present a bold front of indifference to Parisian affairs, while they bought up whatever meager supplies remained, had it milled, and sent the flour to the capital.[79]

The merchants' persistent, if duplicitous, complaints notwithstanding, there was a more distressing reason for the Reserve to question the wisdom of *rayon* purchases in early March. The agents' activities were stirring up local consumers. In Le Havre, for instance, the wives and children of sailors joined dock workers in attacks on "the principal bakeries in the city, many of which were pillaged and their windows shattered." In Elbeuf, buyers dragged sacks of grain from wagons and forced merchants to sell them for only about half their price. The most chilling and widespread assaults took place at the farms themselves. On January 26, there was gunfire at a farm near Dreux (Eure-et-Loir), and two weeks later, the same hungry laborers laid siege to the farm a second time. On February 20, the mayor of Bellême

[78] Since October 1811, many flour merchants from Etampes and Chartres had directed their wares to Orléans, Blois, Nantes, and Bordeaux, while those from Picardy and the Vexin opted for markets in Normandy and the Nord. Those in the Seine-et-Oise claimed to have no interest in sending any more flour to the capital. Others refused to extend credit to the poorer bakers, forcing them to pay cash. AN F11* 3051.

[79] Correspondence of 11 December 1811; 4, 29 January; 8, 15, 29 February; and 15 March 1812, AN F7 3640.

(Orne) was shaken by threats to burn down his house if he did not halt grain price rises. On February 29, a farmer near Blangy (Seine-Inférieure) was killed. Night after night, farms across the Eure, the Eure-et-Loir, the Seine-et-Oise, and the Seine-Inférieure were attacked, grain stolen, and owners intimidated.[80]

The Parisian hinterland was not the sole locale of riots and other "movements." There had been troubles in Nantes, Rennes, Saint-Lô, and La Rochelle, all key arrival or distribution points for grain. Réal, reading the bulletins with growing horror, demanded greater vigilance by local officials and harsher sentences from the courts.[81] The reports of unrest were coupled with Maret's realization that there was far less flour available near Paris than hoped. His clandestine buyers could procure only two-thirds of the 100,000 metric quintals ordered, if that. On March 10, 1812, he shifted his agents' activities to farmlands just beyond the traditional *rayon* in search of larger quantities, lower prices, and greater calm. Such calm was not to be found. As the government merchants scattered further across the countryside, the number of disturbances escalated. In the Eure-et-Loir, "wheat [prices rose] at every market because of the competition between outside merchants . . ." and the prefect worried in March that he would be unable to maintain calm much longer. At night, hungry bands, mostly men, roamed the countryside, especially in the Eure and the Seine-et-Oise. As in the earlier *rayon* purchases, mayors who were also large landowners were among the most vulnerable to attack. The farm of the mayor of Orcement (Seine-et-Oise), for example, was attacked by thirty or forty men who grabbed food and other household supplies and left, threatening to return because "a revolution [was] about to start." The hungry army moved on to the area around Dourdan and broke into ten more farms. Identical reports arrived from every other department where Maret's merchants appeared. There was no safe point along the road from farm to city. In mid-March, a Colonel Henry led 400 troops to the Seine-Inférieure to quell disputes at markets during the day and to protect farms at night. Yet the violence only subsided in early April, when Maret abandoned his attempts to buy grain in northern France.[82]

Although the countryside became somewhat more pacified with the end of Reserve buying in April, the capital's operations broke down. The reports from the government's purchasing agents were alarming: Little imported

[80] The records of the disturbances and throughout AN F7 3619, 3624, 3626, 3634, 3639, 3640.

[81] AN F11* 3052; Jean-Paul Bertaud, *Histoire du Consulat et de l'Empire: Chronologie commentée, 1799–1815* (Paris, 1992), pp. 167–9.

[82] AN F7 3619, 3624, 3626, 3634, 3639.

grain would reach Paris before mid-summer.[83] The efforts to hold grain prices down through strategic Reserve sales were faltering. The government had eased the burden on the Parisian bakers, but had increased the pressure on its own funds to an unbearable degree. The Parisian bakers had decided to sell their remaining first-class flour to bakers in nearby towns. They pocketed the profits and then had the temerity to demand further Reserve supplies. This threw off the delicate balance between Parisian prices and those of the surrounding towns. For months, the ceilings had pitted Paris against the suburbs, especially Saint Denis and Saint Marcel. The ministers knew that whenever Parisian prices were more than a few centimes below area prices, buyers from the outlying towns smuggled bread hidden in laundry bundles or carts out of the capital.[84]

There was no alternative save more extreme measures. To ease the wrenching burden on the Reserve's supplies, the Emperor authorized a daily distribution of free soup to the poor of Paris. He allocated funds to help cities set up public works. Various experiments with other cereals, including the preparation of bread made from potato or rice flour, were carried out with minimal success. In late March, he ordered cities to take up collections from their wealthiest inhabitants and to enlist merchants to buy grain in Belgium and Amsterdam. Nonetheless, anxieties over Parisian supplies and unrest elsewhere were driving Napoleon toward more sweeping measures. Each decision in April 1812 brought the May Maximums nearer. On April 14, the Council urged officials in departments around the capital "to make every effort to enlist the *cultivateurs* to bring all the grain possi-

[83] As of 30 June 1812, the Treasury's purchases stood at:

Date	Funding (F)	Authorized (mq)	Bought (mq)
August 26, 1811	500,000	136,000	136,487
November 3, 1811	500,000	100,000	128,824
February 3, 1812	?	100,000	61,188*
May 12, 1812	?	50,000	27,894**
32rd Military Div.	?	100,000	100,000***
Total	?	486,000	454,393

* The Paris *rayon* purchases.
** May requisitions.
*** The Left bank of the Rhine.

AN F4 2153; AN F11* 3053. [The missing figures are not extant in the archives.]

[84] Throughout the winter, the Council reluctantly allowed suburban bakers to draw on the Reserve's supplies. After March 19, however, bakers from the suburbs were barred from the Reserve flour. In return, the Council allowed them to raise bread prices, only further jeopardizing the capital's supplies. Since the early autumn, the Ministers of Commerce and the Interior had kept tight control over area bread prices, forcing prefects to get ministerial approval before allowing any increases. AN F11* 3051–3.

ble to market," setting the price at 33 francs per hectoliter. Heated meetings between prefects and farmers ensued, although the word "requisition" was avoided. The Prefect of the Seine-et-Marne, in fact, described his orders for owners to supply the Melun market as "persuasion alone." Equally "persuasively," he gave an order that any grain brought to market but not sold be confiscated at the set price of 33 francs per hectoliter.[85] These measures were not limited to the six departments closest to Paris. In the Calvados, Prefect Méchin met personally with local grain owners. Instead of using the words "*taxe*" or "*maximum*," he pressured them make an "*arrangement de famille*" that included setting price ceilings and promising to supply markets.[86]

The orders for "voluntary" shipments from the departments surrounding the capital were accompanied by even more unsettling news of grain censuses. These surveys immediately ignited rumors of a possible return to Revolutionary controls. While the six departments proceeded with their censuses, they placed temporary embargoes on grain leaving their jurisdictions, infuriating would-be recipients. Proprietors throughout the Paris basin hurried to sell their grain before lower ceilings were set, or surreptitiously sent what they could to departments outside the census-takers' reach.[87] By late April, it was believed universally that the government would no longer hesitate to use whatever methods it deemed necessary to provision Paris. Then came requisitions in all but name. During the last week of April, the Emperor ordered grain seized from the six departments at the price of 33 francs per hectoliter.[88]

Those emergency measures threw the provisioning networks into complete disarray. Throughout late April and early May, grain owners and flour merchants rushed to sell whatever sacks they could before the anticipated price ceilings slashed their profits. Markets across northern France bore witness to overall confusion as grain poured into halls or escaped to more distant departments during late April.[89] The greatest reverberations were

[85] If anyone challenged the policy, the administrators were to "explain" it was in the grain owners' "best interest, because . . . in such circumstances, [there could be] unfortunate individual incidents and need might carry *le peuple* . . . to commit terrible excesses against those men who hold out for high prices . . ." Once the grain was at the market, every Seine-et-Marne transaction was scrutinized, and preference was given to local bakers and to townspeople armed with ration cards distributed by the mayors. AN F11* 3053; AN F7 3640.

[86] Vidalenc, "Crise des subsistances," pp. 348–9.

[87] AN F11* 3053.

[88] Prefect of the Seine-et-Marne to Rovigo, May 22, 1812, AN F7 3640.

[89] Some millers in the Seine-et-Marne sent their flour to Paris, "which must have produced an extraordinary abundance at the hall," according to one prefect. That surplus did not lower prices, however. With the exception of one day of heavy sales on 29 April 1812, Parisian buyers, convinced that price ceilings were on the horizon, refused to purchase the high-priced commercial sacks. Seine-et-Marne Dossier, AN F7 3640; AN F11* 3053.

felt in cities at the outskirts of the capital's provisioning zone. There, merchants sped their wares away from Paris and its 33-franc ceiling.[90] Exhausted supplies, disrupted provisioning networks, prices escalating beyond control, and riots sure to resume if *rayon* buying recommenced: Those were the reports littering the Emperor's desk in the first week of May 1812.

The May Decrees

Napoleon's reaction was abrupt and deliberate. Grain prices had reached levels he could term only "artificial and out of proportion with the true price at which this foodstuff should be valued." On May 4, 1812, an Imperial decree restricted all sales to public marketplaces and required buyers and sellers to register transactions with prefects. The Emperor insisted that *libre circulation* be upheld and that anyone caught hoarding grain would have his supplies seized and sold.[91] On May 8, a second decree established a series of maximum prices for grain through September 1, 1812. In Paris, Napoleon announced, grain could not be sold for more than 51.5 francs per setier, a drop of 40 percent from the April 15 peak of 85 francs. In the six departments surrounding the capital, including the Seine department, the ceiling remained the 33 francs per hectoliter of the April requisitions, almost 34 percent below the going rate.[92] Napoleon sent similar instructions to every department normally drawing provisions from outside its territory. The prefects of those departments were to calculate appropriate ceilings, taking into consideration transportation and other costs, and then to uphold them until the harvest arrived.[93]

The 1812 Maximums and requisitions proved such a disaster that they convinced administrators never again to try solving provisioning crises with radical measures. Problems emerged at every turn, from the decrees' definition to their implementation and their eventual impact. They soon became wholly untenable. There was great confusion over which grains were subject to the ceilings: all of them, came the Council's repeated

[90] During the last week of April 1812, the heaviest purchases at the Paris hall were for Rouen, where prices had reached 150 francs per sack. Merchants in the Eure-et-Loire and the Seine-et-Oise sent their flour to the neighboring departments of the Eure and the Orne, where prices had remained "exhorbitant" and, more important, the grain was safely out of the reach of the first round of late April requisitions. 30 April 1812, AN F11* 3053; Prefect Eure-et-Loire to Réal, 1 May 1812, AN F7 3640.

[91] ADSM 6M 1191.

[92] In some outlying departments, the May 8 decree cut prices by almost half. AN F7 3639.

[93] "Arrêté de M. le Préfet . . . Seine-Inférieure," 15 May 1812, AN F11 717–8.

answer.[94] The status of prior agreements, appropriate price levels, and how consistently transactions were to be once again limited to the marketplace all came into question. The issue of confining sales to the marketplace revealed how uncomfortable many administrators were with the profound changes underway in the grain trade. The May 4 decree ran counter to several decades of commercial transformation in northern France. By the early nineteenth century, the marketplace was relinquishing its role as the primary site for the trade in those departments. Most sales were done by sample in the large markets that supplied Paris. In those towns, transactions between farmers and millers or flour merchants were completed in nearby *auberges* or taverns. Townspeople themselves increasingly abandoned markets and bought loaves from bakers. Only when grain prices rose, placing bread prices out of reach, did many families economize by purchasing grain and preparing bread themselves. Thus, the public market hall in larger towns was the emergency depot for many consumers in northern France. This only further heightened the tensions of the marketplace in times of shortages. As grain prices rose, an influx of townspeople competed angrily with bakers and merchant-millers for scarce supplies. These were issues that Napoleon had failed to consider in his haste to reimpose the prohibitions against off-market sales.

Responding to the Emperor's order, Pasquier's staff uneasily set the Paris-area ceiling at 36 francs per hectoliter and flour at 105 francs per sack.[95] There was another flurry of chaotic shipments and sales, as owners sought to garner whatever profits they could before the ceilings were imposed in their departments. That there was no uniform starting date for the Maximums, the decision being the prefect's, only added to the disarray. Commercial agents rushed in, grabbing any remaining grain in areas where the Maximums had not yet taken effect, and rendered departments even more vulnerable than before.[96] The Council labored on without much hope. It had not anticipated the Emperor's decrees, and probably would not have supported them if asked, and so the members found themselves the unhappy defenders of untenable policies.

Almost immediately, the decrees set off price wars as the prefects sought ceilings attractive to sellers but not offensive to buyers. In the Seine-Inférieure, Prefect Girardin proposed a Rouen grain Maximum of 40 francs

[94] Durosiers to Réal, 9 May 1812, AN F7 3640; Prefect of the Seine-et-Marne to Rovigo, 12 May 1812, AN F7 3640; The Minister of Manufacturing and Commerce, 20 May 1812, ADSM 6M 1191; ADSM 6M 1194; AN F11* 3053.

[95] The Council shrewdly exempted any imported grain from several ports, to ensure the continued arrival of anticipated shipments. 12 May 1812, AN F11* 3053.

[96] "Etats des Approvisionnements . . ." and Girardin to Réal, 10 May 1812, AN F7 3639; Prefect of the Seine-et-Marne to Rovigo, 22 May 1812, AN F7 3640; inspecteur général adjoint to Rovigo, 10 May 1812, AN F7 3640.

per hectoliter and 120 francs per sack of flour, a sharp reduction from the 180 francs some of the finest quality flour had reached. His price ceilings were nonetheless well above Parisian levels, and reflected his desire to draw grain from surrounding departments, even those that supplied the capital.[97] The ploy did not go unnoticed by the Subsistence Council. Within only a few days, it had received complaints from all over the region that the Seine-Inférieure's shameful prices threatened to drain neighboring departments. The Minister of Manufacturing and Commerce instantly reprimanded Girardin for his exaggerated stance. The Calvados prefect, perhaps still uneasy after violent riots during March in Caen, took the opposite tack. He set his department's Maximums quite low, near 33 francs. The stark difference in the prices of the two departments sparked more conflicts. The Council spent the late spring adjudicating an infinite number of such disputes between the departments.[98] The members, exhausted, finally judged that price discrepancies "were a necessary consequence of the emergency (*situation forcée*) in which we find ourselves."[99]

There was widespread local resistance to every aspect of the 1812 regulations. Many mayors refused to enforce the policies. After all, as landlords themselves, they had little interest in slashing their own revenues. A letter from the Seine-Inférieure alleged that the many landlord-mayors wanted prices to remain high through the autumn so that they could rewrite their leases during the November period of renewals.[100] Seeing the same interests at work in the Seine-et-Marne, one census-taker finally decided that at least half of the grain had escaped the survey, an accusation that brought only more emphatic denials from local officials.[101] The prefect of the Eure

[97] Girardin justified such high Maximums by claiming that Rouen drew most of its supplies from the Paris hall and that transportation was expensive. (The prefect was ignoring the fact that the Seine-Inférieure drew quantities from other departments.) The prefects of surrounding departments lodged sharp complaints. "Arrêté de M. le Préfet . . . Seine-Inférieure," 15 May 1812 and Prefect S-I to MI, 6 April 1812, AN F11 717–8; 26 May 1812, Oise dossier, AN F11 715; AN F7 3639; ADSM 6M 1194; Vidalenc, "Crise des subsistances," pp. 349–50.

[98] Prefect Méchin in the Calvados also complained that his efforts to set a 33 franc ceiling had been sabotaged by the much higher ceiling (44 francs) set in the arrondissement of Argentan in the Orne. The Emperor sent word to Collin de Sussy to overturn them. Prefect of the Calvados to the Minister of Police, 22 May 1812, AN F7 3624; Napoléon, *Correspondance officielle*, p. 266.

[99] AN F11* 3053.

[100] "Friend of Guilbert" to the Prefect S-I, 23 June 1812, ADSM 6M 1193.

[101] There were strong suspicions of fraud in the grain census done in the Seine-et-Marne department, for instance, and a second one was ordered. The inspector sent to oversee the new survey found it difficult to unravel the deceptions he knew the local officials were carrying out, and he believed mayors were hoarding grain and encouraging their tenants to do so, in order to reap profits when the Maximum ended. Inspector général adjoint

concluded that censuses did not work, and so he simply imposed "arbitrary requisitions," hoping they would bring enough grain to market. He was well aware that many of his mayors ignored illegal trasactions, deciding it was wiser to allow grain owners to achieve some level of profit.[102] There were intermittent efforts to enforce the Maximum in the Seine-Inférieure and elsewhere.[103] Like the Jacobin measures, the Napoleonic requisitions also carried with them the threat of billeting troops in towns to speed the threshing and transport of the grain to market. The Ministry of the Interior repeatedly counseled prefects to make use of the laws permitting them to send troops, and to avail themselves of the judicial system.[104] While there were few soldiers to spare in the midst of war, a handful were mustered to encourage recalcitrant proprietors to part with their grain, but with limited results.[105]

The Council was presiding over a disaster. Markets collapsed and grain disappeared. After the May 19 application of the Maximum in the Seine-Inférieure, shipments to Rouen dropped to less than 3,000 hectoliters.[106] The markets of Dieppe, Fécamp, and Saint Valéry were "almost completely lacking [in grain]."[107] Whether due to hoarding or a genuine shortage, only meager amounts of grain were available for sale in the halls of the Calvados throughout May, and bakery shelves were bare.[108] In the Oise, markets were "null" and the situation was "disastrous." The prefect blamed the 33-franc ceiling and the subsequent Maximum for the grain owners' boycott of markets. The heavy requisitions for the Paris hall had placed further

to Rovigo, 10 and 17 May 1812, AN F7 3640; Subprefect of Meaux to the Prefect of the Seine-et-Marne, 30 May 1812, AN F7 3640.

[102] 12 July 1812, Eure dossier, AN F11 709.

[103] A handful of merchants who had tried to sell grain above the ceiling were given "severe sentences" in order to "intimidate the greed of the speculators a bit." The worsening situation of the spring also brought harsh sentences for two grain owners who had tried to raise grain prices to 210 francs per sack at the Goderville hall on April 28. They went to prison for a year and paid fines of 1,000 francs. Other offenders received sentences of two months in prison and fines of between 1,000 and 4,000 francs. AN F7 3639.

[104] MI to the Prefect of the Oise, 12 April 1812, Oise dossier, AN F11 715.

[105] The mayor of Blennes (Seine-et-Marne) and his assistant, for instance, both *cultivateurs*, had brought few or no sacks to fulfill their requisitions. In mid-May, troops were sent to their farms. Inspecteur général adjoint to Rovigo, 13 May 1812, AN F7 3640.

[106] "Etats des Approvisionnements . . ." and Girardin to Réal, 10 May 1812, AN F7 3639.

[107] Cotteau to Réal, 3 June 1812, AN F7 3639.

[108] There were many suspicions that the Calvados census had been undermined by the mayors' tendency to under-report wheat supplies. Information from the census showed 15,000 hectoliters of barley and *métiel* combined, 27,500 hectoliters of buckwheat, and 13,500 hectoliters of oats, causing Réal to conclude that there were additional stores to be requisitioned, despite Méchin's insistence to the contrary. Prefect of the Calvados to the Minister of the Police, 22 May 1812, AN F7 3624; Vidalenc, "Crise des subsistances," pp. 353–4.

burdens on the department's insufficient supplies.[109] The worst reports came from Paris: On one day during the first week of July, only twenty setiers of grain arrived for sale at the hall, and a few days later, only one. The Reserve supplies alone were the capital's salvation.[110] The empty halls led to more disturbances, and Réal could not keep up with the bulletins piling up on his desk from prefectures across the country.[111]

By June 30, the Council itself had abandoned any hope of enforcing the May Decrees, and urged their repeal. "It is clear today that we no longer receive any good effect from the *taxe*," the members concluded, "and it would be advantageous to repeal it at the first opportunity that presents itself." With the exception of Réal and Dubois, they believed that the beginning of the rye harvest in mid-July offered such an occasion. And, indeed, the Maximum on rye simply "fell by the wayside" as the new grain arrived at market and prices dropped.[112] By late late July 1812, still frustrated by Napoleon's opposition to lifting the price ceilings, the Council suggested the Minister of Trade and Manufacturing write secretly to the nearby prefects, telling "them that they should follow the example of other prefects who had repealed their *arrêtés* at the start of the harvest." It would be, the Council concluded, "a very simple way to reach the desired goal without any fuss."[113] Throughout July and August, if not before, prefects abolished the Maximum, generally declaring it to be a failure.[114] When September 1 arrived, the Maximum simply disappeared. "The transition from one regime to the next occurred as unnoticeably and as smoothly as one could desire," the Council reported.[115]

The Prefects' Consensus

The late summer weeks offered a chance for extended administrative reflection on the impact of the extreme policies. Within the Council and among the prefects, there were some who believed the harsh controls had been appropriate, even if they had rattled commercial networks. For instance, Dubois held out for restricting sales to marketplaces and levying requisitions "as a general principle . . . for all time." According to his arguments, the May decrees' instruction to prefects "to supply their markets, is not only appropriate in difficult circumstances," but as part of their regular

[109] 26 May and 8 August, 1812, Oise dossier, AN F11 715.
[110] 7 July 1812, AN F11* 3053.
[111] AN F7 3619, 3624, 3626, 3634, 3639.
[112] AN F11* 3053.
[113] 28 July 1812, AN F11* 3053.
[114] AN F11* 3053–4; AN F7 3640.
[115] 1 September 1812, AN F11* 3054.

powers.[116] His views were seconded by a handful of the prefects who replied to a lengthy questionnaire from the Ministry of the Interior during the summer of 1812. There were even a few prefects who expressed great enthusiasm for the May decrees and who claimed, as did the prefect of the Eure-et-Loir, that they had "saved the department." There, this prefect explained, with an eagerness that seems almost too convincing, yet characteristic of much Imperial administrative correspondence, that the "measures were extremely appropriate to the situation of [my] department."[117] The Prefect of the Seine-et-Marne wanted continued authority "to order grain to market, and to see that the bakers of the principal cities are well-supplied." If possible, he would also have supported regular price ceilings.[118]

The majority of the prefects who responded to the Ministry of the Interior's survey professed a more moderate evaluation of the decrees' success, however. They found some form of regulation valuable, but questioned whether the forcefulness of the May decrees had been wise. The Oise prefect, for instance, reported that requisitions had compelled owners to part with their grain, although he expressed strong doubts about the efficacy of price ceilings. The May 4 decree had produced a "marked improvement" in his department, but the May 8 ceilings had "entirely ruined" any good the first decree had yielded.[119] Another reply, from the Eure, also applauded the first decree, but criticized the Maximum. In fact, the ceilings had driven owners from the markets and caused rampant black market trading in his department.[120]

Over the course of the summer, more prefectoral appraisals of the May decrees arrived at the Ministry of the Interior. They agreed that the grain and flour trades required careful regulation. The detailed assessment presented by the prefect of the Seine-inférieure spoke for many. Prefect Girardin described a moderate, but still interventionist, approach. Once a firm proponent of free trade, his beliefs had shifted in response to the crisis. "The experience of the past . . . should lead to precautions for the future," he counseled. The trade could not be left to commerce alone. The government, Girardin wrote, should "determine the price levels of the halls . . . and ruin, through a competition that cannot be sustained, the companies or individuals that would seek to raise grain prices." It should do so through "carefully calculated regulation and supervision," destined to limit merchants' profits without driving them from the trade.[121]

116 11 August 1812, AN F11* 3054.
117 31 July 1812, Eure-et-Loir dossier, AN F11 709.
118 Prefect of the Seine-et-Marne to Rovigo, 19 June 1812, AN F7 3640.
119 "Réponses aux questions . . . ," n.d., and 5 August 1812, Oise dossier, AN F11 715.
120 21 July 1812, AN F11 709.
121 Prefect Girardin to [Réal], 20 September [1812], AN F7 3639.

His support of intervention was a marked change from his April and July complaints that the government's actions had only worsened the situation.[122] By the autumn, however, he had become an advocate of carefully planned measures to prevent another such crisis.[123] In fact, on his own initiative, he met with farmers throughout the early autumn of 1812, assigning informal requisitions for his department.[124] Furthermore, he wanted the central government's efforts to extend well beyond Paris. He pointed out that the many sacks of grain and flour in warehouses along the Seine, as far north as Le Havre, could have been employed throughout northern France to counter price increases. The government had had the means to be the "master [of the price]" throughout the region, and had let the opportunity escape by concentrating on Paris. It was critical that future efforts be better coordinated and planned well in advance of potential crises, he argued.[125]

Girardin's moderate interventionist beliefs were wholly representative of administrators in the aftermath of the 1812 crisis. The most solid administrative consensus on the need for action by the state emerged on the issue of tempering free trade when shortages loomed. The Eure prefect spelled out conclusions that were widely held. "Complete free trade in grain will always be preferable to any other system," he wrote. "However, this year the startling rise in prices which could have troubled order and paralyzed provisioning, necessitated government measures." The appropriate measures, however, were "gentle" and included the May 4 decree limiting sales the market. They did not include the May 8 Maximum. And certainly, he complained, the prefects should have been consulted at least two weeks before any such policies were enacted.[126] Several reports suggested that the state should have intervened at the first sign of trouble and that the May decrees would have had a more salutory effect if they had been imposed in February, before the situation had gotten out of hand.[127] The prefect of the Somme presented a very lengthy analysis of the impact of the decrees. He criticized uncoordinated buying for the military and Paris early in the year and the way in which the many government purchasing agents had been too free with information. "They let too much be seen," he

[122] "Examen des questions adressées . . . ," Prefect S-I to MI, n.d., ca. 21 July 1812, Seine-Inférieure dossier, AN F11 717–8.

[123] Girardin to [unknown], 20 September [1812], AN F7 3639.

[124] 24 September 1812, ADSM 6M 1193.

[125] "Examen des questions adressées . . . ," Prefect S-I to MI, n.d., ca. 21 July 1812, Seine-Inférieure dossier, AN F11 717–8.

[126] 21 July 1812, Eure dossier, AN F11 709.

[127] 21 July 1812, Eure dossier, AN F11 709; n.d., Somme prefect, Somme dossier, AN F11 717–8.

asserted.[128] As for the price ceilings, there also the government had erred by allowing every department to have a different Maximum. His solution – noting how poorly price ceilings had worked during both the Revolution and the present shortages – was to limit the number of merchants involved in the grain and flour trades. In June 1812, he revoked most licenses for the trade and placed the provisioning of the largest cities in his department in the hands of a small number of merchant associations. Here, too, "gentle" measures, were deemed best.[129] The crisis had forced the prefects to ponder what measures would most effectively feed their departments. By the autumn of 1812, they had concluded that well-organized, finely tuned strategies, complemented by government grain, promised the best results.

There had been other attempts to find more grain and funds to carry the country through the crisis, although these had stirred resentments. At the Emperor's instructions, prefects had tried to supplement government funds by charitable contributions from the wealthy. These efforts verged on extortion. The Eure prefect dragged 52,600 francs out of the well-to-do in his department, for instance, although he cautioned the Minister of the Interior that such monies would not be forthcoming again. The experience of such measures in the Seine-Inférieure offers a glimpse of the pressures that departmental administrators were prepared to employ against the wealthy – and of the dismal timing of these attempts. When few contributions were forthcoming, Prefect Girardin published the names of donors in the *Journal de Rouen*. His implied threat was that anyone who would not pledge funds would be identified and exposed to the public's hostility. Slowly, Girardin and the other subprefects in cities in the department's north accumulated enough money to commission the omnipresent Vanlerberghe to make purchases. The efforts did not end happily, however, for the grain did not arrive until August. Then it became a costly burden. Vanlerberghe tried to convince the Girardin that the department would need the imported grain in the future, if only to mix with the fresh grain from the harvest. He urged the city to do as Orléans had, which was to leave the bread *taxe* high for several more weeks in order to encourage the city's bakers to buy the expensive imported grain. Such measures would spare the city and the Reserve further losses on the operations. In the long run, his advice prevailed.[130]

[128] Girardin expressed the same concerns in July 1812, blaming the many government buyers for exacerbating fears of shortages in his "Examen des questions adressées . . . ," Prefect S-I to MI, n.d., ca. 21 July 1812, Seine-Inférieure dossier, AN F11 717–8.

[129] Somme prefect, n.d., Somme dossier, AN F11 717–8.

[130] *Journal de Rouen*, 28 June 28 1812; "Rapport," Le Havre, 19 August 1812; and Vanlerberghe to Girardin, 9 August 1812, AN F7 3639; 18 August 1812, Eure-et-Loir dossier, AN F11 709; Seine-Inférieure dossier, AN F11 717–8; AN F11 715; AN F11* 3052–3.

The Emperor's insistence that local elites play a substantial role in provisioning of their communities nonetheless sparked disagreements. The prefect of the Eure-et-Loir, for instance, objected to these so-called "voluntary contributions" and argued that provisioning was the responsibility of the Ministry of the Interior. The Minister, of course, declined to see the matter in that light.[131] He responded by urging every prefect and subprefect to persuade local elites to organize road-building projects and other public works through the winter of 1813 so that the poor would have year-round incomes.[132] The government could not provide all that was needed, and so charity, publics works, and pressure on the wealthy would have to be employed.

The Reserve at the Close of the Empire

Reserve purchases continued, however, although the Emperor's designs kept the Council off balance during the last half of the year. In the mid-summer of 1812, for instance, he almost threw the system into chaos by demanding that the Council put together an immediate Reserve stock of 30,000 sacks of flour to supply the hall throughout the autumn, and that it set up an additional ready surplus of 30,000 sacks by December 1. Such efforts were unfeasible, the Council protested.[133] As the Emperor's orders became public knowledge, prices rose, forcing the reluctant Council to decide to buy what it would close to Paris. At least it would have a steady supply if the bakers demanded more Reserve flour.[134] To ignore the bakers' potential distress at the continued high price levels, the Council judged, would lead to "malicious acts" on their part. It was far more important to maintain peace, even if nearby grain was expensive and subsidizing it was a heavy burden for the Treasury. Reserve sacks appeared at the hall through January 1813. The Reserve itself received a new round of imports via Amsterdam from contracts written in July 1812, although that news once again stirred up merchants, and rumors circulated that the regime planned to provision the city henceforth entirely through its own resources. None of these

[131] 10 September and 12 October 1812, Eure dossier, AN F11 709.

[132] 28 November 1812, Eure dossier, AN F11 709.

[133] If such purchases were made, they might then have to be dumped at a great loss if flour prices fell during the autumn. Moreover, any large purchases would have to be undertaken outside the Paris area, meaning delays in transit that would outrun the December deadline. 4 August and 1 September 1812, AN F11* 3053; AN F11* 3054.

[134] In the meantime, the news of the Reserve's new orders had reached millers in the Paris basin. They spread rumors of further shortages and high prices, while they quickly bought up grain for their mills. AN F11* 3053.

problems caused the Council to doubt the worth of its goal to generate more stable supplies for the city.[135]

Overall, the extremes of the May decrees had been scuttled, confirming the growing sense that the state could not bludgeon either consumers or producers into obedience. It was that simple. What had remained to be learned after the experiments of the Jacobins and the Directory was now etched indelibly in Napoleonic experience: Strong-arm tactics in support of either suppliers or consumers only made matters worse. The legacy of the Napoleonic provisioning policies was not the Emperor's dramatic recourse to Maximums and requisitions. Instead, it lay in the more covert but workable pressures administrators believed could be brought to bear on the supply networks through the government's warehouses.

[135] Initially, the Council had hoped that it could wrap up the Reserve sales as soon as the harvest was complete. Instead, prices dropped somewhat more gradually, and commercial supplies were uneven, so that the best the Council could hope for was simply a slow retreat from the hall. The July rumors that the city's provisioning "be undertaken forever by the Reserve" stirred commercial anxieties, especially in the Seine-et-Marne. AN F11* 3053–4; Circulaire, Prefect of the Seine-et-Marne to the subprefects, 25 August 1812, AN F7 3640.

CHAPTER NINE

The Routines of the Restoration

Among the symbols that Louis XVIII, the restored Bourbon monarch, chose for his reign was a horn of plenty. Yet despite the outward promise of this choice, abundance remained elusive. In 1816–17, with Louis' regime barely secure, European-wide shortages spurred riots throughout the kingdom. In 1829, at the end of the reign of his brother Charles X, rumors of a poor harvest again raised concerns in ministerial corridors. While the two monarchs of the Bourbon Restoration held utterly opposing political beliefs – Louis being a moderate royalist who vowed to uphold the Charter, and Charles an ultraroyalist who would eventually annul it – their provisioning policies were remarkably similar. Each one rejected sweeping intervention, citing practical reasons in particular. Financial concerns predominated. The Treasury could never again fund the vast measures of 1812, they concluded, and their ministers agreed. In the first years of the Restoration, especially, financial matters could not be set aside. The remaining costs of the war and the reparations were distressing, and foreign armies were still housed on French soil. Those Allied soldiers would remain until late November 1818, when the last of the French payments was made, albeit with borrowed money. Even if the Restoration monarchs had wanted to repeat the Reserve efforts of 1812, they were unwilling to exhaust the Treasury to do so. Furthermore, their ministers doubted that energetic actions by the state could assure provisions, believing instead that the state's intervention had too often undermined the market and only generated greater difficulties. Accordingly, both monarchs and ministers counseled a greater reliance on free trade.

Experience and Expertise

Thus, it fell to the bureaucrats in the various subsistence committees to make good on the royal promises of plenty, without adding to the state's debts. Despite the last burdens of wartime mobilization and reparations, circumstances were nonetheless more favorable than those that had faced the committees' Imperial predecessors. After nearly twenty-five years, peace soon would prevail. Soldiers would trudge home, and some would trade their muskets for plows. The Treasury would undertake enormous reforms – auctioning off bonds to cover debts, completing the introduction of double-entry bookkeeping in its bureaus, and streamlining revenue collection and disbursements. Under Villèle's ministry of the mid-1820s, especially, the state's ledger books would be put into order. Eventually, banking and credit mechanisms would permit the accumulation of funds needed by both the state and merchants for grain purchases. Thus, the tools that would allow the government to relinquish its control would come into place during the Restoration. In fact, counter to the general historical view that the Restoration was in all ways mediocre, the energies and capabilities of the administrators who ran the provisioning operations belie that image. They drew on their enormous knowledge of past successes and failures and transformed that knowledge into innovative policies that could feed urban citizens without draining the Treasury – no mean feat. The full transition would take at least thirty years and would provoke many internal debates about the state's goals, but the resources that would confront the 1853 shortages were underway. For that reason, the Restoration represented a signficant moment in the maturation of provisioning policies.

Several groups were involved in these activities during the Restoration. The Subsistence Committee – roughly a continuation of the Napoleonic Subsistence Council – was established on September 5, 1816. As a division of the Ministry of the Interior, its job was to coordinate national measures and to oversee the Reserve, among other duties. In November 1816, as a new provisioning crisis developed, the Reserve's daily operations were transferred to a body that had been created by the Paris municipal government specifically for that task, the *Caisse syndicale des boulangers de Paris*. The members of these two bodies worked together, and in some cases, both their personnel and their duties overlapped.[1] Well aware that the royal government was reluctant to supply Paris, the Subsistence Commission had to find new methods that would be more economical and that also could be based on the capital's existing resources.[2] The Commission, joined later by the *Caisse*, chose measures that built on the Napoleonic strategies, such as bread distributions, public works, grain imports, and Reserve sales, while

[1] The minutes of the Subsistence Commission and the *Caisse syndicale des boulangers de Paris* are in F11 209, 301, F11 2799–2806 and F11* 3056–68. This account is drawn primarily from those exceptionally detailed sources.

[2] The autumn and winter 1816–17 correspondence in AN F11 209, 301, 306–8, and AN F11* 3058–3061 follows these decisions.

hoping nonetheless to limit the scope of its operations as time went along. To do so, however, the members of the two groups realized they would have to create more viable commercial networks. They began by incorporating greater levels of free trade in their strategies than had earlier regimes, and then struggled to maintain those policies during the dearth of 1816–17. After 1821, these authorities began steadily and consciously to transfer the responsibility for provisioning cities from the state to commercial networks and the bakers. These commercial actors, however, cannot be understood as entirely free agents in a free market. Instead, the many participants in the grain trade of the 1820s – in particular, the bakers – benefited from the state's efforts to stabilize the trade and were subject to considerable constraints.

The Fate of the Reserve

From the first meetings of the Subsistence Commission in September 1815, the question of continuing the Reserve weighed on the members' minds. The Reserve had been maintained after the 1812 crisis, and after Maret's dismissal, his successors were directed to keep it at about 250,000 metric quintals. In addition, numerous northern French cities had created their own emergency supplies immediately after the dearth of 1812. Rouen, under Prefect Girardin's guidance, for instance, had begun massive purchases. Even lesser centers such as Elbeuf and Dieppe had tried to buy enough grain to last for six or nine months in order to feed their impoverished citizenry.[3] Despite these efforts, however, the emerging consensus of the Restoration ministers was that the Reserve's 1812 intervention had destabilized the grain trade and that merchants and landowners now had far too many reasons to distrust the government and its operations. Instead, Restoration administrators generally wanted to strengthen commercial ties, while diminishing the role of state-run emergency supplies.

Given such doubts about the value of intervention, the continuation of the Paris Reserve was not a foregone conclusion in the summer of 1815 after Napoleon was finally captured for the second time, bringing his Hundred Days to an end and reestablishing the monarchy. In fact, because the last of the Napoleonic suppliers had been told to liquidate the stocks during May 1815, the Restoration administrators at first could not figure out whom to place in charge, or whether there were any new shipments expected. Louis XVIII nevertheless charged the Commission with the main-

[3] Those cities looked to the Eure for such supplies, and in the process, endangered that department's reserves. Prefect of the Eure to Minister of Police Rovigo, 7 December 1812, AN F7 3626.

tenance of a stable Reserve of 250,000 metric quintals for the time being, and to coordinate measures taken by prefects.

The key to many of the Restoration's overall policies was the emphasis placed on local, not national, initiatives. For instance, the ordinance establishing the Subsistence Commission instructed it to work with prefects to find the means to match loans from the royal treasury with funds secured locally.[4] Primarily the Commission urged municipal authorities to use their own tax revenues to purchase grain and to encourage merchants to organize joint ventures. Once the probability of shortages became known in the autumn of 1816, the Minister of the Interior gave more specific orders to local authorities. For example, he directed the Paris Prefect of Police to persuade landlords and tenants to thresh their grain in order to pay their taxes and rents. He even suggested encouraging landlords to demand payment of rent in full in November 1816, which would force tenants to market their crops early.[5] Charity bureaus and public works projects, developed and maintained by municipal and departmental money, were suggested to prefects and mayors as the appropriate means to address the shortages. Despite the abundance that royal symbolism implied, then, it was clear that the king's intentions were far more modest. Local energies would have to replace many of the enormous programs of the Empire in 1812, or so it appeared in the autumn of 1816.

The pressure on local institutions to assure local supplies was accompanied by long discussions within the Subsistence Commission about the policies for Paris. The Minister of the Interior, for instance, had opposed the practice of keeping bread prices below those of flour during shortages, and wanted Paris to abandon such policies. He believed that the Imperial treasury had lost nearly one-quarter of the 48 million francs it had spent on provisions for Paris alone in 1811–12. When the first hint of future shortages was heard in December 1815, he instructed the Commission to avoid such sacrifices and the members concurred. Concerned about the political impact of an abrupt change, though, the members decided to intervene if the price of a two-kilogram loaf reached 90 centimes, a very high price, and one that Napoleonic officials had deemed dangerous. They also considered placing a ceiling on the price of bread if necessary. Beyond that point, the Commission planned to aid the poor through charity bureaus and to set up work houses. In the meantime, they commissioned Vanlerberghe to restock the Reserve with 250,000 metric quintals of grain as a temporary measure. This was a wise choice, for by the summer of 1816 the rumors of bad harvests across Europe and even in the United States had been confirmed. The

4 6 September 1816, AN F11* 3056.

5 MI to Prefect of Police, n.d., ca. 5 October 1816, AN F11 2801; MI to Prefect of the Somme, January 22, 1816, AN F11 535.

royal government lifted import duties in August 1816 and then offered bonuses for incoming grain.[6] While the Ministry of the Interior made it clear that the government had no intention of resuming the immense 1812 activities, he energetically proposed strategies that could be carried out by local officals.

It is likely that the extensive administrative experience of the men who staffed the Commission, and later the *Caisse*, contributed to their reluctance to perpetuate the Reserve's problematic legacy of 1811–12. The Commission consisted of Prefect of Police Anglès, Seine Prefect Chabrol, and others from Parisian government. All of these men had access to nearly twenty years of ministerial and local records. They consulted them frequently, and used examples from previous shortages or their own memories to support their points during arguments. Their discussions were littered with information: price series, expectations for flour and bread yields from grain, details of contracts with former provisioners, and comparisons of practices in numerous cities. They had witnessed first-hand the disastrous impact of the Napoleonic Maximum and requisitions and vowed never to proceed down that path again.[7] Taboureau, who chaired the Subsistence Commission, ensured that the committees would have access to more reliable calculations than their predecessors had. He hired several skilled employees to handle the broad range of duties he expected the Commission to oversee, and when he moved on to head the soon-to-be-created *Caisse syndicale des boulangers de Paris* in November 1816, he demanded adequate salaries to hire six trained accountants. In both cases, his candidates had served in the offices of the Ministry of Commerce or in military and municipal provisioning bureaucracies.[8] The new appointees went straight to the registers and began correcting the errors they found in Reserve operations since 1812. Their vigilance continued through the crises of 1816–17, to the frustration of the many merchants and millers whose accounts were found wanting.[9] Armed with their extensive knowledge, the committees were in a solid position to rework the system that the Empire had left them. While the finer points of any plans were vigorously contested, the committees were never at a loss for information. They had reams of records – from the Hall Controller, bakers' requests for reimbursements,

[6] The end of the wars finally permitted safe passage for imports across the Baltic and through the Channel, so that merchants were more willing to undertake such missions. The records from ports and several commercial companies are in AN F11 301, 1356, 1512, 1578.

[7] See the analysis of 1812 by the Minister of the Interior and the Commission, 3 January 1816, AN F11* 3058.

[8] Chef de la 3e Division du Ministère de l'Intérieur to Interim MI, March 31, 1817; and "Rapport," presented to the MI, February 28, 1817, AN F11 203.

[9] The rectification of the Reserve accounts was not completed until the early 1820s. The lengthy correspondence of these efforts is in AN F11* 50.

reports on grain prices across the kingdom, and of the revenue from municipal bonds and credits from the Treasury. They gathered additional evidence from earlier shortages, especially that of 1811–12. In fact, the 1812 records were one of their constant points of reference. So frightening had been that crisis, and so far-reaching its Maximum and Reserve operations, that none of the 1816–17 policies could be framed without reference to 1812.

Early Attempts to Rely on Free Trade

The plan the administrators created was known as the "*système mixte*," or the dual system. The goal of this system was to accustom the Parisian population to accepting the idea that the price of bread would float up and down with that of commercial wheat and flour. Holding bread prices artificially low could not work over the long run, and only exacerbated shortages, they reasoned. Yet the committee members were experienced enough to realize that any transition toward freer trade had to proceed slowly and that the onset of a provisioning crisis was not the moment to make dramatic changes. They began outlining the dual system between August 1816 and January 1817.[10] Eventually, its components linked the modest sales of Reserve grain with a two-tiered strategy of bread pricing, overseen by the *Caisse*.[11] Its final form would not emerge until 1821, however, when the many authorities involved would have a chance to review their choices and take advantage of the relatively more abundant years of the mid-1820s.

The dual system thus developed in the throes of deepening shortages and unrest, and the system's initial form revealed numerous reversals and misgivings. The system's broad goals – mainly to move Paris toward greater reliance on commercial networks – were ever-present in debates over strategies, but establishing a new system while being buffeted by shortages was no easy task for the authorities. By October 1816, it was clear that not only French harvests but farmlands throughout the northern hemisphere had been damaged by frigid temperatures.[12] At each turn, the Commission dis-

[10] In addition to the Commission and Caisse registers, see the additional minutes in AN F11 301.

[11] 2 May 1816, AN F 11* 3058.

[12] John D. Post, *The Last Great Subsistence Crisis In the Western World* (Baltimore and London, 1977); Robert Marjolin, "Troubles provoqués en France par la disette de 1816–1817," *Revue d'histoire moderne*, new ser., 10 (November–December 1933), pp. 425–48; Maurice Vergnaud, "Agitation politique et crise de subsistances à Lyon de septembre 1816 à juin 1817," *Cahiers d'histoire* 2 (1957), pp. 163–178; Louis Guéneau, "La disette de 1816–1817 dans une région productrice de blé, la Brie," *Revue d'histoire moderne* 9 (Jan–Feb 1919), pp. 18–46; Roger Marlin, *La crise des subsistances dans le Doubs*

cussed the experience of 1812, rejecting its aggressive policies and working on more moderate solutions. While it wanted to reassure public opinion that the government was taking precautions, it hesitated to do anything bold.[13] The Commission instead defined its responsibilities modestly: It planned to "give every opportunity to *circulation* so that prices will level out, and to call upon only such imports as will maintain the price of grain at moderate levels." In particular, the members did not want to disturb merchants "nor to give them any reason to fear competition" from the government's operations.[14]

Their choices showed an increasing sensitivity to the needs of merchants and revealed the member's commitment to keeping intervention inobtrusive and inexpensive – both of which were features of the nascent dual system and built on Old Regime and Imperial efforts. In the late summer of 1816, the Commission enlisted a number of merchant houses to organize imports that would arrive in the spring, and also to bring regular shipments of flour to the Paris hall during the autumn to be sold at the market price. Among them were the Watin and Thuret companies – the same Watin who had handled the 1812 third-party sales, and Thuret of Le Havre finance fame. Watin was praised effusively by the Commission for moving surreptitiously, thus "entering well into the spirit of his instructions."[15] Complementing the Reserve activities, the Ministry of the Interior and the Commission instituted subsidies for bakers who bought flour at the hall. Any baker who purchased flour for more than 77 francs was repaid 10 francs per sack.[16] The authorities hoped that such subsidies would make it possible for the bakers to rely on millers and *facteurs*, and thus strengthen market ties despite rumors of shortages. Government money, in small amounts, would solidify and protect commercial structures against the ravages of shortages.

In early October 1816, though, the Commission grew apprehensive about reports of rapidly rising prices and the insufficiency of bakers'

(Besançon, 1960); Colette C. Girard, "La disette de 1816–1817 dans la Meuthe," *Annales de l'Est* 6 (1955), pp. 333–62; Pierre-Paul Viard, "La disette de 1816–1817, particulièrement en côte d'Or," *Revue historique* 159 (Sept–Oct 1928), pp. 95–117. For an account of the effect of the famine on the United States, see Barrows Mussey and Syvester L. Vigilante, "'Eighteen-Hundred and Froze to Death': The Cold Summer of 1816 and Western Migration From New England," *Bulletin of the New York Public Library* 52 (September 1948), pp. 454–57.

[13] The Commission and the ministries even tried to maintain publicly that the only reason for shortages was that the farmers were engaged in their normal post-harvest operations and had not yet brought grain to market. 14 September 1816, AN F11 301.

[14] 14 November 1816, AN F11 301.

[15] See especially 10 October 1817, AN F11 301.

[16] There were also some sales of Reserve sacks for 77 francs in the early autumn. 15 October 1817, AN F11 301.

finances, even if subsidized. Worse, the Ministry of the Interior, which was providing funds for the subsidies until November 1, refused to continue beyond that date. Still hopeful of combining state intervention with commercial practices, though, the Commission suggested that its agents prepare batches of 200 and 300 sacks of Reserve flour. The agents would mix various qualities, thus disguising them as commercial arrivals. Selling through discreet intermediaries, the agents would adjust the price of the Reserve sacks "in such a manner that the resulting price could be lowered in line with the decrease . . . that usually takes place at this time of year." Such secrecy and the clear attempts to follow the routines of the market were of course a continuation of the "simulated" and third-party sales of earlier regimes. The Minister of the Interior refused to allow the sales, however, wanting the bakers to rely solely on commercial flour. His prohibition remained in effect until the following April, and caused the Commission immense frustration.[17] Both the Minister and the Commission wanted market networks to prevail, but differed sharply on the extent of state resources that would be necessary to sustain them.

Not only did the Commission's efforts in Paris attempt to imitate the workings of the market, but its plans for imports followed similar strategies. Discretion continued to be the watchword. Thus, when it became clear that government agents buying in the *rayon* had raised prices once again and stirred chilling memories of 1812, the Commission limited the purchases, and finally abandoned them altogether in October 1816.[18] Making arrangements to bring grain from more distant areas, the Commission again tried to mimic commercial practices and prices. When the members heard that its contracts held agents to price levels that were unrealistically low, for example, they revised the next agreements. They offered their merchants a 2 percent commission on any grain purchased and placed no limits on its price, even though the members worried about the debts that might accrue. Moreover, they constructed an elaborate scheme to conceal their buyers' activities in Baltic ports, and barred government grain inspectors from accompanying them. "We must hurry slowly," explained the merchant conferring with the Commission. He meant that the Commission should send buyers scurrying to numerous ports and markets throughout the Baltic, but instruct them not to be noticed. They were to wait for respectable houses to contact them with grain for sale. They should each buy small amounts, feigning disinterest, and do so on the same day, allowing the whole operation to be wrapped up by the time the news spread. Several equally respectable French merchants were asked to organize the grain's transport. The Commission, on the merchant's advice, judged it to be less disruptive

[17] 3 and 5 October 1816, AN F11 301.

[18] 29 October 1816, AN F11 301.

in the long run to imitate commercial practices quietly rather than to thwart them openly.[19]

Organizing the Dual System

The difficulties of financing further purchases and of assuming responsibility for the reimbursements to the bakers, however, forced the Commission and the Paris Municipal Council to redefine the scope of their mission in October 1816. The resulting decisions underscored the need to establish the dual system that they had been sketching out. For many months, the Minister of the Interior had been pressing the Commission "to come to the aid of artisans, the *petits bourgeois* and the indigent" and to leave the rich to their own devices. Ideological and budgetary concerns ran together for him.[20] When the price of commercial flour reached 140 francs in October 1816, the Commission itself began to question the value of the system of subsidies – even as it battled the Ministry of the Interior for an extension of the funds for the program. The Commission had to agree that subsidies, which drew on municipal money as well, were expensive and did not cover the bakers' losses. As the city fell further in arrears, bakers were slowing their flour purchases and liquidating their personal stocks. Soon they might be unable (or, at least, unwilling) to continue the trade. During the same weeks, anxious consumers from the Parisian suburbs had been flooding into the capital to buy the low-priced, subsidized loaves.[21]

Reviewing the reports, the Commission reluctantly conceded that the Minister of the Interior was correct: It was neither fair nor feasible to continue the subsidies. The wealthy could no longer be allowed to benefit from policies that – arguably – were only needed by the poor. The Commission hoped to raise the legal price of bread to match that of grain and flour, a primary goal of the emerging dual system, while using Reserve wares and revenues to supply inexpensive loaves to the truly needy. Thus, bakers would be able to buy flour from the multitude of millers and flour merchants in the region. Vital trade links would be preserved at the expense of the well-to-do, while the indigent would receive essential assistance from

19 One buyer even pretended to be a gentleman-scholar who was touring ports, for instance. 16 October, 9, and 21 November 1816, AN F11 301. For evidence that these plans succeeded, see the correspondence of Thuret's agents in AN F11 306–8.

20 Letter of December 30, 1815, AN F11* 3058.

21 The 1816 discussions were complicated by the knowledge that the Ministry of War owed Paris money, which it could not repay. The Minister of the Interior criticized the Paris Prefect and the city's officials for not securing adequate funds in advance. See the extensive correspondence between Seine Prefect Chabrol and the Ministry of the Interior on late 1816 provisioning concerns in AN F11 209.

the Reserve. The Commission finally ceded to the Minister's demands in late 1816 and directed its limited efforts toward the capital's poor. After further negotiation, the Minister of the Interior, having consulted with the king, told the Commission to draw up a list of the Parisian inhabitants who deserved cheap bread. To handle the daily operations of the Reserve and the bakers' subsidies, the Parisian authorities and the Commission's members drew up plans for the new institution, the *Caisse syndicale des boulangers de Paris*.

By mid-November 1816, then, the Committee had determined that it was essential to dissolve the costly system of reimbursements to the bakers and get the dual system underway quickly. The members brought the Municipal Council into the plans, seeking advice and institutional structures to carry the project forward speedily. Very quickly, such assistance appeared. The mayors' offices and the charity bureaus volunteered that they might have adequate information to identify those in need – and thus eligible for subsidized loaves – although they still insisted that the royal treasury ultimately had to assume more financial responsibility for the capital's provisioning. The estimates of how many people would qualify for help was deeply distressing. Somewhere between 80,000 and 200,000 might legitimately lay claim to the subsidized loaves. To pay for the program, the Commission and the Ministry pressured the Municipal Council to levy heavier taxes.

In the midst of the rapid-fire planning of the late autumn, bitter disputes surfaced within the Commission itself. With shortages confirmed and the subsidy program well established, it seemed to several on the Commission that the vast changes under way would be profoundly unsettling to the city's population. Prefect of Police Anglès, apprehensive about both public order and the well-being of his bakers, argued to continue the bread subsidy for the entire city. The Commission could figure out how to pay for it later, he explained. Still another member pointed out that Napoleonic officials had never allowed Parisian bread to cost more than 90 centimes, causing the Commission to consider nervously the detrimental impact of higher prices, even if only the wealthier of the city would have to pay them. There followed a torrent of reservations about all of the programs under discussion. The members realized that subsidies generally encouraged merchants to raise their flour and grain prices and even to refuse credit to bakers. Thus, the Commission considered aiding bakers directly, discussing whether such aid would have the same worrisome effect. Finally, though, the looming financial crisis drove the Commission to decide that ration cards issued through the capital's mayoral offices was the best of the admittedly problematic range of options. Until the ration cards could be printed and distributed, though, the city would place a ceiling of 1 franc on two-kilogram loaves. As soon as the dual system was in place, the 1-franc limit would be lifted and well-to-do buyers would pay for bread at its full cost, however high. The "indigent" and "*mal aisés*" on the official lists, spared by the new program, could present the cards at the bakeries and buy a loaf for 90 centimes. Bakers could turn the cards in for the corresponding reimbursements. The dual system could protect the poor somewhat, while pushing the rest of the city to depend solely on the forces of supply and demand.[22]

While the plans were under discussion in November and December, Seine Prefect Chabrol waged an unrelenting campaign to wrest more money from the royal government. He received repeated rebuffs from the Minister of the Interior, which only exacerbated the tensions between the institutions trying to feed the city, such as the Subsistence Commission, and the Ministry. The Minister even informed the committees that both the king and the Minister of Finance were uneasy with the plans for the dual system, especially because as many as 200,000 people might be identified as poor. They feared riots at bakeries between those with and without cards.[23] Yet, despite Louis' insistence that the Commission continue to supply the city and hold bread prices down, he absolutely refused to give the Commission any more money. In fact, he told the capital's Municipal Council, Chabrol, and the Commission to stop harassing the Minister of the Interior about a purely Parisian matter. Furious, the Commission accused the restored monarch of the most blatant hypocrisy. "All of France has an interest in maintaining the order and tranquility of the city," it argued, "and consequently . . . it is obligated, out of political solidarity, to help find the necessary funds to achieve that goal." It was not a question of the king's preferences, the Commission insisted, but rather "*celle de la bourse ou de la vie* . . . ," of balanced budgets or urban riots. They knew which they would choose.[24]

The Creation of the *Caisse syndicale des boulangers*

Stunned that the capital was abandoned to its own resources, the Paris Municipal Council passed a 4-million franc tax in December 1816, and dedicated the revenue to the *Caisse*.[25] Formally established in February 1817,

[22] Prefect Seine to MI, n.d., ca. 10 December 1816, AN F11 209. See also the winter 1817 records of the Caisse, AN F11 3061.

[23] The Commission eventually determined that 93,889 people deserved the ration cards, although estimates continued to fluctuate. Seine Prefect to MI, n.d., ca. 10 December 1816, AN F11 209.

[24] [MI] to Prefect Seine, 12 December 1816, AN F11 209; Chabrol to MI, 7 January 1817, AN F11 209, p. 29 and weekly meetings in AN F11* 3058–60. For the larger context, see AN F4 2153, F11 209 and 301.

[25] "Extrait du régistre des procès verbaux du Conseil général du Département de la Seine," 22 December 1816, AN F11 209, p. 16.

the *Caisse* was to oversee daily Reserve operations and to supervise reimbursements to the bakers.[26] Despite the new levy, the Commission knew the money would never meet that year's debt.[27] To overcome the deficits, the *Caisse* was allowed to take possession of the Reserve flour and claim the receipts from the sales at the set price of 77 francs per sack. It was widely acknowledged, however, that the *Caisse* would soon go bankrupt. The Paris Prefect of Police estimated that by February 1817, the reimbursements to bakers would cost almost 500,000 francs per week.

The Commission's dire predictions of escalating unrest soon proved to be true. Crowds rose up in marketplaces and along roadsides across the country and bloodshed followed.[28] Already in October 1816, buyers in Vienne in the Isère had forced local authorities to set price maximums on the grain in their market. In Toulouse, the export of grain was blocked the following month, and dragoons were needed to restore order. During the third week of January 1817, unrest spread across the Norman countryside, where the rural poor stopped wagons and distributed their cargo for 3 to 4 sols per pound. The attacks targeted critical supply lines leading from farmlands and ports to the cities of Rouen and Paris. Crowds in Bondeville stopped wagons along the road from Dieppe to Rouen. The following day, 100 people confronted a *cultivateur* in Pavilly, just north of Rouen, demanding grain at 15 francs per *mine*. Playing along with the crowd, the mayor promised to carry out their wishes. He held them off until the *garde royale* arrived, arresting fifty "*mutins*" and taking them to prison in Rouen. In Doudeville, troops tried first to shoot blanks into a market crowd. When the townspeople realized the deception, they became even more incensed, and pelted the captain with rocks. Scrambling to safety, he ordered the troops to shoot bullets at the rioters, killing at least two men and injuring one other. The *gendarmerie*, the most heavy-handed of the military and police forces available to departmental authorities, had to be called out to quiet market disturbances in Tôtes, Cany, Yerville, and Déville, all in the Seine-Inférieure. The prefect lost no time in complaining that the source of his department's problems lay in its never-ending competition with Paris for grain. He wanted any remaining buyers for the capital to vacate the region.[29]

The disturbances continued through the spring and summer of 1817, forcefully reminding the ministries and the Commission that they dared not

[26] Prefect Seine to MI, 24 November 1816, AN F11 209, p. 36. "Rapport au Roi," 27 December 1817, AN F11* 3060.

[27] The Seine prefect set aside an additional 500,000 francs from municipal bonds. 29 November 1816, An F11 301; Seine Prefect to MI, 24 Movember 1816, AN F11 209.

[28] Report of the riots in northern France are in AN F7 9888–9.

[29] Seine-Inférieure dossier, AN F11 733.

let down their guard. As many as 8,000 frustrated consumers in the Brie, outraged that grain-laden wagons were leaving their towns, attacked merchants and warehouses in May 1817. Frustrations peaked in the Calvados during July 1817, when buyers surrounded merchants and bakers, demanding lower prices and greater access to grain. When news of unrest in Lyons reached Paris, ultra-royalist leaders, remembering the previous year's pro-Bonapartist conspiracy, cried out for harsh repression against those who would destroy the Bourbon Restoration.[30] Unlike the 1812 riots, however, troops were on hand in 1817 and were deployed more aggressively than before. The Ministry of the Interior had instructed prefects to request soldiers in advance of any disturbances. Yet even the presence of troops who were willing to fire on angry buyers could not guarantee order. By June 1817, the Minister of the Interior would have outlined how local authorities were to handle riots and their suppression afterward. Until then, during the tense winter and spring, crowds still had their way, if only for the few hours before soldiers with gunpowder and bayonets arrived.

Reserve Sales Resume

Anxious about the spread of disturbances, and under fierce pressure from the Commission, Louis finally sanctioned ration cards in early March 1817, and ordered the Minister of the Interior to prepare the lists of the "indigent" and the "*mal aisés*."[31] The funding of the new operation, which was to begin on 1 April, remained uncertain, however, and Louis postponed its implementation at least once. As the lists were drawn up, the various provisioning committees turned to the question of the Reserve grain itself, prompted by the expected arrival of grain and flour imports from the December operations in the Baltic. The Commission begged the Minister to allow it to put the wares to use. Several of the merchants who had been selected to act as intermediaries planned to take small lots to the Paris hall, and to sell them for the going rate, thus reviving the plans the Minister of the Interior had scuttled the previous autumn. At first, the only flour in question were some sacks that risked spoiling in the coming months.[32] After 1 April, however, the Subsistence Commission expected to have approxi-

[30] 22 January 1817, AN F11 301; Calvados dossier, AN F11 723; Post, *The Last Great Subsistence Crisis*; Marjolin, "Troubles provoqués en France par la disette de 1816–1817;" Vergnaud, "Agitation politique et crise de subsistances à Lyon;" Guéneau, "La disette de 1816–1817;" Marlin, *La crise des subsistances*; Girard, "La disette de 1816–1817 dans la Meuthe;" Viard, "La disette de 1816–1817."

[31] Royal Ordinance, 3 March 1817, AN F11* 3061.

[32] 4 March 1817, AN F11 301.

mately 14,000 sacks of flour arriving per month through July, and wanted to begin distributing the wares as soon as possible.[33]

For all the minister's opposition to Reserve sales, however, he was willing to consider the Commission's proposals in late March. When that news reached the *Caisse*, its members threw themselves into planning the sales and almost immediately became wholly divided on how to use the supplies. Count Anglès, the Prefect of Police, wanted extensive sales to stave off unrest. Others resisted such sales, fearing they would drive merchants away. The members arrived at a carefully worded compromise which made evident their understanding of the appropriate role for the state. First, any sales had to be made "at commercial rates." Second, sales were "to help *regulate* prices, which would be useful, but not to *lower* them."[34] Although such distinctions may seem confusing, their intent was to slow, or even halt, further price increases. The amounts of flour released could be adjusted each day to counter such rises. The members contrasted this approach with that of the Reserve in 1812, when an enormous number of poorly disguised sacks had made the government's role all too apparent. If the Reserve were to be involved, they counseled, the government would have to realize that its moves would be discovered, and it should plan accordingly. Any attempts to shape prices needed to be gradual and inoffensive to merchants. To meet those challenges, the *Caisse* was allowed to channel the first flour distributions directly to the bakers as part of the arrears for the flour subsidies the city still owed.[35] Quickly, though, bakers demanded more than the 10-franc subsidies that had been promised the previous fall. They insisted that they be reimbursed for every centime above 77 francs that they had paid for commercial flour, which the city had promised them in January, but not paid.[36]

The mounting problems with the bakers propelled the Commission toward precisely the kinds of wide-ranging intervention it had hoped to avoid. The legal price for bread could not be allowed to follow flour prices, and Louis, in any event, had forbidden Prefect of Police Anglès to raise it further. Thus, half of the dual system – the plan to allow bread prices to match those of flour – was supposed to be abandoned. The *Caisse* members were at loggerheads over whether to demand permission to supply more Reserve flour to the bakers – and if so, at what prices and amounts, through what avenues, and with what effect – or to renew their efforts to push bakers to find commercial sources. The *Caisse* launched into a long discussion of the possibilities before it, generally agreeing that Reserve flour

[33] 28 March 1817, AN F11 301.
[34] 29 March 1817, AN F11* 3061. Italics mine.
[35] April minutes, AN F11* 3061.
[36] 16 April 1817, AN F11* 3061.

should be used to repay the bakers for their losses on the sales of bread. In addition, Count Anglès and another member wanted to raise the price of bread somewhat, despite the Minister of the Interior's injunctions. The bakers' distress could not safely be ignored. Moreover, they argued, there could even be political benefits to such a plan. If done "firmly," price increases could show the people that the government was not to be disobeyed. Finally, the other members yielded. The *Caisse* petitioned the king for permission to raise bread prices, and received royal approval. In that way, it attempted to provide relief to both the bakers and to the municipal coffers, which were saddled – unfairly, it believed – with the subsidies.[37]

The matter of Reserve sales was still unsettled, however, and the Commission and the *Caisse* were running up against the intransigence of the Minister of the Interior. Since the autumn, he had opposed any expanded distribution of Reserve supplies, demanding that bakers be forced to find commercial wares at reasonable prices. Throughout May 1817, the *Caisse* contested the Minister's limits with renewed vigor; it needed more flour to repay bakers for their losses. When the *Caisse* recalculated the capital's daily requirements, it found that the old formulas, based on 1,492 sacks of flour per day, were insufficient. A more accurate estimate was 1,600, or even 1,700, sacks per day, which lent a great deal of credence to the bakers' complaints that the subsidies were falling short. Whether the Treasury provided more credits for reimbursements in flour or in cash, or allowed the *Caisse* to take full control of the Reserve flour and distribute it as it pleased, mattered little by then. The *Caisse* members simply wanted to be able to do more, and to have its measures sufficiently underwritten by the government. It could not continue to operate in the red.[38] With or without permission, it began to organize distribution of increasing amounts of flour throughout the summer.[39]

By the last week of May, flour reached 168 francs per sack at the Paris hall. The Subsistence Commission, sharing the *Caisse*'s anxiety, sanctioned disbursements from the Reserve warehouses in amounts just below those the *Caisse* had requested.[40] The Prefect of Police, still worried, reported that "a spirit of rebellion" was manifesting itself among "the people" and that "crowds" had begun gathering in front of bakeries. The situation demanded "serious attention, and could worsen suddenly," he declared. Unexpectedly, the Minster of the Interior gave in. Within hours, he arranged further credits

[37] 3–10 May 1817, AN F11* 3061.

[38] See, especially, the sessions of 17 and 21 May 1817, AN F11* 3061.

[39] From 1–10 June, the Reserve would release 600 sacks per day. After 11 June, it would release 800 per day, and from 21 June to 1 September, 1,000 sacks per day. 24 May 1817, AN F11* 3061.

[40] From 1–10 June, the *Caisse* could claim 600 sacks per day, and from 21–30 June, 800 sacks per day.

from the Minister of Finance, and allowed Reserve flour to be sold to bakers at reduced rates.[41]

Expanding Operations: June 1817

The *Caisse* began finalizing plans for the flour's distribution to the bakers, expecting to provide about a third of the city's needs. On 9 June, however, those designs escalated beyond anything that even the boldest members of the *Caisse* could have imagined. That day, Louis ordered the Reserve to undertake the city's entire supply starting on 12 June. The *Caisse*'s startled members formally received word on 11 June, giving them only one day's notice. Joined by the Subsistence Commission, the *Caisse* immediately put an expanded system into motion. The Reserve would send hundreds of sacks of flour per day to the hall. Bakers, holding certificates issued by the *Caisse*, would go to the hall to purchase their allotted amount of Reserve flour at the subsidized rate of 92.6 francs per sack. The royal ordinance further specified that the Reserve would take possession of any private flour supplies that the bakers had, and also of any flour they had bought on credit. The Reserve would pay them for the flour at the going rate on 9 June, and would sell their flour back to them as needed. Only one kind of bread – a medium-quality loaf that used mixtures of Reserve and the bakers' former wares – could be produced.[42] Those last instructions would help stretch supplies by prohibiting the production of the finest quality loaves, while lifting the nearly impossible financial burden of the city's provisioning from the bakers.

While the scale of these 1817 measures matched, and briefly exceeded, those of the 1812 Reserve, the *Caisse* and the Subsistence Commission were at pains to avoid the "abuses" of the earlier crisis. The greatest crime of the 1812 system, they believed, had been that merchants had been chased from the city, and the Reserve had had to supply the city for over a year. The effect had devastated merchants. The 1817 administrators distanced themselves entirely from the Empire's activities and carefully avoided the 1812 pitfalls as they understood them. Clarifying the issue, Prefect of Police Anglès presented a lengthy analysis of the advantages of system he and his colleagues were creating. Because the Reserve had absorbed all of the bakers' supplies and paid the bakers a good price, the bakers could not complain about potential losses. Moreover, the 1812 supplies had originated with the government alone; a substantial portion of the 1817 wares

[41] 31 May 1817, AN F11* 3061.

[42] "Ordonnance concernant la Boulangerie de Paris," 9 June 1817; *Caisse* session, 9, 11 June 1817, AN F11* 3061.

came from the private stocks that bakers had amassed independently. Unlike the 1812 measures, then, the 1817 Reserve would not compete directly with merchants. Instead, it would assume the losses between the prices merchants demanded for flour and the prices Parisian consumers could afford to pay. It would sustain, not undermine, commerical connections. Better, the *Caisse*'s forty-eight inspectors would visit bakeries to ensure that the flour was used to produce the greatest number of loaves possible. Prefect of Police Anglès calculated bread prices that would guarantee a small but steady profit for the bakers. In addition, the Reserve would continue to have cheap loaves sold to the very poor at even lower prices. The *Caisse* was ready to shoulder this responsibility, the Prefect of Police declared. Every inquiry, every discussion of the preceding months had prepared the members for this day. They knew the bakers in every quarter, how much bread a sack of flour would yield, and where the troubled neighborhoods were located. With their knowledge of the trade, the *Caisse*'s June 1817 measures would meet the city's needs temporarily without destroying the trade, the Prefect insisted.[43]

Between the Reserve's flour, the incoming imports, the supplies in bakers' stores, and additional credits from the Treasury, the *Caisse* was able to pull together the resources it needed. One particularly fortunate auction of *Caisse* bonds at the end of July sold well, to the members' increasing relief.[44] Thus, the *Caisse* was able to maintain the necessary level of flour distributions throughout the summer, to almost universal approval. Within days of the June 11 announcement of the Reserve's expanded duties, *Caisse* members were reporting great calm among buyers and steady bread production in bakeries.[45] That fortunate situation continued for many weeks.

By August 1817, the *Caisse* felt ready to ease the capital's supply back to commercial resources. Once again, it weighed vast amounts of information and discussed the possibilities at length. Finally, it decided to gradually reduce the number of low-priced sacks that the bakers could buy with certificates, and to urge the bakers to begin restocking the private supplies that they were required to maintain. If they wanted, bakers could buy any leftover Reserve flour to meet those obligations before the 1 November deadline the *Caisse* had decided to impose. By combining those two strategies, and by setting gradual and well-publicized expectations for the bakers' return to commercial wares, the *Caisse* smoothed the transition from the

[43] 9 June 1817, AN F11* 3061.

[44] The bonds were auctioned off successfully for between 5 and 6 percent and backed by city tax receipts. July 1817, AN F11* 3061.

[45] June–August 1817 sessions, AN F11* 3061.

emergency measures of that summer.[46] Despite a few persistent but minor problems, the Reserve's sales tapered off uneventfully by the late autumn, and the bakers' supplies regained their required levels.

The Commission ambitions for shaping commercial activities extended far beyond Parisian borders, however. By the 1816–17 crisis, the Commission had determined that only a regional program could address the shortages and protect cities, especially Paris. Unlike the 1812 efforts, then, the 1816–17 Commission scattered its sacks of imported grain across the Seine basin and throughout northern France. Such plans had surfaced in 1812, when Frochot had suggested distributing some incoming Reserve cereal at trouble spots along supply lines. Seine-Inférieure Prefect Girardin had urged similar tactics when he had assessed that year's policies. It was only fair – and certainly practical – to share some of the government's grain with the hungry people who lived in towns along its route to Paris, his 1812 conclusions ran. In 1817, the Commission agreed, judging that its grain could be used wisely to keep calm along the transportation lines. Riots and other market disturbances in Upper Normandy and in the Somme had obstructed shipments throughout the winter of 1817, and so the Commission and the Ministry of the Interior focused on those regions. Also, given the success of the Commission's efforts to amass grain supplies, it could envision such ambitious measures. Both the capital and the people of northern France, it reasoned, would benefit from the distribution of government grain in the areas north of Paris.

The Veuve Le Couteulx and Sales in Northern France

The merchants that engineered these shipments were Merian and Company of Le Havre and the Watin and Thuret companies.[47] Thuret's assistant in Rouen, the Widow Le Couteulx, handled the movement of grain and flour from ports to cities and was omnipresent in the activities of that year.[48] She based her operation in Rouen. Starting in the autumn of 1816, she positioned the price of her grain and flour imports at 25 percent below the going rate in that city's market, and sent reports of her success.[49] Throughout the

[46] August sessions, AN F11* 3061.

[47] Merian and Company brought grain from the Mediterranean, while Thuret and Watin operated in the Baltic. On the operations by Merian and Company, see AN F11 1604. For Thuret and Watin, see AN F11 301, 1578, and 1605–1606.

[48] On these business, political, and banking dynasties, see Bertrand Gille, *La Banque en France au XIXe Siècle, Travaux de droit, d'économie, de sociologie et de sciences politiques*, 81 (Geneva, 1970), pp. 19–21; Bergeron, *Banquiers, négociants et manufacturiers*, pp. 59–60.

[49] 5 October 1816, AN F11 301.

next months, her agents asked 10 to 15 francs less than Rouen's market prices, steadily decreasing their rates week by week.[50] Her tactics were sometimes wanting in subtlety, though, causing her trouble with Rouen authorities. In May 1817, for instance, she placed small amounts of imported grains at a very low price with an unfamiliar seller in the Rouen hall, and reported to Thuret that local sellers were quite frightened. She expected, nonetheless, that her sales "would have some effect."[51]

Not surprisingly, the Rouen officials protested. Since the autumn, heeding the government's instructions that local needs would have to be met by local activities, they had been struggling to help the estimated 14,500 poor of their city. They had raised taxes, secured the promise of a loan from the Treasury (but not the money itself), and coerced prominent citizens to contribute to the city's charitable funds.[52] They did not want Le Couteulx to do anything that would destabilize the volatile situation. They wanted her to distribute supplies directly to bakers or to ask prices closer to the going rates. Objecting to the unexpected and deep price decreases she had effected, they asked what she planned to do if her methods drove merchants from the city entirely. Would she be able to supply the city herself if that happened? The department's general council doubted it. Overall, the council insisted that "the only way" to lower prices was "to demand sales at our market of a large enough quantity of grain to bring about a decrease in prices."[53] Initially, Le Couteulx had not seen her purpose in such a light. She had wanted to intimidate grain owners with a few very low-priced sacks, force prices down quickly, and keep lines to Paris open. By later that spring, however, she seemed to have yielded to the municipal authorities. Her sales began to conform more closely to the prices of the city's market, and she took care to place the sacks with intermediaries who would not draw attention to themselves.[54] Thus, she adapted her activities to prevailing practices, and moved more cautiously.

Her plans for outlying towns were less refined, but produced favorable results. The Commission was well aware that the winter's riots had savaged grain shipments moving from the ports to Paris. Local farmers were

[50] The records of her extensive activities are in AN F11 301, 1578, 1605–1606; and ADSM 4M 114.

[51] The wares may have been of poor quality, and some of her letters indicate that she was anxious to be rid of them. The issue of the cereal's quality may have angered the Rouen officials further. 14, 16, and 19 May 1817, AN F11 1605.

[52] *Ville de Rouen. Conseil municipal. Analyse des Procès verbaux des séances du 22 décembre au 20 novembre 1874*, 2 vols. (Rouen, 1982), 1: 52–7. The city channelled money to the needy through the municipal *bureaux de bienfaisance*, poorhouses, parishes, the Consistoire, and, finally, through public works projects.

[53] Procès verbaux, General Council, 1817, AN F1c V, Seine-Inférieure, 2.

[54] Le Couteulx to the Commission, 14 and 16 May 1817, AN F11 1605.

demanding excessive prices in small markets, and buyers were hungry and restless. Working with local administrators, Le Couteulx sought to force grain owners to lower their asking prices, while calming the townspeople. She used the grain from Thuret's long-awaited shipments, which were arriving from the Baltic and the United States. Le Couteulx began distributing unanticipated sacks of grain at markets where grain owners were insisting on prices that neither local consumers nor Parisian agents could afford.[55] Her actions in the former Jacobin stronghold, Yvetot, a town of small textile mills along the route between Le Havre and Rouen, were typical of her tactics in other remote markets. Since February, the *cultivateurs* who brought grain to Yvetot had disregarded the pleas of buyers and the advice of officials. The threat of violence did not move them; they appeared "neither humane nor frightened, focusing on only their objective, more money," a police official reported. On 5 June, one seller demanded 180 francs per sack of grain, outraging the handful of buyers still left at the hall. The next morning, Le Couteulx's assistant delivered eighty-three sacks, stood back to survey the scene, and then announced a price of 125 francs. The *cultivateurs* fought to keep prices high for an hour, then lowered them to 140 francs. Triumphantly, the police official noted the *cultivateurs*' chagrin: "Seeing the wheat arrive, they sensed the end was coming for their gluttony."

Throughout June, prices in Yvetot stayed low as long as Le Couteulx could make good on the government's threat to send imports. By early July, the price had fallen to 95 francs.[56] During the same weeks, she dispatched forty sacks of grain to Saint Romain, also along the Le Havre-Paris axis. Local officials had signaled the place to her as a "site of speculation." When the *laboureurs* began to demand 200 francs per sack, Le Couteulx's assistant unloaded his sacks at the hall and in turn demanded a price of 127 francs. The astonished *laboureurs* lowered their prices to 155 francs. The agent declared that "this form of *police* had a good result."[57]

In most of these sales, Le Couteulx and her assistants asked about one-third below what local farmers wanted. In some cases, when they could make repeated dispatches, they worked over the course of several weeks to decrease price levels. Her missions in Saint Romain, for instance, drove the price of a sack of grain from 172 francs to 115 francs, and from 175 francs to 117 francs in Montivilliers. When Thuret sent grain to Bolbec, in order

[55] Her daily reports are in AN F11 1605–6.

[56] See the June and July reports from the mayor of Yvetot and Le Couteulx in AN F11 1578 and ADSM 4M 114.

[57] 15 June 1817, AN F11 1578. The strategy was not without its risks, however. When news spread that there was cheap grain in Montivilliers, the market town was invaded by impoverished buyers seeking a few measures for themselves. 6 June 1817, AN F11 1578.

to "reinforce the greater softness in prices," the mayor even used the resulting rates for the bread *taxe*, bringing unexpected relief to his community.[58] Le Couteulx's daily reports in the late spring and summer of 1817 reveal at least ninety-three forays to markets in the former Upper Norman region alone. On market days in Le Havre, Saint-Saëns, Dieppe, Cailly, Pavilly, Montivilliers, Elbeuf, Forges, Ry, Cailly, and Duclair, among other towns, buyers witnessed the unexpected arrival of dozens of hectoliters of grain.[59] She sent additional sacks to the Somme in April 1817, where the Prefect oversaw their use.[60] The operations continued through August. While she could not send grain to every market or monitor every price, she used the element of surprise to her advantage, and kept many grain owners in check.[61] She guaranteed restless villagers access to afforable grain, while securing the safe passage of other shipments to Rouen and Paris during the crucial late spring and summer *soudure* of 1817.

There rural forays contrasted markedly with the sales Le Couteulx oversaw in cities. While she had eased the crisis in Rouen by reshaping her sales to follow market practies, her actions in remote towns had not needed such finesse. Instead, she struck fast and hard. What was the impact of these unanticipated shipments, fifteen and twenty sacks at a time? They forced prices down briefly and lulled anxious consumers with the illusion of adequate supplies. For grain owners, the message was less reassuring. They might welcome calmer markets, but not the drop in prices, nor the government's implied threat that any inclination to hoard grain and demand high prices would be challenged. Best to sell now, for less, the administrators cautioned, before prices dipped lower. In short, the government's shipments to outlying markets liberated local supplies, letting buyers get on with the business of obtaining supplies for their households or carting grain to the cities. They met the needs of that day, as perceived by buyers from those towns or elsewhere, and left the owners cowed and resentful. Thuret sent other shipments to Charleville and Sedan in eastern France, where the supplies were to be used "with as much wisdom as economy," suggesting that these methods were characteristic of other regions.[62] Peace in Paris – and in Rouen and the other northern cities – was being bought for the price of a few dozen sacks of grain dumped strategically in restless market towns throughout the spring and summer of 1817.

[58] 12 May, 6 and 22 June, and 14 August 1817, AN F11 1578; 19 May 1817, AN F11 306–8.

[59] AN F11 1606.

[60] She sent at least 998 sacks of grain to that department. "Etat des blés expédiés . . . [au] dep^{t} de la Somme," 29 April–9 May, AN F11 1606.

[61] See her records of May–July 1817, AN F11 1605 and 1606; and the correspondence by local officials who praised her methods in ADSM 4M 114.

[62] 10 December 1816, AN F11 301.

CHAPTER TEN

Relinquishing Control: Bakers and the End of the Paris Reserve

After the grain imports from Thuret's operations, among many others, had reached their destinations, and the harvest of 1817 had made it to barns, the crisis eased. Reviewing the events of the past year, the Commission and the officers of the *Caisse* began to make changes in the overall operations of the Reserve. While they had had many successes, there had been some worrisome problems. They all recognized that the *Caisse* could not be expected to fund such extensive operations on a regular basis. Furthermore, the members agreed that even the most covert purchases by the government had been discovered quickly, and that merchants continued to be infuriated by the government's willingness to undersell them. Thus, the members wanted to reconsider the uses of the Reserve's supplies and how they influenced commercial activities.[1] During the next year, the Commission, with a reorganized *Caisse* on better financial footing, returned to its initial goal – that is, allowing the price of bread to match that of flour, so that the mechanisms of supply and demand alone might provision the city.[2] The outlays for the 1816–17 payments to the bakers had been immense, despite the Commission's determined efforts to work cheaply, and all agreed that the only long-term solution to Paris's supply problems lay with commerce.

Reorienting the Reserve

For the next three years, the members of the Commission and *Caisse* board explored better means to supply the city.[3] They found their solution in a

[1] See the late summer discussions of the problems in AN F11* 3062.

[2] 19 November 1818, AN F11* 3062.

[3] This account of the reorganization of the Reserve and of the policies toward bakers is taken primarily from the *Caisse*'s minutes in AN F11* 3062–8.

revamped dual system. Neither the Commission nor the Minister of the Interior was willing to disband the Reserve. By the Commission's calculations, a shortage of some level could be expected every four to five years. The Reserve was to continue to hold a supply of 250,000 metric quintals of grain, some of it milled.[4] The funds were occasionally lacking, however, and the Minister of the Interior wrote regular letters criticizing the Commission and the *Caisse* for their frequent deficits and delays. While the minister refused to let the Commission reduce the amounts kept in the Reserve warehouses, he unfortunately offered no money. Moreover, the Director of the Reserve insisted that he had to be able to send 600 sacks of flour per day to the Paris hall if necessary, and so he too opposed any decreases in the supplies. The two committees were thus confronted with demands that they maintain the Reserve's supplies despite the absence of any guaranteed source of money.

The administrators' post-1817 goal was to restructure the entire operation and to make its inadequate funds meet multiple needs. They would have to limit the Reserve's activities so as to avoid further antagonizing merchants and millers, and to continue to seek strategies that would render market forces acceptable to large segments of the capital's growing population. There were the expected financial matters to consider, matters that pressed more heavily on the local Parisian authorities after a royal ordinance transferred the Reserve to the *Caisse*, and thus to the Municipal Council. In short, this meant that Louis was renouncing all further government responsibility for the Reserve and Parisian provisioning, even though the Minister of the Interior would retain the right to decide the pace of any sales of Reserve supplies. The Treasury would provide some start-up money. After that, the Municipal Council would have to fend for itself.[5] In the next year, officials from the Subsistence Commission, the departmental administration, and a variety of municipal groups worked together in the absence of any further funds from the Treasury. In September 1818, the Commission and the *Caisse* began outlining a series of reforms, while several municipal and ministerial bodies scrutinized their conclusions.[6] Throughout these months, the Reserve continued its normal course of purchases and sales, sometimes sending several hundred sacks of flour per week to the city's market hall, and drafting contracts with the many merchants and millers that it had come to depend on.

Finally, the months of deliberations yielded a new system. A January 1819 proposal linked Reserve sales with the activities of bakers and the levels of bread prices in a complicated fashion. The end result, the com-

[4] AN F11* 3061.

[5] 26 November 1817, AN F11* 3061.

[6] The discussion of the methods is extraordinarily detailed. AN F11* 3062.

mittees hoped, was a means to ensure that Parisians would bear the cost of maintaining the all-essential Reserve supplies. The new plan depended on a change in the ways that Parisian bread prices were set. For some time, the Prefect of Police had been making unofficial but accepted adjustments to the bread price ceilings. If grain prices were low, he kept bread at a slightly higher price. This tactic allowed bakers slightly greater profits when harvests were good, he reasoned, in ways that the urban population might not notice or that it might not resent.[7] He hoped that the extra profits would allow bakers to pay more for flour so that merchants and landowners would not become discouraged whenever a bountiful crop sent grain prices downward. During shortages, of course, the Prefect of Police was faced with the opposite problem, that of holding bread prices artificially low so that buyers would not revolt. This practice was merely the continuation of Old Regime strategies, and had always been an important means of offsetting, or sometimes smoothing, fluctuations in grain prices. By 1819, however, the practice no longer responded to the problems facing Parisian administrators. It gave the appearance of inexacitude, and was vulnerable to charges of capriciousness, arbitrary judgments, and favoritism from either bakers or consumers. Moreover, these practices did nothing to bring emergency supplies to the city. The Prefect of Police and the other members of the Commission desired a more certain means by which to use bread prices and Reserve sales to influence the commercial market for grain and flour, and also to fund any further purchases for the Reserve.[8]

Complicating the matter was the Restoration's promulgation of the French corn laws in 1819. Landowners had become accustomed to the high grain prices that the war and periodic shortages had brought them. Once the Revolutionary and Napoleonic wars had ended, the prospect of a torrent of grain from the Russian steppes led the panicked landowners to petition the king to block cereal imports. Peace and the possibility of good harvests and imports would bankrupt them, they argued. Louis acceded, and placed a sliding scale of heavy import duties on grain. Thus the Commission and the *Caisse* had to respond not only to the shortages that nature might inflict, but also to the high prices that a nervous and entrenched elite was determined to impose.

[7] In 1818, the board of the *Caisse* discussed having the bakers declare the amount of their extra profits during periods of low grain prices, just as they had declared their losses during the period of high prices and subsidies in 1816–1817. The baker syndic present informed the others that any inquiry into the bakers' profits would cause the board to realize that the bakers constantly lost money. This argument increased the board's desire to reorganize the system. 17 September 1817, AN F11* 3062.

[8] The *Caisse* board criticized the tactics, and believed the bakers made far too much in years of abundance. The board wanted the bakers to turn the extra revenue over to it, so that the money could be used to buy Reserve grain. 22 September 1818, AN F11* 3062.

The proposed reforms for a dual system attempted to address all of those concerns. The plans outlined a series of surcharges on cereals brought to Paris that would generate revenue to fund the Reserve. For many months, the committees debated the best way to collect such a surcharge and how much it should be. The members considered imposing a fee on every sack that entered the city, and to lower the amount of that surcharge each time the price of flour rose. When the price reached uncomfortable levels, the surcharge would be suspended. The *Caisse* would use the receipts to purchase supplies for the Reserve. The bakers would be allotted a small increase in the price ceiling, or *taxe*, for their bread, so that they could pay the surcharge.[9] In that way, the cost of maintaining the Reserve would be passed along to the people of Paris, with the heaviest surcharges collected when grain prices were the lowest. Since the surcharge would decrease as flour prices rose, the system would raise bread prices at a pace that lagged behind flour price increases. Overall, the timing of price increases and decreases would be politically astute, which was a matter of great concern after the shortages and riots of 1817. Additionally, when bread prices reached dangerous levels, Reserve flour would be brought discreetly to the hall in lots of several hundred sacks per day in the new unmarked sacks that the *Caisse* had designed. The *Caisse* also would commission several bakers to use Reserve supplies to make low-quality loaves to be sold in markets or through charitable institutions so that the very poor could afford bread.

The discussions about the best way to organize the new system – including how and where to collect the surcharge, how to calculate the bakers' expenses, how to estimate the average daily needs of Paris, and what prices were to be considered dangerous – lasted for many months. The members displayed the expected detailed knowledge of the activities of bakers and flour merchants, although their facts frequently conflicted. In particular, they argued about how each of the parties might react to the surcharges and Reserve sales, and whether more pressure could be placed safely on the bakers.

After many disagreements, the committees decided it would be more practical to impose one flat fee of 3 francs on every incoming sack of flour, rather than to create a sliding scale of surcharges that would be difficult to assess and collect. At the same time, the *Caisse* decided that the capital's bakers and the Reserve should cooperate on a project to increase the city's emergency supplies. Through various transfers of flour, credits, and subsidies, the bakers could purchase some of the Reserve supplies and then store them at their own expense – a plan that had been followed in 1817. Since all of the capital's bakers had to keep some private emergency sup-

[9] Anglès proposed an increase of 5 centimes per two-kilogram loaf, as long as the price of bread was below 75 centimes. 12 January 1819, AN F11* 3063.

plies, the *Caisse* thought that the Reserve flour could be forced on the bakers to bring their private stores back up to the required amounts. The *Caisse* could allow them to buy the flour on credit, or perhaps the flour could count as part of the reimbursements that the *Caisse* still owed the bakers from 1816–17. However the transfers came about – and the committee debated many schemes – the result would be larger stocks under the bakers' responsibility and less money from the *Caisse* for cereals, warehouses, and personnel.[10]

The capital's Municipal Council had other plans, however. Since the Council underwrote many of the Reserve expenses, it had the final say on the reorganization of the *Caisse*. As the various municipal insitutions scrutinized the proposed reforms, they raised troubling doubts. They wanted more ambitious changes, and maintained that the Treasury ultimately had to accept responsibility for the city's well-being. The Municipal Council was weary of shouldering this burden alone, and resented the obligation. The royal orders that the Reserve hold 250,000 metric quintals seemed impossible to finance year in and year out, despite the success of the *Caisse* bonds, which were selling well. By 1820, tight accounting methods and steady revenues from *Caisse* bonds had put the *Caisse* in the black. Its good fortune did not mean that the Municipal Council would be willing to squander that modest bonus unwisely, however.[11] The Minister of the Interior, too, hesitated to sanction the revisions. He noted that leases for some of the Reserve warehouses were due for renewal, which strengthened the feeling that even more massive changes could be carried out if needed.

Recalculating Bread Prices

These objections spurred additional questions within the *Caisse* itself. Over the next year, several of its members admitted that they too were uneasy with the surcharge plan.[12] Their doubts centered on the formulas that underlay bread prices. Since 1801, bakers had complained that the imple-

[10] The committees were especially anxious to rid their warehouses of any wares that had been damaged by water or that had spoiled en route. They hoped that the bakers could be forced to accept the flour for a low price, and then mix it with better flour for their loaves. In this way, the *Caisse* would not have to purchase first-quality flour to mix with its own stocks and then resell at higher prices. The bakers were not pleased with this proposition, however, and had to be forced to accept the flour. See AN F11* 3061–3602, and in particular the discussions on 9 September and 19 November 1818, AN F11* 3062.

[11] See, especially, 5 January 1819, AN F11* 3063.

[12] At least one expert on the *Caisse* board realized that bakers and merchants would arrange fraudulent transactions in order to push flour rates just high enough to lift the surcharge and thus bread price ceilings. 2 February 1819, AN F11* 3063.

mentation of the metric system had calculated their traditional 325-pound sacks of flour and their 4-pound loaves of bread incorrectly; they claimed they were losing several centimes per loaf. The Prefect of Police agreed, finally, and a series of assays proved that the bakers' frustration was warranted. To be fair, bread prices should be raised overall, some concluded. Moreover, samples of the grain, flour, and bread yields and qualities of 1819 and 1820, along with records from previous assays, showed that there were wide variations from year to year. Somehow, the price schedules had to begin taking those fluctations into account in a more systematic fashion, or bakers or consumers would suffer. And given that bakers bought flour, not grain, and that many of their supplies were bought on credit and not purchased through the Paris hall, the entire mechanism of the city's bread price setting – which was based on the price of grain at the hall – was never going to be accurate. The bakers had alleged this for many years, and now the *Caisse* and the Municipal Council had to listen. The stakes were enormous. Jettisoning the price schedules completely would jeopardize public order, a fact that became more clear with the upheaval that followed the assasination of the Duc de Berry in February 1820. But if the Prefect of Police continued to use price schedules that harmed the bakers and that failed any mathematical test, the consequences would be equally grave.[13]

If those concerns were not sufficient, there was also the administrators' commitment to finding workable means to allow limited free trade to feed the city. There was general agreement that the price schedules could not be tossed out, but also that they could not continue in their present, highly objectionable, form. Within Paris, the provisioning committees wanted the price of bread to fluctuate with that of flour, and for Paris consumers, save the poorest among them, to follow the rhythms of the market. Thus, the 1818–21 efforts aimed to find more accurate procedures and formulas for drafting the price schedules. Those goals received substantial support from the offices of Elie Decazes, the former Minister of Police who became the Minister of the Interior in December 1818, combining the two posts. He was ambitious, exacting, and as Louis' confidant – the king called him "my dearest son" – had the monarch's full support. During the fourteen months of his ministry, Decazes was determined to bring an unassailable level of accuracy to pricing practices. He demanded yearly inquiries into the weight and yields of each type of cereal. The information was to be sent also the Ministry of Agriculture so that the figures could be compared. Mayors who wanted to set bread prices would have to refigure the weight of the wheat each year and then draft a scrupulously reliable price schedule.[14] Many

[13] AN F11* 3064.

[14] Circulaire No. 78, September 16, 1819, ADSM 6M 1215.

officials had voiced a desire for such practices for several years, and the methods, once instituted, continued into the mid-century.[15]

The decision to construct more accurate bread price schedules was a forceful move toward an increased reliance on free trade. The October 1820 meetings of the *Caisse* acknowledged that the drive toward greater precision was breaking with past policies. In an earlier time, one member explained, "when political concerns influenced bread prices as much as the rates of the hall," the government poured immense sums of money into provisioning. That former government – which could have been any of the regimes since 1789, but most likely that of Napoleon – had operated without any pretense of legitimacy. It used funds as it wished, and felt no obligation to account for them or to repay the debts it accrued. The Restoration had "inherited that ruinous system," the *Caisse* agreed, but it did not have to continue it. It had a responsibility to all its people to balance its budgets and not to favor one group over another. Thus, even though the *Caisse* members conceded that the poor had to be helped, they were not inclined to burden the municipal finances with subsidized bread for the entire city. The *Caisse* and the Subsistence Commission had wanted to implement such distinctions in its spring 1817 lists of the 200,000 or so poor of the capital, but had abandoned the plan when the Reserve was ordered to supply the entire city that June. According to the *Caisse's* reckoning, Parisians had paid perhaps an average of 10 centimes less per loaf of bread than it cost the bakers to produce them. Even with the other buffers that worked in the bakers' favor, the price ceilings could not continue to be used so greatly to the advantage of the entire city.[16]

Thus the Municipal Council, along with the Commission, the *Caisse*, and a committee from the Ministry of the Interior (now headed by De Serre), imposed the reforms of 1821. The new rules outlined formulas for the price of bread and revised methods for distributing the Reserve's supplies. In general, the price of bread was to follow that of grain.[17] During shortages, the poor could buy low-quality loaves baked with Reserve flour. In fact, according to the 1821 reforms, the Reserve was to limit itself to such activities and to refrain from selling its supplies at the hall. This part of the plan encountered immediate objections. The mills around Paris only produced fine wheaten flour, and Parisians, even the poorest, accepted only high-quality loaves by the 1820s. Thus, finding mills that could churn out lesser

[15] Minister of Agriculture to Prefects, November 26, 1850, ADSM 6M 1216; Circulaire, MI to Prefects, September 13, 1817, Archives du Finistère 6M 951.

[16] 29 October 1820, AN F11* 3064.

[17] The issue of whether to use grain or flour prices remained unresolved.

flour qualities for the Reserve would be difficult.[18] If the Reserve stocks could not be used to produce bread for the poor alone, then the issue once again was what to do with those supplies. The provisioning authorities agreed that they did not want to chase merchants out of the city with threats of competition from the government grain. Thus, in the 1821 accords, the Municipal Council promised to wait until bread had reached the price of 1 franc per 2-kilogram loaf before ordering the Reserve to become involved. Once bread prices had risen, however, the Reserve agents were to place sacks for sale at the going rate of the hall, while the mayors' offices distributed ration cards to the needy. After the 1821 reorganization, Reserve supplies grew to the equivalent of almost 90,000 sacks of flour, or an amount that could feed the capital for over fifty days (Figure 10.1). While the Reserve sales occasionally upset flour merchants, some on the Commission countered with the argument that the merchants received healthy profits overall, and that the price of bread generally followed that of grain. Thus, they concluded, the merchants had little to complain about when shortages occurred. The Reserve wares were not going to be used to compete directly with the merchants, they added, but instead would be placed at the going rate and could do no harm. Several members dissented, and pointed out that any increase in the supply at the hall was bound to lower prices. Further negotiations with the Ministry of the Interior finally earned full approval for the plan in April 1823.[19]

The Shortages of 1828–29

The April 1823 plans worked smoothly for the next few years. Good harvests allowed the Reserve to buy up the supplies it needed at moderate prices, and its flour was used sparingly. The shortages at the end of the Restoration, though, tried the resolve of the *Caisse* administrators and the resources of the Reserve. The scarcity began to be felt in late 1826, when the first of several poor harvests led to an overall economic contraction. By 1827, tight credit and a plunge in demand for manufactured goods resulted in runs of bankruptcies and commercial failures. Moreover, the political strife that would bring down the Restoration in July 1830 was intensifying.

[18] Despite that information, the various provisioning administrators held out the hope that such milling could be arranged. Later that decade, for instance, when the Municipal Council again announced that it would use the Reserve grain only for low-quality bread, a quarrel broke out between it and the other provisioning administrators. 19 January 1828, AN F11 2806.

[19] 26 April 1823, AN F11* 3067. See also the analysis of past measures, 19 January 1828, AN F11 2806.

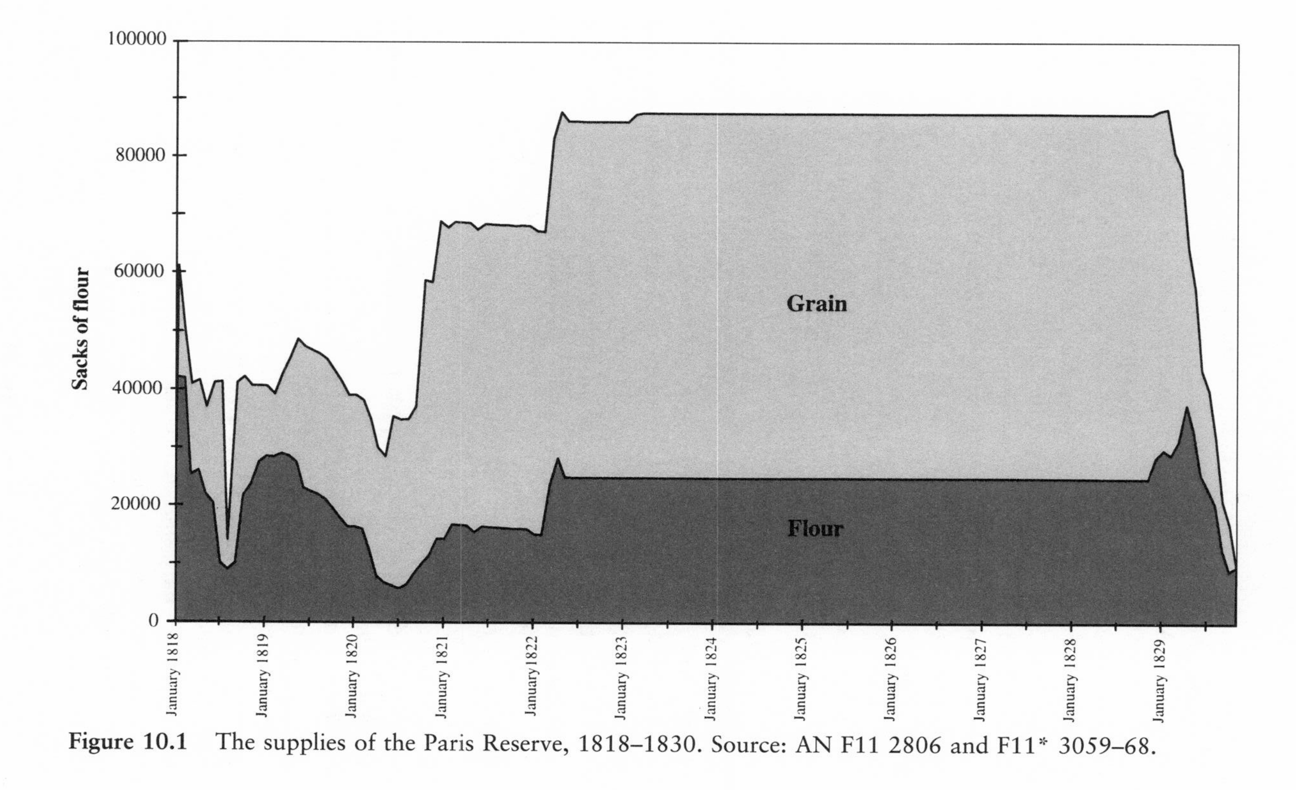

Figure 10.1 The supplies of the Paris Reserve, 1818–1830. Source: AN F11 2806 and F11* 3059–68.

Frustration among liberals, including those from the business and professional classes, had mobilized opposition to Charles X and his ultraconservative policies. More liberal, bourgeois candidates prevailed in the elections of the autumn of 1827, voting out the ministry of the Comte de Villèle that had endured for nearly six years.

The moderate 1828–29 ministry of the Vicomte de Martignac dismantled some of the most-hated laws of the Villèle years, loosening press restrictions and decreasing clerical involvement in schooling. Martignac could not navigate successfully between the competing demands of liberals, who wanted much more far-reaching change, and the ultraconservatives, who still hoped to restore the monarchy to its full powers. Charles X dismissed Martignac in November 1829 and placed the ultraconservative former émigré, the Prince de Polignac, at the head of the government. In the midst of these embittered electoral and ministerial battles, the members of the *Caisse* and the Paris municipal government turned with anxiety to the matter of steadily rising grain prices and possible urban unrest.

The initial test of the dual system involved the consensus confirmed in the 1823 plans that Reserve supplies would not be used until bread prices in the capital had exceeded 1 franc per 2-kilogram loaf.[20] The *Caisse* officials hoped that such delays would encourage commercial suppliers to bring grain to the capital in search of high prices, and that any Reserve intervention could be avoided altogether. During 1828, these administrators monitored flour and bread prices and kept a careful eye on bakers. The rising prices of that autumn demolished the hope that the shortages would end without any Reserve sales. By 3 December 1828, batches of 250 sacks of Reserve sacks per day began to appear at the Paris hall, and the officials made plans to distribute ration cards to the poor so that they could buy bread at subsidized prices.

The familiar debates commenced within the *Caisse* offices over the perennial matter of how to deploy the wares – whether directly to bakers or through third-party sales at the Paris hall, for instance, and if so, at what prices and in what amounts. Hour after hour was spent assessing price data, flour yields, and the success of previous ventures. What was striking, however, was the sharp shift that the Reserve strategies again took to imitate commercial practices even more fully. Each of the previous shortages had brought about greater certainty that the government would act with discretion in its sales and rely on methods that mimicked those of the trade.

The shortages of 1828–29 proved to be no different in that respect. Once again, the *Caisse* authorities crafted yet more refined strategies for using its wares. Pressures from the Paris Municipal Council, which had the ultimate

[20] This account is drawn primarily from AN F11 2806, the minutes of the *Caisse syndicale des boulangers* in 1827–29.

authority over the Reserve and the *Caisse*'s budget, probably sharpened the *Caisse* board's desire to use its wares as frugally and unobtrusively as possible. The Municipal Council was still frustrated by the expectation that the city had to fund the Reserve, even if sometimes the entire department benefited from low-cost loaves smuggled out of the city, and even if the monarchy itself benefited from public order in the capital. Thus, anticipating deep shortages, the Council had told the *Caisse* in January 1828 to limit its activities to selling subsidized loaves to "*la classe mal aisée*," and not to sell any flour through the hall. The Municipal Council feared Reserve sales "would bring about low prices in the hall . . . which would violate the agreements made with merchants, because one would be initiating competition with them."[21] The *Caisse* members rejected these anxieties, informing the Council that merchants knew full well that when bread prices reached that 1-franc limit, the Reserve would step in. In fact, the members sensed that this knowledge helped to keep prices in check precisely *because* flour merchants did not want the Reserve to begin selling supplies. Furthermore, since the *Caisse* had subcontracted the storage and milling of Reserve wares, the merchants and millers involved had even made profits from the Reserve's activities.[22] Thus, the *Caisse* members dismissed any fears that the merchants would be hurt by Reserve sales, and argued instead that merchants and the Reserve had entered into a mutually beneficial, if not ultimately cordial, relationship. Based on that sense of matters, the *Caisse* board held to its understanding of its mission to sell its sacks at the hall at long as it did not undersell merchants. That the Municipal Council was being more forceful on the matter only strengthened the *Caisse* members' already strong intention to avoid upsetting commercial suppliers as Reserve sales began.

The *Caisse*, in fact, was already discussing some unexpected questions that had forced it to reaffirm the 1823 decision on pricing strategies. The policies it adopted point to new insights into how the Reserve flour could be used to imitate commercial sales. In December 1828, the first sacks of flour the Reserve had sent to the hall through several *facteurs* – the official hall brokers – had been of high quality. The *Caisse* members had wanted to sell the flour for 92 francs, even though most commercial flour cost about 90 francs. Few sacks had been sold, and the next week, the *Caisse* members discussed lowering its price. Prefect of Police De Belleyme, in

[21] The Municipal Council pointed to one article of its deliberation of 15 December 1821 that stipulated that "none of the supplies can be used to reimburse bakers [for their losses] nor to lower the price of flour, but only to supply bread to the indigent. . . ." The Caisse members disputed this understanding of the Reserve's use. 19 January 1828, AN F11 2806. See also 17 March 1823, F11 2805.

[22] 19 January 1828, AN F11 2806.

particular, was angry and feared that the high-priced Reserve flour might raise the hall's average price – and thus the bread *taxe*. The head of the Reserve, Busche explained to De Belleyme that the matter of pricing the flour was far more complicated than the Prefect of Police wanted to believe. First, Busche noted, the quality of the flour was exceptionally good, and thus worth every franc. Second, he argued, such high-quality flour was precisely the means by which the *Caisse* could counter the constant rumors (often spread by merchants) that Reserve flour was generally poor, and possibly spoiled. What better way, Busche suggested, to make Reserve sacks invisible and untroubling to buyers and sellers alike than to have it blend in with the best-quality wares when the Reserve was fortunate enough to have such fine flour? De Belleyme protested in vain that the Reserve "would thus be ceding its supplies to speculation and not to the consumption of the city." There, too, the *Caisse* members differed with De Belleyme. They reminded him of the overall strategy they had been developing for over a year. Through ration cards that were being printed up even as they spoke, they planned to ensure that about 200,000 indigent Parisians would receive subsidized loaves of bread. They had developed a way to fund the bread rations by selling roughly enough flour through the hall *facteurs* to produce the subsidized loaves. (This plan had the additional advantage of feeding the poor without disrupting one of the normal paths of flour from the mill to the hall and then to the bakery.)

Given these strategies, the Reserve had every reason to want to get a good price for its flour. The 92-franc price, entirely appropriate considering the flour's quality, would help the poor by ensuring that the Reserve could afford to give them ration cards, while the wealthier would pay the higher going rate for bread. Although Busche could make no promises about the exact amounts of flour that would be sold any given week, the *Caisse* charged him to continue to "put himself in precisely the position of a merchant, to follow price movements higher and lower by obeying the stimuli given by commerce." Thus, when he had very good flour, he could even place its price above that of commercial flour, which he did throughout December 1828.[23] Busche's arguments would have seemed to run counter to Reserve activities since the warehouses had first been established, but the logic of the sales was certainly intact. The Reserve intervened best when following commercial practices and prices, even if in this instance it meant selling high-quality wares for a high price.

The political consequences of any government intervention weighed heavily on these discussions and forced the *Caisse* board to proceed with even greater care. Such concerns had been among the first issues raised when the board began formulating its options in late December 1827. The

[23] 13 December 1828, AN F11 2806.

members had expected that at some point the Reserve might need to supply some portion of the city's population for five or six months. Initially, they discussed expanding one of the Reserve's normal operations, the commissioning of several bakers to use Reserve flour to produce low-quality loaves for sale at several markets or charity bureaus in working class areas. It was important not to raise alarms that a serious shortage was on the horizon. But if, as the *Caisse* board calculated, there might be 150,000 or even 200,000 people within the capital who legitimately needed bread, what would be the effect on "opinion?" Was it in the monarchy's best interest to have bread given away free to the poor, or to find a way for the charity bureaus to distribute ration cards to known "indigents?" Should each arrondissement's notables be called on to help draw up lists of those deserving aid? Certainly, the notables were in a position to verify needs and establish qualifications. (They were to be careful not the mix the 140,000 "already identified indigents" with the 60,000 merely "*mal aisés*," for the latter would take offense.) In short, the *Caisse* members wanted to find an unobtrusive and inexpensive means for distributing the supplies. If, as the members generally agreed, it was best to limit its help to subsidies worth a few centimes per loaf, and if they wanted that aid limited to only a portion of the capital's population, then ration cards were the best way to proceed. While the administrative tasks associated with the system of cards was daunting, the authorities preferred to follow that path. Despite the decision, there was the chilling "fear that this measure will awaken unfortunate memories that the Revolution has left us; that [this system] will give rise to *rassemblements* and then to disorders."[24] Thus, the *Caisse* board decided to structure the distribution of the ration cards through the city's twelve arrondissements, and to supply only one card per household, in five-day batches. The board hoped that such plans would reduce the number of consumers arriving each day looking for ration cards and thus check the potential for riots.[25]

In mid-November 1828, with little apparent notice, the Minister of the Interior instructed the *Caisse* to begin distributing cards that would allow the poor to buy bread for only 80 centimes per 2-kilogram loaf on December 15. (His decision had resulted from the news from the *Caisse* that on December 1, the city's legal bread price would have to be increased to 95 centimes.) The orders forced the *Caisse* into rapid motion, printing the cards, contacting the charity bureaus for recent lists of the poor, and verifying lists produced by mayoral offices. The ration card distributions had to be suspended initially because the charity bureaus were not fully prepared for the throngs of needy that appeared in the opening days. By late

[24] 22 and 29 December 1827, 5 and 19 January 1828, AN F11 2806.

[25] 19 January 1828, AN F11 2806. Requests for ration cards are in AN F11 2808/B.

December 1828, however, the ration cards made their way from the distribution points into the hands of the poor and then to the bakers, who later redeemed them for reimbursements from the *Caisse*. Between December 1828 and September 1829, when the system of ration cards ended, the *Caisse* distributed 6.5 million cards, worth over 1.1 million francs in reimbursements to the bakers.[26]

How does one evaluate the 1828–29 efforts? First, it is significant that the *Caisse* stuck to its 1823 decision to limit its involvement in the grain trade. Even though the members had grave fears about the possible political implications of a shortage, they honored the agreement the *Caisse* had with merchants to send no Reserve flour to the hall until bread prices reached genuinely high levels. Even then, the *Caisse* had scaled back its expectations that Reserve flour would bring about significant price reductions, and indeed, it saw any deep reductions as dangerous. It even sanctioned selling its flour at prices above the going commercial rate if the flour's quality merited such levels. Thus the 1828 sales conformed even more fully to the long-stated desires of several generations of administrators to use simulated sales – the covert, carefully priced, and market-mimicking sales – to shape but not destroy commercial practices. The distribution of bread to the poor through ration cards, although difficult to disguise, was deemed an economical means for satisfying the poor while not disturbing the normal pattern of flour and bread sales in the capital. Bakers received every centime of the legal price of bread for every loaf they sold. The wealthy paid in full and the poor paid with cash or promises and a ration card, which the bakers then turned in for reimbursements from the *Caisse*. In other words, while the Prefect of Police and the *Caisse* board worried about the consequences of raising bread prices to follow those of flour, they had found a way to channel extra funds to the poor. This tactic meant that flour prices could rise and fall, and bread prices along with them, but that neither bakers nor merchants would bemoan that any demands by the Prefect of Police that bread prices stay artificially low. Only 3 Parisian bakers out of 600 went bankrupt during the shortages – a very good sign.[27] Moreover, the system of Reserve sales and ration cards put the help – albeit still insufficient help – into the hands of the most needy and compelled the more affluent to pay the full commercial cost of producing a loaf of bread. Thus the plans of 1828–29 fulfilled the *Caisse* members' best sense of how to proceed, and especially of how to maintain commercial networks even when shortages and political tensions loomed. They did so by tying the Reserve's activities ever more closely to prevailing commercial practices and costs, and by directing the *Caisse* funds to a much smaller proportion of the capital's population.

[26] See the figures in AN F11 2806. [27] 14 November 1829, AN F11 2806.

In late 1829, the *Caisse* board, following orders from the Municipal Council and the Ministry of the Interior, began liquidating its warehouses and dissolved the Reserve. The plans to dismantle the system that had served the capital so well had actually been under way for several years. In general, the Municipal Council really did resent having to shoulder the responsibility for supplying the city, and feared that it might suddenly accrue enormous debts if a truly grave crisis occurred. Most important, though, was the growing optimism that commercial ties were almost strong enough to supply the city, except in the worst crises. Thus, the 1828–29 shortages actually had interrupted the process of terminating the Reserve, and although the Reserve and the ration cards had been necessary that one last time, the overall consensus in late 1829 was that the moment had come to end the practice. That November, the Comte de Chabrol, the Seine Prefect, could announce at a special meeting of the *Caisse* board that the Reserve had been a "*service extraordinaire*" – unusual or exceptional – but one no longer needed. The board responded by describing its hopes for the capital's future to the Minister of the Interior, who was also present. The members believed "that it will be especially in commercial operations that any significant help will be found; that it is only necessary perhaps to allow that type of speculation to take its full development . . ." and went on to support greater international free trade in grain, so that imports could be found easily when needed.[28] The Reserve sold off its last sacks of flour (some of it forced into the hands of bakers in lieu of cash reimbursements for the ration cards), and closed it warehouses and its books.

Thus the broad lines of the Restoration provisioning policies encouraged a greater acceptance of market forces as the appropriate source of prices. The Parisian population would buy loaves whose prices had been calculated with great precision to reflect the going rate of flour. The trade would not be turned loose altogether, but the government would further the growth of commercial networks through its more restrained activities. The poor, who could not afford to follow the price increases that accompanied shortages, would be spared the worst through subsidizied loaves. The rest of the population would feel the forces of supply and demand each time they entered a neighborhood bakery.

Restructuring the Baking Trades

The eventual dissolution of the Reserve did not leave the capital without stable emergency stores of grain, however. Every administrative discussion

[28] 11 November 1829, AN F11 2806.

that took up the question of the Reserve – its expense, the way it might scare away commerce, the possible low quality of its wares – came back to the problem of how to feed cities, especially Paris, if such supplies were disbanded. No administrator, not even the most ardent proponent of free trade, advocated allowing the market to supply the capital entirely. Thus, other sources had to be organized. For this, the ministries and many others involved in urban provisioning turned to the bakers. Each time one of the provisioning committees lamented the cost of running the Reserve, one of the members would recommend instructing bakers to amass supplies. The idea of forcing bakers to increase their supplies gained momentum after 1817, when it became clear that the royal treasury would not underwrite supplies for Paris or any other city. In fact, the 1818 and 1829 attempts to pressure the Parisian bakers to accept Reserve grain as part of the remaining reimbursements owed them were part of such schemes. In the 1820s, the Municipal Council, ever more anxious to relieve itself of the financial burdens of the Reserve, pointed repeatedly to the bakers as the solution to the entire problem of emergency provisions.

These plans focused on the supplies that bakers were required to own as part of the requirements of their permission to practice the trade. Gradually, bakers were pushed to assume the responsibility for stable emergency supplies. As their resources increased, the need for a government-sponsored Reserve diminished. The bakers' willingness to maintain those all-too-necessary emergency supplies was not a result of their concern for their communities, or even for their own shops' financial stability, however. Instead, the growth of these stocks came about through the state's steady demands on the bakers that had begun well before the Restoration.

The government's ability to exert such pressure originated in the statutes that reorganized the baking trades under the Consulate. The shortage of Year X (1801–02) had brought about not only the creation of the Reserve, but a broader move to stabilize urban provisioning in general. The desire to restore the baking guilds was one facet of an uneven return to economic regulation under Napoleon. By the beginning of the Empire, there were a number of proposals for regrouping many trades.[29] Thus, starting on 19

[29] Mutual aid societies, corps of *huissiers* and notaries, Conseils de Commerce, and Chambers of Commerce were part of this shift away from the interdictions on associative life that the Revolutionary ideology had imposed. As Jean-Pierre Hirsch explains, "the wish for regulation proved more powerful than the rejection of corporations." Trading companies, purged of the taint of Old Regime "privilege," also reemerged as the protectors of professional standards and orderly business conduct. Jean-Pierre Hirsch, *Les deux rêves du Commerce: Entreprise et institution dans la région lilloise (1780–1860)*, *Civilisations et Sociétés* 82 (Paris, 1991); Tulard, *Révolutions*, 226; Michael David Sibalis, "Corporatism after the Corporations: The Debate on Restoring the Guilds under Napoleon I and the Restoration," *FHS* 15 (1988): 718–30; Christian Bonnet, "L'encadrement institutionel

Vendémiaire Year X (11 October 1801) in Paris, bakers guilds were reestablished, and in the next years Bordeaux, Rouen, Marseilles, Troyes, Amiens, Toulon, and many other cities followed suit.[30] Like many of his colleagues, Seine-Inférieure Prefect Girardin's resistance to controls had been overcome in 1812. "I know everything that can be argued against the reestablishment of *corporations*," he wrote to Réal, "but in political economy especially there are no rules without exceptions. Everything that has to do with the food supply is outside ordinary principles." Bakers had to be brought to order because, where bread was concerned, "the need to sell will never equal the need to buy." The government's correct role was to "intervene in order to establish an equilibrium," wrote this former advocate of complete free trade.[31]

Like many of the pre-Revolutionary regulations, the post-1801 ordinances served multiple goals, not the least of which was to control the unruly membership of the bakers' guilds.[32] In Paris, for instance, municipal government exerted direct authority over the trade through the four corps officers, or syndics, who were effectively hand-picked by the Prefect of Police. Unlike the semi-autonomy with which many Old Regime guilds had operated, the Napoleonic societies were forbidden to call any independently organized meeting of the collective membership, and in Paris, as elsewhere, their cramped meeting room held only a few seats. Whenever the bakers needed to be apprised of a change in policy or prices, the syndics were to spread the word personally.[33] While the number of baking permits was in theory unlimited, the Paris Prefect of Police was under strict orders to make them difficult to obtain. There, as in other cities, bakers had to pay a fee and a security deposit and were subject to unannounced visits from the police. Bakers who failed to follow the statutes could be forced out of the

et juridique du commerce marseillais de l'Ancien Régime à la Restauration," in *La Révolution française et le développement du capitalisme*, pp. 201–2, 205, 252–3; Etienne Martin de Saint-Léon, *Histoire des corporations de métiers*, Reprint (New York, 1975), pp. 620–9.

[30] "Reglements des Boulangers de la ville de Bordeaux," 27 March 1809, AN F11 209; Mayor of Rouen, "Ordonnances concernant le commerce de la Boulangerie," 22 July 1809, ADSM 6M 1195; "Rapport à Sa Majesté," Direction générale de l'Agriculture, du Commerce . . . , 11 July 1814, AN F11 209. The report summarized the organization of the baking trade during the Empire.

[31] Girardin to Réal, 16 September 1812, AN F7 3639. See also the comments of the Mayor of Dieppe, 10 August 1812, ADSM 6M 1193.

[32] See the list of regulations republished in the *Tableau des boulangers de Paris, pour l'exercice de l'an 1831* (Paris, 1831).

[33] The government's control extended inside the bakery, in ways that the bakers did not welcome. The searches provided municipal officials information on the level of the capital's flour supplies, a powerful negotiating tool. Prefect of Police to the Minister of the Interior, 26 September 1810, ASVP 4 AZ 795; and [2 October 1810], AN F11 209, p. 6.

trade, and any baker who decided to leave the profession without giving at least six months' notice forfeited his or her sacks in the depot.[34]

Stabilizing Bakeries

One way of stabilizing the baking trade was to get rid of insolvent bakers. In general, local officials decided that the abolition of the Old Regime guilds in 1791 had allowed too many newcomers to rush into the trade. Few could hope to earn enough from the meager and ever-changing profits on bread to guarantee the steady supply of the city. These efforts may have hit female bakers, who tended to have smaller establishments, hardest.[35] More affluent Parisian bakers, with the encouragement of the Ministry of the Interior and the prefect, provided funds to buy up and close lesser shops. In 1807, there had been 689 bakers in the capital. By 1822, there were only 562. During the 1820s, another 40 shops were closed.[36] Other cities pursued the same policies, anticipating that more substantial bakeries, with a greater clientele and possibly more efficient production, would yield a more stable profession.[37]

Throughout the century, the police and the bakers' syndics devised complicated formulas based on a city's population, the expected per capita bread consumption, and the profits to be made from each loaf in order to establish the appropriate number of bakers for each city. During the July Monarchy, a subcommittee of the Seine Department estimated that ideally there would be one baker for approximately every 1,800 citizens – or a total of 585. In smaller cities and towns, where rents and the cost of help were less, it was believed that one baker for every 1,000 or 1,200 customers created a suitable balance. Most towns had far more bakers than they needed, and struggled to drive the least sound operations into retirement.[38] In the smallest *bourgs*, however, some administrators suggested allowing

[34] The First Consul issued an *arrêté* ordering all bakers in the capital to apply for permits from the city's Prefect of Police. After 1811, the *facteurs* at the hall had the first claims on flour left in the depot. AN F11 209; Paris Prefecture of Police, Commission des subsistances, 1854, ADSM 6M 1299; Cambacérès, pp. 112–15.

[35] See the summer 1812 protests of Femme Caruel, a baker in Rouen. ADSM 6M 1195.

[36] Although ten more with special equipment were allowed to open in 1829. See the 1854 printed report of the Paris Prefect of Police and the Commission des Subsistances in ADSM 6M 1299.

[37] See the policies in Fécamp, Rouen, and elsewhere in the Seine-Inférieure in ADSM 6M 1299.

[38] See the Fécamp mayor's report. There were twice as many bakers as he felt his city could support. ADSM 6M 1299.

more bakers. It was important to give their villagers a range of possible bakers from whom to buy on credit.[39]

The efforts to stabilize the baking trade in the Seine-Inférieure were typical of those undertaken throughout the country. Under Prefect Girardin's supervision, attempts to force insolvent bakers out of the business heightened after 1812. The new rules limited the number of bakers, and required them to place ten sacks of flour in a municipal depot or in private warehouses. In addition, they had to provide between 2,000 and 5,000 francs apiece to purchase an emergency grain supply. To buy out insolvent shops, every municipal baker had to contribute 24 francs to a common fund.[40] At first, Prefect Girardin, still skeptical of controls, objected to the plan, even though Mayor Hellot reminded him that 20 of the city's 135 bakers had closed their shops during the shortage, and now faced bankruptcy. These more stringent requirements, Hellot admitted, "were more proposed to the bakers than proposed by them."[41] The Minister of Manufacturing and Commerce overruled Hellot's plan, justifying the decision in a way wholly consistent with his commitment to routine and an appearance of neutrality: No city could limit the number of bakers arbitrarily. However, the Minister deemed it appropriate to place greater pressure on the poorest bakers by increasing the number of sacks of flour required for the municipal and private stores and then encouraging wealthier bakers to buy them out. In other words, as long as the Prefect and city hall did not appear to play a direct or capricious role, the reduction of the number of bakeries could continue apace. While occasional skirmishes broke out over which shops should be closed, the practice of eliminating insolvent shops persisted for many decades.[42]

Increasing Bakers' Supplies

As for the bakers themselves, throughout the country they were grouped into three categories, depending on how much bread they produced per day. Since the Empire, membership in the bakers' *syndicat* had required a commitment on the part of the baker to store sacks in a community warehouse and to keep a private supply of flour in his or her shop. After 1819, Minister of the Interior Decazes required prefects to send in monthly lists, naming the bakers in every town and attesting to the exact number of sacks each

[39] See the 1855 report on conditions in the Seine-Inférieure, ADSM 6M 1299.
[40] "Réglement concernant la boulangerie," ca. 16 October 1812, ADSM 6M 1295.
[41] ADSM 6M 1295.
[42] ADSM 6M 1295.

one had placed in the designated storage areas.[43] By 1828, one of Decazes's successors judged this supervision so successful that he only needed annual lists of bakers, categories, and stocks.[44] Here, then, were the supplies that would make up for the termination of the Reserve and of other municipal granaries across France.

Essentially, the state was forcing bakers to use commercial cereals to provide the back-up supplies their cities needed. Each decade, the regulations for municipal bakers' *syndicats* stipulated more stringent prerequisites for entry into the profession. Paris led, and others followed quickly in the move to bolster supplies. In the capital, for instance, the 1801 statutes specified that each baker had to deliver fifteen sacks of flour to the Saint Elizabeth warehouse, and keep a private supply of between fifteen and sixty sacks for emergencies, depending on the size of his business and what category the baker belonged to.[45] Those amounts were increased in 1807, and again in 1818, reaching the equivalent of thirty-five to forty days of flour. To raise the amounts more fairly, the 1818 Paris statutes redivided the bakers into four categories, based again on the numbers of sacks of flour they used each day. The new category consisted of the bakers with very large establishments, and they were held to very high standards – 140 sacks of flour each in their private stores alone.[46]

Statutes in other cities followed suit, whether at the urging of ministers or through the stubborn efforts of prefects and mayors. As early as 1815, Versailles regulations had pushed standards higher by ordering bakers to keep only first-quality flour in their depots.[47] In Rouen in 1809, the best the Prefect could do was to require his impoverished bakers to put ten days of grain (maybe seven or eight sacks, depending on the calculations) in city

[43] If a baker did not keep a sufficient private supply, he or she could lose the right to practice the trade. In periods of shortage, the government used such failings as a means to threaten bakers and increase local supplies. Those warnings were only rarely carried out during the crisis of 1817. See the letter from Prefect of Police Anglès to Minister of Police Decazes, 5 December 1816, AN F7 9889^{A}.

[44] MI to Prefects, 25 August 1828, AD Eure 6M 701.

[45] There is a great deal of correspondence on the reorganization of the Seine-Inférieure bakers, and in particular on the post-1812 attempts of that department's prefects to force poor bakers out of business, in AN F11 717–8; AN F7 3639; ADSM 6M 1193 and 1295; *Du commerce des grains et farines*, pp. 29–33. See also the discussion of the September 1807 law in "Réglement concernant la boulangerie," n. d. [October, 1812], ADSM 6M 1295; AM Le Havre FM F^4 9.

[46] AN F11 717–8 and AN F7 3639; "Réglement concernant la boulangerie," n. d. [October, 1812], ADSM 6M 1295; AM Le Havre FM F^4 9; Prefect of Police, Paris Commission of Subsistence, 1854, ADSM 6M 1299; and *Du commerce des grains et farines*, pp. 29–33.

[47] *Tableau des boulangers de Versailles autorisés par le Maire de Versailles pour l'an 1832* (Versailles, 1832), AN F11 2796.

magazines, and to place fifteen more sacks in their personal magazines.[48] By 1812, though, those amounts rose to between thirty-five and sixty-five sacks of flour in several forms of depots.[49] Additional supplies were wrested from *forain* bakers who wished to bring bread into the city; after 1812, each one had to deposit eight sacks of grain in a Rouen depot as a bond against any possibility that the *forain* baker would abandon the city during a crisis.[50] By 1824, the Rouen bakers regularly stored between thirty-two sacks apiece – a slight decrease for the poorest bakers – and eight-four sacks apiece – a substantial increase for the largest bakeries.[51]

In Le Havre, the 1812 mayor had resisted the idea of obligating his bakers to create such supplies, and criticized the supplies in force in Rouen and elsewhere, because they generally "last only a few weeks." Those amounts were hardly enough to bring a city through a crisis.[52] In 1816, the subprefect ordered the mayor's successor to draw up more appropriate statutes. The resulting Le Havre ordinances of February 1817 instructed bakers to keep between approximately fourteen and thirty-nine sacks of grain in their various warehouses, a modest amount that might have offered two to four weeks of relief.[53] By 1858, the bakers of Le Havre had to offer guarantees of three months of flour in order to enter the trade. Furthermore, every new baker in that city had to qualify as a second-class baker, not a third-class baker, which raised supplies further.[54] The prescription that only candidates who had the wherewithal to enter as second-class bakers and not as third-class bakers could qualify was added in other towns. Elbeuf, for instance, upheld a new minimum requirement to enter as a second-class baker in 1826.[55] Similar transformations in the department of the Eure ratcheted up the requirements there also. In 1820, for instance, the Minister of the Interior directed the prefect to reclassify a number of the bakers in Evreux in order to increase supplies. He deemed the mere twelve days of stocks that the departmental *chef-lieu* had on hand insufficient.[56]

[48] Mayor of Rouen, 22 July 1809, AM Le Havre FM F4 9.

[49] "Réglement concernant la boulangerie," ca. 16 October 1812, ADSM 6M 1295.

[50] Rouen bakers had fought the presence of *forain* bakers for many years, viewing them as threats to their monopoly. By placing these requirements on the *forain* bakers, the Rouen administration was able to halt the worst of the disputes between the two groups. See the lengthy correspondence, and especially "Réglement concernant la boulangerie," ca. 16 October 1812, ADSM 1295.

[51] AN F11 562–3.

[52] Mayor of Le Havre to the Mayor of Dieppe, 24 September 1812, AM Le Havre FM F4 9.

[53] "Ordonnance du Roi," 26 February 1817, AM Le Havre FM F4 8; and bakers' reply to the mayor's orders, n.d., AM Le Havre FM F4 9.

[54] AM Le Havre FM F4 8.

[55] "Avis au Maire," 4 July 1826, ADSM 6M 1297.

[56] MI to the Prefect of the Eure, 29 February 1820, AD Eure 6M 764.

There were thresholds, however, beyond which it made little sense to press the bakers. Even with the closing of insolvent shops and the corresponding stabilization of the trade, bakers successfully resisted the more substantial pressures that were placed on them in the middle of the century. If the number of sacks each baker were to supply rose too high, the bakers would not be able to afford to buy them, much less store them. Furthermore, the stocks would be susceptible to spoilage. Several times, the ministries tried to oblige bakers to amass several months of grain and flour. Those orders, though, ran straight into the problem of storage. Not only were the bakers unable to afford to lease such massive warehouses, but many sacks were certain to spoil if not used month after month. The general consensus grew that any more than three months of supplies might well rot before they were used. Gradually, the goverment allowed the bakers to follow a yearly cycle: In the autumn and winter, the bakers bought up as much as three months' supply, and in the summer they sold it. The summer depletions required special permissions at first, but year by year, these practices gained greater acceptance from the provisioning authorities. To ensure that sufficient supplies were available for emergencies, the state encouraged cities to extend the statutes to bakers in nearby towns. An important example of these efforts was carried out by a specially appointed commission set up by the Paris Prefect of Police in 1854. The commission examined ways to reorganize the bakery trade near Paris, and made several strategic recommendations. One of its most convincing proposals was to extend the Parisian standards to the bakers of the eighty-four communes in the Seine department. This measure would alleviate the need to push Parisian bakers to accumulate still more sacks. If one is to believe the figures in a prefectural committee's report, the 1854 revisions would require Parisian bakers to have a total of 213,120 sacks on hand – an increase of 131,840. The *banlieu* bakers would supply another 98,010 sacks – all in all, a very substantial increase.[57]

The *Tarif* Reforms of 1819

The shift from government-funded reserves to those owned and tended by bakers was accompanied by corresponding efforts to revive the *taxe du pain* of the Old Regime.[58] After all, if bakers were going to be forced to main-

[57] Préfecture de Police, Commission des Subsistances, "Réorganisation de la boulangerie . . ." 1854, ADSM 6M 1299.

[58] During the 1812 Maximum, most towns had held the price of bread in line with the ceilings for grain and flour. Once the Maximum had been lifted, however, there was a relentless move toward *tarifs*. Circulaire, Prefect S-I to the Subprefects, 24 September 1812, ADSM 6M 1192; Girardin to Réal, 16 September 1812, AN F7 3639.

tain private supplies bought through commercial networks, administrators wanted to make sure that bread prices conformed as well as possible to the forces of the market. At first, local officials used the price schedules in the familiar ways – as a means for limiting bread price increases. In the Seine-Inférieure, the prefect steadily extended the use of *tarifs* to smaller towns by early 1814, and even announced a desire to craft a uniform schedule for the entire department. To hold bread prices slightly lower than the *tarif* itself indicated, he changed the ways in which the average market price of grain was calculated. By using only the market's opening prices – generally lower – he achieved that aim.[59] His measures were as much a public relations campaign as an attempt to police the bakers. The bakers were powerless to challenge these efforts, although they petitioned almost yearly for better rates and a more accurate *tarif*.[60] The Old Regime was no more, nor were there any parlements or jurisdictional rivalries for them to exploit with their appeals. The bakers' only audience was the unsympathetic prefect, subprefect, or Minister of the Interior.[61]

By the Restoration, however, there was growing administrative concern that holding bread prices too low would have a detrimental effect over the long run. These concerns had led to the 1817–21 reforms in Paris. They continued, encouraged in large part by Decazes, who, as Minister of the Interior, issued orders establishing baking guilds in every town and requiring bakers to store specified amounts of flour for emergencies. The plans did not end there, however, and the Restoration's ministers even considered imposing a kingdomwide bread price schedule. Frustrated by piecemeal attempts to re-create bakers' societies and rewrite *tarifs*, Decazes demanded greater uniformity in his *tarif* reforms of 1819.[62] The *tarifs* were to be forwarded to his Ministry, so that his assistants could compare price levels across France. Statisticians in the two ministries would spot discrepancies and demand explanations. Decazes wanted not only to coordinate *tarifs*, but also for the price schedules to reflect annual changes in the yields of flour from grain. The resulting *tarifs* would follow the market prices more closely.

[59] Cotteau to Réal, 20 March 1812, AN F7 3639; ADSM 6M 1301. On the problem of establishing a single *tarif* for the entire department, see the comments of the Subprefect of the Rouen arrondissement, 10 February 1814. See also the efforts of the Subprefect of Dieppe in ADSM 6M 1193.

[60] See their many petitions and treatises, including one drafted as part of a country-wide movement by bakers to have their trade reorganized in the 1850s. ADSM 6M 1301 and 1308.

[61] See Michael Sonenscher, *Work and Wages*, for a fuller discussion of the impact on the trades of the parlements' dissolution.

[62] Circulaire No. 78, 16 September 1819, ADSM 6M 1215; Minister of Agriculture to Prefects, 26 November 1850, ADSM 6M 1216; Circulaire, MI to Prefects, 13 September 1817, Archives du Finistère 6M 951.

A critical part of Decazes's reforms of the baking trade lay in the timing of bread price changes. For several centuries, local administrators had manipulated the process of raising and lowering bread prices. The careful timing of those changes helped offset fluctuations in the price of grain. Decazes sought a more uniform method, one that would have the appearance of neutrality and regularity. Adopting proposals that had originated under the Empire, he ordered municipal officials to devise monthly schedules. In Paris, for example, the price was to be announced on the 15th and 30th of each month. Such measures would silence bakers and consumers, who demanded new prices every time the price of flour rose or fell at the Paris hall.[63] The Paris prefect of police added a further measure to regularize the process. In October 1819, he proposed that the preceding month's average flour price be taken as the basis for any changes. His suggestion would seem to be an insignificant detail. By using average prices, however, as he explained to another police official, the price of bread would conform more accurately to the price that bakers had paid for flour. He knew that bakers frequently bought in advance of their needs, and concluded that the *taxe* had to take those purchases into account. Also, in periods of rising prices, using the previous months' levels would slow bread price increases and encourage bakers to seek more reasonably priced supplies. Once prices came down, the delay would offer bakers compensation for the earlier losses. Most important, using the previous month's prices would silence the "public's conjectures, almost always false, on the causes of rapid price rises."[64]

These measures were adopted quickly, often with little need for ministerial directives, in other cities.[65] Regularizing price changes, and bringing the resulting prices more closely in line with those of flour and grain, when possible, helped strengthen the ties to commercial networks. They would also put an end to the generation of rumors and, the police hoped, make urban provisioning a smoother process.

The routines established by the Restoration did not follow a wholly smooth course, however. Successive regimes seem to have embraced several conflicting visions of the state's appropriate role in the food supply. Primarily, it seems that neither the July Days of 1830 nor the Revolution of 1848 caused any grand rethinking of the controls on the baking trades. Instead, authorities seem to have reinforced the practices established by the Restoration. Officials between 1830 and 1860 engaged in increasingly detailed explorations of the relationships between wheat and flour prices,

[63] AN F7 9889^{A}.

[64] Prefect of Police to Mounier, 24 May 1820, AN F7 9889^{A}.

[65] Officials in Rouen chose to announce price changes at exactly ten o'clock every Sunday. Prefect S-I to Subprefect of Neufchâtel, 28 October 1829, ADSM 6M 1301.

their yields, and milling practices. They used that information to structure bread price schedules that followed commercial prices more exactly, and gave even fuller consideration to determining the optimal ratio between a city's population and the number of bakeries it could support. Thus, throughout mild shortages in 1838–39 and a more severe harvest failure in 1846–47, ministers, prefects, and mayors did not stray too far from the plans that the Restoration had established.

The continued commitment to controls can be found in the many inquiries, some sponsored by the Ministry of Agriculture and Commerce, into practices of the baking trade during the mid-nineteenth century.[66] In 1831–32, for instance, the pressing question within the ministry was how the price of the bakers' actual purchases was to be included in the monthly bread price ceilings. Was it necessary to include off-market sales, a ministerial survey asked in 1831.[67] How should purchases made on credit or in advance of delivery be considered when it came time to set the *taxe*, one circulaire inquired. Twenty years later, a new set of ministers posed the same question to the prefects of the early Second Empire. Some of these questions may have reflected the desire of each incoming regime to get its bearings, to find out what practices were being followed, and whether there was any need for significant reform. Yet, the constant preoccupation of the surveys and the prefectural replies was that of adapting regulations to changing commercial practices rather than removing them.

A significant question, for instance, was whether flour or grain should be the basis of a market's average price. The matter had only recently been resolved in favor of flour prices in Paris, after many years of discussion by the Napoleonic and Restoration provisioning bureaus. Either choice offered disturbing possibilities for collusion and fraud among bakers and their suppliers, and prefects and mayors of the July Monarchy wanted to make the best choices for their *administrés*. A related set of discussions led administrators to consider whether to allow bakers to sell by weight, rather than by the loaf, in the 1840s. Their offices, and those of the police, had been filled with several decades of allegations against bakers who had short-weighted their customers. In fairness, the bakers could offer somewhat credible excuses, especially that their loaves weighed less as they cooled and dried. Thus, even a perfectly prepared loaf could be the target of complaints a few hours after being placed on a shelf. The solution, some administrators thought, was to allow bakers to sell their bread by weight, like meat

[66] See the extensive discussions on each of these issues in AN F11 2757, 2799, 2801, and ADSM 6M 1301.

[67] See the replies to a 13 January 1832 circulaire by the Ministry of Commerce and Public Works, AN F11 2755.

and other foodstuffs. Indeed, in 1840, sales by weight and not by the loaf, were introduced in Paris.[68]

Each of these discussions could be described as a set of practices, and even a discourse, that sought to ground policies in the unassailable logic of mathematical calculations. These refinements, however, remained within the bounds established under the Restoration. By and large, these officials were only tinkering with what the Restoration had given them. The prefectural replies generated a great deal of information on the practices in effect, and sometimes listed the policies' shortcomings, but rarely suggested that regulations be abandoned. Instead, the general concern seemed to be how to craft formulas and practices with the greatest precision, and to make sure that neither bakers nor consumers would suffer when a shortage occurred. Urban provisions had been tied roughly to the workings of the market, and the rules in force helped to keep the bakers, and thus the population, connected to a reputedly commercial system.

Loosening Controls During the July Monarchy

Despite the general obsession with bread price schedules, there is evidence that support for greater liberalism in the baking trades was growing during the July Monarchy, and that not every prefect and mayor was willing to uphold the regulations in their entirety. In fact, during that regime, the practices of the Restoration appear to have lost some of their coherence – a coherence that was regained only after the Revolution of 1848. The reasons for the July Monarchy's often contradictory stances on provisioning policies are not completely explicable from the correspondence and debates of the period. It is possible that the unrest that had accompanied the shortages of 1838–39 had prompted doubts among some members of the Chamber of Deputies and the ministries. During the food riots of those years, the Minister of the Interior demanded that *libre circulation* be enforced, and ordered the heavy-handed suppression of riots. His vision of the solution to provisioning riots seems to lie in troops and free trade, and not in the further fine-tuning of regulations. His actions were entirely consonant with the stark policies of the government in those years. Strikes and labor unrest in 1832 and 1834 had provoked bloody repression, and by the late 1830s, the government, having gotten the upper hand, was intent on keeping it. The disturbances may have even hardened the convictions of those most committed to laissez faire. In the Chamber of Deputies, representatives engaged in embittered wrangling over a wide range of issues con-

[68] "Ordonnance concernant la vente et la taxe du pain dans Paris," 2 November 1840, AN F11 2801.

cerning working-class poverty, charity, and labor laws. Several deputies argued that any intervention in the economy was contrary to the rules of nature, and opposed any aid to the poor. The turbulence of the Chambers and the distress of the poor revealed the tensions and uncertainties of the French economy during the last decade of the July Monarchy.[69]

The increasing acceptance of free trade that developed during the July Monarchy may help to explain the contradictions of the policies that successive ministers applied to the baking trade. Despite the ongoing discussions of the finer points of price schedules and supplies, the administrators of the July Monarchy seem to have experimented with relaxing some of the controls on the baking trade, or even to have taken steps to undo them. It is possible that the period's incessant and detailed examination of bread pricing strategies had generated concerns about whether any calculations could ever yield reliable results. In other cases, bakers seemed to have been able to amass warehouses that went well beyond the requirements, so prefects felt that regulations were no longer needed. Commercial networks and the bakers' own initiative had finally brought the much-desired stability to some towns.[70] Whatever the source of the skepticism about regulation, by the late 1830s and early 1840s, some cities had ceased obligating the bakers to keep their supplies at the required amounts. The uncertainty reached new levels in the mid-1840s, when the Minister of Agriculture and Commerce expressed his opposition to the statutes governing the trade. In a 19 August 1843 circulaire, he suggested that the laws requiring bakers to maintain supplies "were contrary to free trade and to personal liberty."[71] An 1844 letter to the prefect of the Eure pointed out the minister's belief that back-

[69] See, especially, H.A.C. Collingham, *The July Monarchy: A Political History of France, 1830–1848*, with R.S. Alexander (London and New York, 1988), pp. 214–19, 347–64; and Giovanni Procacci, *Gouverner la misère: La question sociale en France (1789–1848)* (Paris, 1993).

[70] The general well-being of the baking trades seems to have occurred first in cities that served as important distribution points between the wheat fields of the Beauce and Paris, and where commercial ties were well-developed by the late eighteenth century. In Versailles, for example, the shops of all the insolvent bakers had been bought out and closed by 1819, and the remaining bakers had exceeded the supply requirements by 1823. During the 1820s and 1830s, their warehouses held almost double the number of sacks required. See the correspondence in AN F11 2796. Other cities, such as some in the Seine-Inféreirue, struggled throughout the period to enforce the requirements. AN F11 2757.

[71] The prefect of the Eure-et-Loir reported the 1843 comments to the Second Republic's Minister of Agriculture and Commerce and asked for clarification. Minister of Agriculture and Commerce to the Prefect of the Eure-et-Loir, 7 September 1849, AN F11 2757. The prefect of the Eure also commented on the circulaire in his December 1848 discussions of the problems facing the ministry. Prefect of the Eure to the Minister of Agriculture and Commerce, 22 December 1848, AN F11 2757. Other departments related similar instructions, and mentioned that they had relaxed supply requirements somewhat in the early 1840s. See the many letters in AN F11 2757 and 2796.

up supplies "sequestered, in a certain sense, for the benefit of an isolated commune, a quantity of grain or flour, [wares] that otherwise would have been part of the [country's] general supply." Perhaps, the minister implied, the time had come to reconsider the utility of the supply requirements. Despite those sentiments, the minister nonetheless insisted that bakers maintain the provisions. He could not, of his own accord, undo the imperial decrees, the royal ordinances, and the other legislation that had shaped the baking trades for almost forty years.[72]

In the meantime, the Chamber of Deputies had created a commission to examine the question of baking trade regulations. In response, bakers began establishing national contacts and discussing the need for some of the requirements to be lifted. As part of the Chamber's investigation, it allowed a group of bakers from Strasbourg to address the deputies in March 1843 on that subject.[73] The commission announced its findings on 18 March 1844. The statutes and legislation needed to be reestablished and enforced, it concluded. The order, legality, and stability that regulations guaranteed were essential to the smooth functioning of the trade. More significantly, it feared the suffering that might ensue if bakers were allowed to deviate from their supply requirements, a fear that the shortages of 1846–47 bore out.[74] Thus, on the eve of the crises of 1846–48, the loosening of controls that had developed in the early 1840s had come under sharp scrutiny, and legislation was pending. Bakers, mayors, and prefects from across the country awaited the outcome, and the flow of resulting ministerial directives, that would end the contradictions and uncertainty that had arisen in the practices since 1840.

They waited in vain, however, for the July Monarchy's Chamber of Deputies to settle "this *vive polémique*," as the prefect of the Eure described it. The July Monarchy collapsed before any reforms emerged from the Chambers. No doubt, the shortages and unrest of 1846–48 did not allow any ambitious reconsideration of policies, and the most the government seemed able to do was distribute a massive number of ration cards to the poor of Paris and the surrounding Department of the Seine.[75]

[72] Quoted at length from a 22 July 1844 letter from the Minister of Agriculture and Commerce to the prefect of the Eure. Prefect of the Eure to the Minister of Agriculture and Commerce, 20 September 1849, AN F7 2757.

[73] A 20 February 1850 letter from the prefect of the Eure to the Minister of Agriculture and Commerce describes the bakers' 1843 political activities briefly. AN F11 2757.

[74] Prefect of the Eure to the Minister of Agriculture and Commerce, 22 December 1848, AN F11 2757.

[75] See the reflections of one administrator in the *arrondissement* of Saint Denis, who wrote in 1853 that, wisely, the 1847 system of ration cards had extended help to the rest of the Seine department. He was arguing for a similar extension in 1853, which was eventually granted. Commune of Boulogne, 4 October 1853, AN F11 2799. On political conflicts in

Strenghtening Regulations After 1848

After the Revolution of 1848, however, both the Republic and the Second Empire hastened to reinstitute supply requirements in any city where they had lapsed. Because the harvests of 1848 and 1849 were relatively good and grain was cheap, the Ministry of Agriculture and Commerce was unyielding in its demands that bakers get their supplies back up to the assigned minimum amounts in 1849 and 1850. The minister may also have believed that some of the hardships of 1846–47 were caused by the lack of adequate back-up supplies, although his letters make no such connection. He was, however, utterly committed to holding bakers to the requirements of their statutes and several decades of legislation. He waged a dogged campaign to force bakers to observe all the former regulations.[76] Could the town of Nogent-le-Rotrou release its bakers from some of its requirements, given that so many of its townspeople baked their own bread, the prefect of the Eure-et-Loir inquired. The minister refused to allow any deviation, and recommended that the prefect's next reports "be much improved."[77] The municipal council of Evreux had determined that a one-month supply of flour would suffice, but the minister would have none of it. The royal ordinances had long required a larger supply, and the bakers of Evreux would have to move quickly, he informed the prefect.[78] And what if the mayor of La Rochelle believed that his bakers could not afford to amass the supplies, and therefore wanted an exemption for his city? the prefect of the Charente-Inférieure was asked. There could be no exceptions, the minister replied. If he had ordered the supplies to be reestablished, it was not merely as an issue of "local provisioning," but of "the complete execution of a general measure whose utility seems . . . to be incontestable." Public order, a respect for the law, and the wisdom of buying ahead when prices were low – all of them demanded immediate compliance with the minister's orders.[79] Thus, while the policies of the July Monarchy presented many contradictions – combining the fine-tuning and loosening of controls on the baking trades – the Second Republic and then the Second Empire moved unflinchingly

Rouen, and especially the disturbances of the July Monarchy and the Revolution of 1848, see Ronald Aminzade, *Ballots and Barricades: Class Formation and Republican Politics in France, 1830–1871* (Princeton, 1993), pp. 174–91.

[76] The correspondence in AN F11 2757 follows those efforts.

[77] Minister of Agriculture and Commerce to the prefect of the Eure-et-Loir, 7 January 1850, AN F11 2757.

[78] Minister of Agriculture and Commerce to the prefect of the Eure, 25 April 1850, AN F11 2757.

[79] Minister of Agriculture and Commerce to the prefect of the Charente-Inférieure, 2 September 1850, AN F11 2757.

toward restoring the back-up supplies that bakers everywhere were supposed to have.

The prefects complied, albeit only reluctantly in some cases, and by 1852 and 1853, they reported that the bakers had once again been brought to order. Ready supplies had been pulled back together. The emergency supplies could not have been more timely. By 1852, the first suspicions that the harvest was short became known. Within a year, the worst fears of the administrators would be realized as they scanned table after table of prices that moved ever upward, and contemplated sheaf after sheaf of reports attesting to the anguish and disorder caused by a second poor harvest.

Energetic Intervention: The *Caisse de service de la boulangerie* and the Crisis of 1853–55

As the crisis unfolded, administrators cast off the uncertainty of the 1840s, and moved swiftly to put their expertise and knowledge to use. Several decades of reports gave the Ministries of the Interior and of Agriculture and Commerce ready information on local practices and viable strategies. Their commitment to maintaining calm in Paris was unshakable. Northern France also deserved particular attention. As early as 1851, the Ministers of the Interior and of Agriculture, Commerce, and Public Works, along with the Paris prefect of police and the prefect of the Seine, monitored flour sales across Europe. They followed commercial activities in the Baltic and in Great Britain, noting any unusually high prices or large purchases with apprehension, and then seeking out explanations.[80] Publicly, they admitted to no worries about the French food supply. Finally, after almost two years of observation and planning, the administrators acted.

In late September 1853, as bread prices exceeded the equivalent of 1 franc per 2-kilogram loaf of first-class bread – the price that Napoleon and then the Restoration had judged hazardous to public order – the Paris Municipal Council imposed a price ceiling of 40 centimes per kilogram on first-class bread and 32 centimes on second-class loaves.[81] The Council, along with the Prefect of Police, expected the bakers to use their back-up supplies to cover any losses in the coming weeks. Indeed, the announcement of the artificially low price ceiling initially caused the bakers to slow their purchases, and they appeared to be waiting out the period of high prices. Millers, however, only increased the prices they wanted, and by November 1853, the city and departmental authorities had concluded that some very bad months lay ahead. Every calculation yielded one sure fact:

[80] See the daily correspondence in AN F11 2802.

[81] The following account is drawn from AN F11 2796, 2799, and 2801–3.

Even if the bakers depleted their entire back-up supply, there would not be enough flour to get the capital through to the next summer's harvest.[82] Also worrisome was the knowledge that bakers could not be expected to shoulder the losses that came with the bread price ceilings. If the city were going to continue to hold first-class bread prices at 40 centimes per pound – which, given the violence that might result if it did not, was the only sensible choice – either the city or the Seine Department would have to fund a system of reimbursements. But neither institution had the money to do so. Thus the municipal and departmental authorities found themselves in the same situation that their Napoleonic and Restoration predecessors had faced repeatedly. Fortunately, the bakers' back-up supplies had been brought up to the required levels before the crisis was under way, which gave the authorities several weeks of leeway in the late autumn of 1853.

Lengthy meetings, ministerial directives, and reams of calculations filled those weeks. Finally, in late December, an ambitious system was unveiled – the *Caisse de service de la boulangerie de Paris*.[83] The *Caisse*, not unlike its 1817–1830 predecessor with a similar name, had been developed to meet a broad range of related provisioning needs, most importantly to subsidize bread prices as economically as possible. Unlike the former *Caisse*, however, the 1853 establishment had no grain of its own to dispense and hence no means of shaping prices or funding its outlays. Thus the ministerial, departmental, and municipal authorities had come up with an ingenious plan to carry the city through the crisis. The 1853 *Caisse* would act as an intermediary between the bakers and their suppliers. Every time a baker arranged to purchase flour, he or she would report the agreement to the *Caisse*, which would pay the supplier on the baker's behalf. In that way, grain and flour prices could continue to fluctuate according to supply and demand – the government would be taking no measures to hold down grain or flour prices. Thus, millers and flour brokers could go on supplying the capital without fear of losses. The bakers would eventually have to repay the *Caisse* for the supplies, but the *Caisse* would offer them lines of credit at 5 percent interest, and would use their back-up supplies as the guarantee for those loans. The bakers would receive *mandats*, or certificates, from the *Caisse*, and would use them to pay their suppliers. The suppliers would turn the *mandats* back to the *Caisse* for payment in full. In the meantime, the *Caisse* would keep track of the difference between the low price at which bread had been set and the higher prices of the bread *tarifs* in effect across the department. A new municipal *octroi* levied on grain entering the capital would subsidize the margin between the two – and it appears that the 5

[82] See the figures and discussion in AN F11 2801.

[83] "Decret . . . qui institue une caisse de service pour la boulangerie de Paris," 27 December 1853, *Bulletin de Lois*.

percent interest that the bakers paid on their lines of credit was intended to help cover those subsidies, at least to an extent.

Once flour prices started to decline, the prefect of police would hold bread prices artificially high. The bakers would step up the repayment of any outstanding balances on their lines of credit, and eventually, the crisis would run its course. Merchants and millers could not complain, for no one was going to meddle with their prices, and there would be no government wares to compete with theirs. Bakers, too, would be protected from losses through the lines of credit and the subsidies. Bakers could also continue to produce the finest loaves – the *pain de luxe* – and sell them at whatever price they chose. Those sales, generally to the wealthiest customers, would keep at least one portion of the department's population tied solely to market forces. Thus, without entering the grain trade on its own, the *Caisse* established a means to tide bakers over. Even better, it was doing so without interfering with commercial flour and grain prices. By 1854, it had extended the *mandats* to the bakers of the entire Seine department, and begun plans to place Parisian and suburban bakers under the same statutes. The shortages lasted for nearly three years, and grain and flour prices did not come down until 1856. During that period, bread prices remained fixed The *Caisse* remained in operation until the 1861, and in that nine-year period, dispensed over 500 million francs of *mandats* to the bakers in the department.[84]

An important consequence of these measures was the forceful resumption of efforts to limit the number of bakers in cities, and to hold them tightly to the back-up supply requirements. While the government in previous shortages had been reluctant to order any great changes for fear of the disruption they might provoke, it showed no such hesitation in the 1850s. Reviewing the policies of the last half of the July Monarchy, and especially the loosening of the controls on the bakers, the authorities of 1853–55 determined that only by reinstituting the controls in their entirety would they provide appropriate and economical safeguards for their cities. In an 1854 report, the Subsistence Commission of the Paris Prefecture of Police announced that it would combine the Parisian corps of bakers with those of the rest of the department. Because the entire corps of bakers would benefit from any subsidies – and it was anticipated that their bread prices would be the same – the bakers outside the city limits would have to begin stocking more substantial private supplies. They could not, as the commission pointed out, continue to profit from Parisian subsidies without bearing the same burdens as their Parisian counterparts. And, returning to the issue of the back-up supplies, the commission explained that the past decades of experience had shown one irrefutable fact. While many systems had been

[84] See the daily tabulations in AN F11 2800.

tried, and few had been fully successful, "the only workable means that has presented itself, and that has been used to a certain extent, is the creation of emergency supplies in the hands of the bakers." Instead of allowing some of those supplies to decrease during the July Monarchy, administrators should have held the bakers to their responsibilities. Yet if the bakers were going to have to find funds for those purchases, even in the midst of this crisis, the government was going to have to encourage departments to place firm limits on the number of bakers each city could have. The two conditions – profits to be used for back-up supplies, and limits on the number of bakers plying their trade – had to go together. These constraints had been neglected in the 1840s, and the time had come to understand that there could be no stability in the baking trades without them. Furthermore, such measures were useful for the entire country.[85]

Thus, in Paris, Rouen, and elsewhere, another energetic round of buy-outs was initiated. Solvent bakers who wanted to continue their trade contributed to municipal funds that were used to purchase and close insolvent shops. More precise calculations offered better estimates of the size of the clientele a baker needed to remain financially secure. By 1855, the prefect of the Seine-Inférieure had set such buy-outs in motion in Rouen, Fécamp, Dieppe, and Eu. The mayor of Fécamp applauded the decision. Keeping bread prices at affordable levels "is of such a high degree of interest to the well-being of the people and the security of the country," he wrote, that the number of bakers had to be decreased and the back-up supplies reorganized (although he did want the state to help pay for the supplies).[86] In Rouen, the reductions were quite stark: 35 of its 115 bakeries would be closed. The 80 bakers chosen to remain in business would have to make annual payments of 200 francs for 15 years in order to cover the estimated 245,000 francs the closures would cost. In Fécamp, the reductions were even more severe: 12 of its 24 bakers were forced out of business.[87]

By 1858, Napoleon III had ordered bakers to put together massive private supplies of three months of flour, and local authorities were pressuring their bakers to comply.[88] The following years, the Emperor added to the Parisian bakers' charge: They would have to handle all the costs of storing their reserves. They were to rent the hall, pay a guard, and insure the grain.[89] Taken together, these requirements and the pattern of buying

[85] "Réorganisation de la boulangerie de la Seine," Préfecture de Police, Commission des Subsistances, 1854. ADSM 6M 1299.

[86] Mayor of Fécamp to the subprefect of Le Havre, 7 October 1854, ADSM 6M 1299.

[87] Draft, prefect to the Minister of Agriculture, [1855], ADSM 6M 1297.

[88] Copy, Ministry of Agriculture to the prefect of the S-I, 22 November 1858, AM Le Havre FM F^4 8.

[89] November 16, 1859, AD Eure 6M 763.

and selling created the long-needed urban emergency supplies. They had the added benefit of forcing the bakers to rely on commercial networks. The state had relinquished its direct control and encouraged the grain and flour trades, while still ensuring the safety of French communities.

The provisioning policies of the early nineteenth century suggest the steady, if not always successful, development of safety nets for urban inhabitants through the government's moderate intervention. The experience of these years persuaded administrators of the need for control, while thoroughly convincing them to reject any extreme measures. If the Jacobin Maximums of 1793–94 had not taught them the risks involved, Napoleon's hasty and unworkable May Decrees were the final lesson. After the searing 1812 experience of rural bloodshed, commercial chaos, and administrative resistance, no minister could argue for a return to radical policies. Instead, the state went underground, formulating routines and structures to guide the grain trade and to reinforce urban supplies. The moderate conclusions of the Subsistence Council of 1812 laid the groundwork for the Restoration's more extensive yet even more standardized practices. Some of the measures, such as the "third-party sales" of Reserve grain, were meant, at least at the outset, to remain covert. Other measures relied on giving the impression of standard practices, such as the routine setting of bread prices, the regular arrival of grain shipments at warehouses, the disciplining of bakers, and the spectacle of calm administrators simply carrying out their duties, quietly warding off trouble by approved and increasingly well-worn methods. The problem of supplies had been resolved also. By first building up the Reserve, and then pressuring bakers to amass their own stocks, the state had created a necessary buffer against shortages. It had transferred that responsibility gradually to the bakers. It had done so not only through complicated ordinances and requirements, but by slowly buying out insolvent bakers and stabilizing profits. Bakers still protested the *taxe*, and claimed they were impoverished, but in good years, they sometimes exceeded their necessary stocks.[90] The state had guided and pressured them and forced them at times to rely on commercial sources.

[90] In 1845, the Paris baker's reserves held 478,475 metric quintals, more than needed. February 1845, F11 2801.

CHAPTER ELEVEN

The Market Mastered

The stability of the French state depended on grain. This had been true in the eighteenth century, and remained so for the nineteenth. The word stability must be used with some caution. The nineteenth-century state had not been able to resolve its political fragmentation. Those fissures seemed only to deepen through the revolutions of 1830 and 1848. Industrialization would unleash further threats to fragile regimes. The apparent silence of the Second Empire was deceptive, and led to the Commune of 1871. Yet, by mid-century, the list of seemingly intractable problems confronting any minister no longer included provisioning. To a reassuring extent, the market had been mastered. This did not mean that the food supply was no longer the object of constant concern and surveillance. It did mean that when shortages loomed, there were well-honed measures to use and sufficient supplies to draw from granaries. Bakery shelves would not be empty, and few markets would feel the justified wrath of hungry crowds. Administrators could monitor their markets from their desks rather than on horseback.

There had been some setbacks under the July Monarchy, when supplies and regulations had been relaxed, and they may have contributed to the misfortunes of 1846–47. The Second Republic and then the Second Empire, however, had regained the lost ground. The *Caisse de service de la boulangerie de Paris* had seen the entire department through the crisis. Despite the gravity and duration of those shortages, commercial networks had not failed. The lines of credit that the *Caisse* extended to the bakers, along with modest bread subsidies, held the ties between bakers and commercial suppliers together. Moreover, since most of the outlays from the *Caisse* would be covered when the bakers repaid the advances on their lines of credit, the entire operation had essentially paid for itself – or, more accurately, the bakers had paid for it out of their own pockets at 5 percent interest. Nonetheless, as they repaid the *Caisse* for the credit it had extended them, they benefited from the increased stability that the subsequent closure of insolvent bakeries brought to their trade. Once stabilized, the bakers' own resources would serve as the emergency supplies every city needed. Credit lines, strict limitations on the number of bakers in any city, and the enforcement of the regulations for back-up supplies – these were the tools of the mid-nineteenth century administrator.

Numerous authorities had lamented the insufficiency and inconstancy of the grain trade. Representative Le Quinio, when he addressed the Convention in 1792, had wished to make the grain trade honorable, and if not honorable, safe. Yet, merchants had continued to flee the trade when they were most needed. Their legitimate fears of violence caused them to halt their purchases and bolt their granary doors at the first whisper of shortage. They even refused to be known as grain merchants. The Napoleonic attempts to establish lists of those dealing in grain brought only the constant reply: There are few in our department who undertake this form of trade. Since they tended to abandon markets during shortages, and provided only the bare minimum of supplies even in better years, the state stepped in. Gradually, the administrators began to consider ways in which their efforts could be used to draw merchants into the trade and to stabilize supply lines.

Their efforts were born of exigencies and constraints. They were forced to seek out supplies and distribute them when circumstances were worst. They had to do so without damaging the ever-fragile commercial connections that the state wished to promote. The authorities moved to the background, and sought methods that were clandestine, simply routine, or both. Moreover, the state provided a structure for easing grain along pathways toward markets. It weighed its efforts against the disruption they caused to merchants and producers. Gradually, the administrators took the imperfect trade of the present and generated the stable networks of the future. Through simulated sales, Reserve shipments, and increasing pressure on bakers, the state guided the provisioning trades toward reliability. Ultimately, it created substantial emergency supplies and then transferred them from its own hands (and treasuries) to those of the bakers and merchants. It had succeeded in creating a self-sustaining form of free trade in grain. This form of free trade did not leave bread prices to chance, at least not before 1863. Nor did it permit bakers to buy, sell, and stock grain as they pleased. But it did guarantee most bakers a modest security and fairly stable profits over the long haul. In short, the nineteenth-century state had traced and smoothed the paths that led to stability.

Repressing Riots

The stabilization of the provisioning trades was one part of a larger picture of growing state power. Certainly, troops, musket-fire, arrest warrants, and prison sentences played a role in quelling disturbances. While this book has focused on the state's efforts to assure urban food supplies by strengthening commercial networks, those networks would not have survived without the presence of soldiers willing to round up rioters and courts willing to prosecute them. The full story of the nineteenth-century repression of food riots has not yet been written. Some preliminary conclusions based on the events of 1817 may nonetheless be sketched out briefly.[1] Until 1817, a crucial problem was how to handle the extreme, but all too familiar, matter of market riots. The national guard had been unreliable in emergencies, and was altogether too likely to sympathize with buyers. By 1817, once soldiers had returned home, the Restoration was able to reconsider the problem of confronting riots. The disturbances of that winter had been acute, and the ministries turned in earnest to the question of public order. In June 1817, the Minister of Police issued a *mémoire* on the responsibilities of prefects and subprefects to handle such disturbances. They were to call out the *troupes de ligne*, the national guard, and the *gendarmerie*, "who have arms, who [can] impose [order]." If that failed, as the Minister feared it might, the prefects' only duty after summoning the crowd to disband, was "to meet force with force." "No hesitation, no weakness," he commanded. These efforts had to be carried out with energy, and the most zealous soldiers would be rewarded. Far better to let a full-scale riot break out in reaction to a show of force than to give in to the crowd in any way, he explained. If the *fonctionnaires* did not have enough troops at their disposal, they were to continue to observe the crowd's actions and to take careful note of the instigators. What ground the administrators initially lost in the marketplace would be regained in the courtroom. To let the townspeople know in advance, the prefects were to reprint and post laws from the Directory that laid down severe fines for any community that pillaged grain.[2]

These were harsh measures. During the Old Regime and the Revolution, too often the most local officials could do was to summon the crowd to desist and then make the distribution of supplies as orderly as possible. When they could, they scanned the crowd for familiar faces to include in the *procès verbaux* they wrote up when everything was over. Rarely did they have troops at hand. There were cases in which the Minister's 1817 instructions were honored. After a riot in Sens, three people were executed by firing squad, inspiring (in the Minister's words) "a salutary terror." He

[1] Bouton's contribution to *The Politics of Provisioning in Britain, France and Germany, 1690s–1850*, forthcoming, will provide an essential analysis of the nature of repression in France.

[2] Circulaire of June 1817, AN F7 6690. Many of the merchants who had supplied grain imports had been urging such policies for several months. Delorme in the Auray and the Widow Le Couteulx had expended much ink demanding armed escorts for their wagons and barges. 17 December 1816 and 16 January 1817, AN F11 301.

suggested similar sentences for a messy disurbance involving 700 or 800 people in Ervy in the Aube.[3] The tenor of those episodes may have been more violent than many Old Regime *attroupements*, and perhaps the presence of troops aggravated any disturbances that escalated to full-blown riots. When the people of Ervy saw the *gendarmes*, for instance, they screamed that they would "kill or be killed," and if the *gendarmes* used their sabres, the mayor and the *juge de paix* would be the first to be hanged.[4] In the Calvados, the confrontations appear to have been less bloody. Soldiers arrested two men who were trying to raise grain prices at one market, and the gendarmerie was present throughout the department, but did not need to fire its rifles or bare its lances.[5] These episodes suggest that the Restoration's markets were more heavily policed, and that those forces included armed troops ready to respond to any threat.[6] Under the July Monarchy, the authorities in Paris continued to press prefects and subprefects to organize the national guard and *troupes de ligne* in advance of market disturbances. In fact, the presence of the military essentially turned these matters into questions of public order and criminality. In 1839, the subprefect of Pont-Audemer, for instance, was advised to stay away from the marketplace and to allow the troops to do their job. Except in "grave cases," he was to remain in his office.[7]

The trials of the 1817 food rioters yield a sense that these disturbances, while troubling, were no longer viewed as containing politically subversive undertones. The Old Regime and Revolutionary equation of shortage, legitimate grievances against the state, and eventually revolution was slowly being dissolved. In the 1820s, when the appeals of many of the 1818 sentences were being heard, the immediate threat of shortages had diminished. The appeals dossiers of the Ministry of Justice include petition after petition from local authorities pleading for pardons. The *procureurs* who had prosecuted the 1817 cases now believed that the guilty should be freed. Guilbert Pingué, sentenced to ten years of hard labor for a violent, premeditated attack on a grain merchant in Riom, merited a pardon because he had only acted on his belief that merchants were draining his community of necessary supplies. At the time, the *procureur* explained, they needed "an example of severity." Now they could risk indulgence for what they presented as logical actions on the part of anxious buyers.[8] A similar plea was lodged for Jean Tavenne, who, armed with a hatchet, had attacked two

[3] 11 June 1817, AN F11 723.

[4] For the correspondence surrounding the Ervy riots, see 4 June 1817, AN F11 723.

[5] 8 July 1817 AN F11 723.

[6] See the numerous references to the presence of the gendarmerie and the military in riots between 1829 and 1832. AN BB24 64–84, 85–99, 116–135 and 136–154, AN F7 6690.

[7] Prefect Eure to Subprefect Pont-Audemer, 23 October 1839, AD Eure 6M 701.

[8] 5 March 1825, AN BB24 18–33, Dossier 2290. The request was granted.

wagons going to Orléans. While the crime, in the *procureur*'s estimate, was "*bien grave*," nonetheless the guilty had only been afraid of "being reduced to the most extreme misery." It was time to forgive the guilty and release them from their chains.[9] The list of excuses grew, but contained the now-acceptable claims for release – drunkeness, fear, and starvation. These all figured prominently in the requests for pardons, and were consistently honored.[10] The howling of marketplace crowds did not constitute the *cris séditieux* of nineteenth-century political culture. Of course, it was in the interest of the petitioners to hide any political dimensions to their actions.[11] Calls for "bread at two sous," while regarded with sympathy and anxiety, no longer evoked the specter of revolution. Once the riot had been repressed, and a few examples made, the state could return to the business of reshaping the grain trade.

The state's steady efforts to shape the trade eventually yielded stable grain supplies based on commercial networks. This was especially true for cities. Where merchants had been unable to create sure mechanisms for the purchase and transport of supplies to metropolitan areas, the state had provided a framework. Once it had strengthened that framework, the state was able to remove itself. It had created and shored up the links between producers, merchant-millers, bakers, and consumers. It had provided special subsidies for the poor – ration cards, workhouses, and *soupes à la Rumpfort* – so that commercial prices would not feel the stark pressure of working-class demands. It had established public acceptance of the fact that bread prices generally would follow the market price of grain, and that most consumers would pay those rates.

The 1863 *Liberté* of the Baking Trade

The final steps in the liberalization of the grain trade occurred in 1863. The measures came at the end of a decade of falling barriers to international free trade, increasing demands for (1) more thorough liberalization of the economy, and (2) the suspension of the sliding scale of import duties on grain.[12] (The sliding scale had been established in 1822, although it was often lifted when shortages threatened, such as from 1853 to 1859.) On 22

[9] AN BB24 18–33, Dossier 2919.

[10] See the many dossiers in BB24 18–33 and 64–84.

[11] On the forms of popular political resistance, in particular, *cris séditieux*, see Sheryl Tracy Kroen, "The cultural politics of revolution and counterrevolution in France, 1815–1830," Doctoral Dissertation (University of California at Berkeley, 1992).

[12] Michael Stephen Smith, *Tariff Reform in France, 1860–1900: The Politics of Economic Interest* (Ithaca and London, 1980), pp. 18–25.

June 1863, Napoleon III announced that the following 1 September, most of the regulations on the baking trades would be annulled. The decree lifted limits on the number of bakers who could enter the trade in any given town, and suspended the requirements for back-up supplies. The only areas of the trade that would still come under the supervision of local authorities were the cleanliness and purity of the bread, and the accuracy of the scales in bakeries. The decree specifically did not mention bread price controls, and instead, through its silence, appeared to sanction the continuation of the practice.[13] Later in the summer of 1863, the Paris Municipal Council pressed successfully for some exceptions for the capital. To replace the surcharge that the 1853 *Caisse* had levied to fund its operations, the city could set other fees on cereals arriving in the capital. Furthermore, if bread prices exceeded 50 centimes per kilogram, the *Caisse* would resume its operations, as long as the Minister of Agriculture, Commerce and Public Works believed it was necessary.[14]

Despite the stark change that the 1863 decrees seemed to usher in, the actual effect of the policies was more muted. Municipal authorities everywhere continued to set bread prices, and many bakers continued to maintain their back-up supplies. If regulation had disappeared in fact, one 1899 critic – a Beauvais lawyer – complained, it had remained "in spirit." More than thirty-five years later, the city of Beauvais, like many others, continued to set an unofficial price on bread and its customers expected bakers to comply.[15] The Minister of Agriculture, Commerce and Public Works had tried to sort out any confusion about the *taxe du pain* in 1863, but clearly to no avail. His 22 August and 10 November circulaires had explained in puzzling terms how the "provisional lifting of the *taxe*" was to work. First, local officials were to continue to calculate the bread *taxe* according to their regular practices. But, he cautioned, the government only wanted that price to be "*officieuse*" – "an internal method of record-keeping (*contrôle*) for the administration." Local officials were not to make the price public, because buyers would expect it to be enforced.[16] Whether the *taxe* became public or not, and whether it was enforced or not, bakers throughout France continued to observe the *taxe* and sold their bread in the loaves familiar to their customers. The appearance of routine and fairness that the *taxe* had come to embody was deemed an essential element of stable urban provi-

[13] "Décret impérial qui abroge diverses dispositions de décrets, ordonnances ou règlements généraux concernant la boulangerie," 22 June 1863, *Bulletin des Lois*.

[14] "Décret relatif à la boulangerie de Paris," 31 August 1863, *Bulletin des Lois*.

[15] Antonin Lefort, *La boulangerie et le décret du 22 juin 1863 proclamant sa liberté* (Paris and Beauvais, 1899), pp. 14, 17–31.

[16] Lefort, *La boulangerie*, pp. 14–15.

sioning.[17] Thus, even in the absence of formal controls, the regulatory practices lived on.

The implications for a historian of France are that the state's halting development can be viewed from the chambers of ministries, prefectures, subprefectures, and mayors just as validly as from the barricades of 1830 and 1848. Certainly, the state's political development was uneven. Many of the tensions that had been aggravated or provoked by the Revolution, and then by industrialization, could not be resolved. The legacy of the Revolution was especially burdensome. But one dimension of those problems – provisioning – had been brought under control. The skills of nearly two centuries of administrators, and especially of the military suppliers, had succeeded in creating a moderate form of free trade. Toiling at their desks, sometimes riding out to confront crowds, and their own lives threatened on occasion, they had removed provisioning from the roster of possible threats to their regimes.

The Broader Context: State and Economy

Further evidence of the state's intermittent effects to structure and shape the economy may be found in other areas. The French government's role in the building of railroads parallels that of the provisioning policies. When commercial companies judged railroads to be too speculative in the 1830s, the government provided enormous subsidies and other aid to get projects started. Once private companies had been reassured by the government's measures, they rushed in to lay kilometer after kilometer of track in the 1840s.[18] Jean-Pierre Hirsch's comprehensive and ground-breaking work on Lille merchants reveals their sometimes contradictory positions regarding free trade, arguing to be permitted to pursue profits, yet consistently requesting institutional protection in those endeavors.[19] Recent work on commercial law, while drawing a dichotomy between free trade and intervention, maintains that merchants and administrators "looked backward to a limited conception of market enterprise more than forward to an expansive one." Here, too, there is documentation of the state's creation of institutions that would allow commercial stability and growth.[20] Still other

[17] In the departments of the Eure and the Seine-Inférieure, bakers seem to have kept their prices at a few centimes above the former *taxe*. Only rarely did bakers exceed those accepted limits. See the reports between 1863 and 1880 in ADSM 6M 1308–9.

[18] Pinkney, *Decisive Years in France*, pp. 34–60.

[19] *Les deux rêves du Commerce*.

[20] James Robert Munson, "Businessmen, Business Conduct and the Civic Organization of

aspects of the state reveal a reliance on the clandestine, or at least routinized, methods that were preferred in provisioning. The Napoleonic shift from direct to indirect taxation, for instance, made the fiscal weight of the regime less apparent, in particular to the land-owning elites that supported it. By 1815, over half of the state's revenues came from indirect, rather than direct, impositions.[21] Further afield, but wholly consistent with this argument, the publishing industry also submitted to increased controls and standardization in these years.[22] Until the workings of the early nineteenth-century regimes have been studied more thoroughly, an overall model for the state's development can only be suggested. But it is clear that one of the main elements of that model is the gradual development of skills and methods that made possible a sustainable form of free trade, especially that of foodstuffs.

I have argued throughout this book that the gradual refinement of the controls – such as the *taxe* that persisted even after its abrogation – on the grain and baking trades was essential to the stability of urban provisioning. Yet there were critics in the nineteenth century who objected to any form of regulation, and who had long demanded the complete liberalization of the provisioning trades. They blamed the government's interference for France's lackluster economic performance – and their comparisons with England almost always showed France to be well behind its rival across the Channel.

How does one address their challenges? What would have happened if many generations of ministers, intendants, prefects, and mayors had not stepped in to guide the provisioning trades toward a sustainable form of free trade? It is always hard to play the historical game of "what if." Comparisons, whether between France and England or further afield, often falter because no two countries offer the same resources or have the same needs. Would France have looked more like England if the likes of Desmarests, Miromesnil, Pasquier, and Decazes had abandoned the market to the forces of supply and demand? It is doubtful that a swifter and more complete adoption of free trade could have overcome harvest shortfalls, insufficient transportation networks, and inadequate sources of credit. In the long run, perhaps the market would have supplied those goods. In the short run, it could not. And where food was concerned, it was the short run that mattered.

Commercial Life under the Directory and Napoleon," Doctoral Dissertation, Columbia University, 1992.

[21] Hincker gives a brief overview of Napoleonic tax policies in *La Révolution française et l'économie*, pp. 197–9.

[22] Carla Hesse, *Publishing and Cultural Politics in Revolutionary Paris, 1789–1810* (Berkeley, 1991), pp. 205–39.

Did the policies that privileged the short run over the long run leave any legacy, beyond that of the more stable provisioning practices of the mid-nineteenth century? It is entirely possible that the state's intervention did in fact reinforce the French people's reliance on wheat and other cereals. Potatoes – except in the far north – and rice generally failed to satisfy French consumers, even in the worst crises. Only the prospect of starvation could convince the Rouennais in 1812 and 1817 to sample loaves of bread made with potato flour. Rice sacks were greeted with hostility and contempt, although their contents were eventually consumed. The state conceded to that hostility, and found sufficient imported grain – albeit at great cost – to feed its famished cities.

By the nineteenth century, the effect of the state's intervention could be found in the whiter loaves on bakery shelves. Even the lower-quality loaves had improved. Administrators frequently noted that the bread prices of the nineteenth century could not be compared easily with those of the eighteenth century. The poor third-class loaf of the 1770s had vanished by 1817. In general, every category of bread had jumped up a notch. The nineteenth-century second-quality loaf was essentially the equivalent of the best loaves of the eighteenth century. The typical first-qualities loaves of the nineteenth century surpassed all but the finest wheaten loaves of the period before the Revolution. Improved milling technologies and increased expectations – both reinforced by the government's provisioning policies – contributed to the French dependence on grain. That dependence may have delayed any shift in diet or agricultural production toward a wider and more productive array of other crops. Thus the government's policies perhaps had hidden costs. Not only did the treasury suffer from the demands of French consumers for bread, but French agriculture may have also paid the price of those demands.

The legacy of the French government's ability to shape the grain trade can be felt each time a visitor to France gets off the plane in Paris and heads immediately for a neighborhood bakery. Strolling back into the crowded street, tearing off the end of a baguette, the visitor sighs with contentment. France and baguettes – the two are ever united. Perhaps it would not have been so if Pasquier, Decazes, and all their *confrères* had not labored so hard and so successfully all those years ago to assure the bread supply to France's urban population.

There is a certain irony in saying that the state created free trade, yet that is what had happened. The Physiocrats had envisioned no significant role for the state in such a process, save that of lifting the endless list of controls and removing its meddlesome servants from the marketplace. Turgot, viewing rising prices in 1774, had argued that no intervention was appropriate. High prices were "the only possible remedy for scarcity." The country would have to wait out the shortages.[23] The National Assembly had taken much the same tack by ordering free trade in August 1789. Even after the riots of the winter and spring of 1817, the Minister of the Interior insisted that free trade would be the only long-range solution to the kingdom's provisioning problems.[24] Yet, by then it was well recognized that an adequate grain trade would not be generated overnight and that the consequences of standing aside were too serious. Thus, where the market had failed, the state eventually succeeded. A fundamental part of the legacy of the food riots of the Old Regime and the political turmoil of the Revolution can be found in the French state's creation of a moderate free trade in the nineteenth century. Free trade in grain was not the result of the market, but of the state's intervention.

[23] "Lettres patentes sur arrêt du conseil," 2 November 1774, ADSM C 103; Turgot, *Oeuvres*, 3:102–10.

[24] 15 May 1817, Minister of the Interior to the Prefect of Calvados, AN F11 723.

Archival Sources

Archives Nationales (France)

AD I 23A
AD XI 9, 14, 23(A), 68
AD+ 856–9, 917, 1260
AFII 44, 101, 149, 177
AFIII 607
3AQ 351
69AQ 1, 3
BB3 119
BB18 297, 806, 807, 830, 1166, 1167, 1170–1173, 1375
BB24 8–99, 116–154, 170–218, 286–368, 409–418, 448–456, 478–488, 494–506
C 356, 536
DXXIX bis 1, 2, 25, 26
DXL 1, 2, 4, 7, 15, 27
DXLI 2
D§I 10, 17, 18, 20
E 2445, 2462, 2512, 2513, 2515, 2523
F^{1c} III Eure 7
F^{1c} III Seine-Inférieure 1, 8, 9
F^{1c} V Eure 1, 2
F^{1c} V Seine-Inférieure 1, 2, 3, 4
F4* 306
F4 1028, 1032, 2096, 2153
F7 3022, 3619, 3624–3626, 3634, 3638–3640, 3701, 3905, 4183, 6690, 6691, 7090, 7092, 8730, 9786, 9787, 9869A, 9888, 9889A
F10 1A, 203A, 242, 291, 339–340
F11* 1–5, 50, 1759, 1766, 1768, 3049–3068
F11 203, 208, 209, 213, 220, 223, 225–226, 229, 231, 242–243, 252, 264, 265, 278A, 300, 301, 306, 308, 312(A), 322, 324, 325, 338, 390,

402, 403, 435–436, 446, 447, 450, 453, 457, 463, 464, 465, 480–481, 514B, 515A, 520, 533, 534A, 535, 562–563, 565, 592, 593, 628, 629, 709, 715, 717–718, 719–720, 723, 733, 741, 743, 750, 1178, 1179, 1191–1192, 1200, 1235, 1356, 1363, 1373A, 1385, 1403, 1504A, 1512, 1548, 1549, 1578, 1594, 1604, 1605, 1606, 2733, 2734, 2739, 2740, 2753–2755, 2757, 2758, 2770, 2796, 2798–2808, 2810, 2813

F12* 6

F12 560, 655A, 936B, 937, 1240, 1251C, 1240, 1269A, 1269B, 1539, 1544(28), 1544(41), 1547B–C, 1966, 2045B, 2056, 4551

F20 101, 256, 282[11] 293, 296, 560, 564, 567, 570, 616, 715

G7 15–6, 1708, 1872, 1903

T 1157

U 1441

X^{ia} 8033

Y 9487, 9499, 9648, 10558, 12827, 15405, 17058

Bibliothèque Nationale (Paris)

Manuscrits français 6877, 8127, 11347, 11348, 21635, 21638, 21642

Collection Joly de Fleury 1095, 1107, 1111, 1159–1165, 1168, 1428, 1742, 1743

Collection Le Senne

28 (5)

Z 30

4Z 2263

8Z 4275, 14259

Archives du Calvados

1B 1665A–B, 1666A, 2060, 2063, 2064, 2074B

C 245

6E 23, 52–55bis

L 15, 17, 22

L ex 10051bis, 10221, 10224, 10266

Non-côté L 11

M 8885

Archives de l'Eure

1B 256–8, 392, 398

6B 207

73B 176, 182

14L 1

17L 1

18L 1
229L 24, 40
238L 84
6M 699, 700, 701, 762, 763
6U3 263

Archives du Finistère
6M 951, 959, 960
8M 49, 52, 71

Archives de la Seine et de la Ville de Paris
4 AZ 794, 795
D^4 B^6 13, 14, 19–29, 31, 34, 36, 38–63, 65–71, 75, 76, 78–80, 82–95, 97–106, 111, 113
D^5 B^6
D^5 Z 9
DQ^{10} 351

Archives de la Seine-Maritime
1B, 276, 277, 290–300, 1626, 1881, 3234, 3256, 3786
4BP 5911–5916, 5918–22, 5926, 5927, 5930, 5931, 5939, 5944, 5949, 5954, 5956, 5958, 5963
4BPL/60–73, 75; 4BPLY 1/7, 5/1
6BP 9, 11, 180, 184, 194
17BP 5135
201BP 22, 23, 530–562, 560, 595, 599, 605, 609, 621, 639, 682
202BP 14bis
C 62, 102–110, 118, 144, 182, 345, 350–357, 360, 362, 364, 366, 368, 372, 375, 378, 381, 384, 515, 518, 520–524, 526–528, 560, 609, 929–932, 934, 935, 945, 1085, 1086, 2200, 2220, and non-côtés
3E 169/49, 169/116
4E 215, 2105, 2106, 2145, 6611
5E 9, 52, 195–6, 423, 431, 512
L 15, 30, 139, 298–301, 307, 309, 320, 328, 342, 343, 361, 378, 379, 381–385, 387–390, 403, 407, 450–454, 456–457, 1387, 1393, 1563, 1559, 1784, 1819, 2172, 2173, 2175, 2286, 2297, 2347, 2375, 2378, 2379, 2381, 2382, 2390, 2391, 2393, 2394, 2409, 2453, 2462, 2734, 2820, 3009, 3010, 3012–3014, 3098, 3143, 3225, 3227, 3277, 3605, 3642, 3643, 3647, 3726, 3727, 3790, 3876, 4053, 4111, 4124, 4169, 4255, 4318, 4324, 4444, 4461, 4512, 4580, 4715, 4771, 5084, 5711, 5794, 5846, 6375
LP 6391, 6425, 6428, 6431, 6432, 6442, 6445, 6459, 6466bis, 6467, 6468, 6476, 6481, 6484–6486, 6488–6495, 6500–6505, 6651, 6652,

6633, 6652*, 6653, 6658, 7020, 7811–7814, 7893, 8092, 8105, 8440, 8442, 8450–8452, 8650, 8703
4M 114
6M 1070, 1118, 1181, 1188–1195, 1202, 1204, 1205, 1216, 1223–1228, 1295–1297, 1299, 1301–1304, 1308, 1309, 1378, 1381
8M 471, 472, 486, 492
*1NP 1
9 UP 2, 3, 6, 37, 79, 80

Archives de la Somme
C 80, 83, 85–88, 91, 99, 100, 103, 104, 105, 458, 459, 483, 484
L 439, 954, 1545
L^a 446
L^i 2446
Archives Départementales des Yvelines
2B 1142, 1144, 1145
2F 17, 18

Archives Municipales du Havre
CC 126
FF 24, 53, 59, 61, 64, 66
HH 25
PR F^4 148, 150, 153, 158, 166
FM F^4 7, 8, 9

Archives Municipales de Rouen
Chartrier, no. 262
A 36–40
B 15–17
F3 2, 4, 5, 7, 8, 10, 14
F4 16
I2 3, 4
I5 6
I9
Ms 1172 U 75
Fonds Leber

Archives Communales de Caen
F* 19, 31

Guildhall Library (London)
7798, 7798^A, 7798^B

Selected Bibliography

Primary Sources

Archives Parlementaires de 1787 à 1860, recueil complet des débats législatifs et politiques des chambres françaises. Première série. 95 vols. Paris, 1862–1988.

Barbier, [E. J. F.]. *Chronique de la Régence et du règne de Louis XV (1718–1763) ou Journal de Barbier*. 8 vols. Paris, 1857.

Code criminel et correctionel. 2 vols. Paris, An XIII-1805.

Collection complète des lois. . . . 2nd ed. X Vols. Ed. J. B. Duvergier. Paris, 1835.

Collection des principaux économistes. Ed. Eugène Daire. Reprint, Osnabrück, 1966.

Correspondance des Contrôleurs Généraux des finances avec les intendants des provinces. Ed. A. M. de Boislisle and P. de Brotonne. Paris, 1874–1897.

Cahiers de doléances du Tiers Etat du Bailliage d'Andely (secondaire de Rouen) pour les Etats Généraux de 1789. Ed. Marc Bouloiseau and Philippe Boudin. Rouen, 1974.

Cahiers de doléances du Tiers Etat du bailliage de Gisors (secondaire de Rouen) pour les Etats Généraux de 1789. Ed. Marc Bouloiseau. Paris, 1971.

Cahiers de doléances du Tiers Etat du bailliage de Rouen pour les Etats Généraux de 1789. Ed. Marc Bouloiseau. 2 vols. Rouen and Paris, 1960.

Cambacérès, [Jacques-Régis de]. *Lettres inédites à Napoléon*. Ed. Jean Tulard. Paris, 1973.

Chaptal, *De l'industrie française*, Ed. Louis Bergeron. Paris, 1993.

[Condorcet, Marie Jean Antoine Caritat de, and Anne Robert Jacques Turgot]. *Correspondance inédite de Condorcet et de Turgot (1770–1779)*. Ed. Charles Henry. Paris, n.d.

Delamare, Nicolas. *Traité de la police*. 4 vols. 2nd ed. Amsterdam and Paris, 1729.

Dupin, Claude. *Oeconomiques*. 2 vols. Ed. Marc Aunay. Reprint. Paris, 1913.

Documents relatifs à l'histoire des subsistances dans le District de Bergues pendant la Révolution (1788-An V). 2 vols. Ed. Georges Lefebvre. Lille, 1914.

Fain, [Agathon-Jean-François]. *Mémoires du Baron Fain, Premier Secrétaire du cabinet de l'Empéreur*. Ed. P[aul] Fain. 3rd ed. Paris, 1909.

Ferrière, Claude-Joseph de. *Dictionnaire de Droit et de Pratique....* 2 vols. Nouvelle édition. Toulouse, 1787.

French Royal and Administrative Acts, 1256–1794. Microfilm. New Haven, 1978.

Herbert, Claudes-Jacques. *Essai sur la police générale des grains, sur leurs prix et sur les effets de l'agriculture.* Paris, [1755].

Les Elections et les Cahiers de Paris en 1789. 4 vols. Ed. Ch.-L. Chassin. Paris, 1888–1889.

Memoires des intendants sur l'état des généralités dressés pour l'instruction du duc de Bourgogne. Ed. Arthur André Boislisle and Michel de Gabriel. Paris, 1881.

Ministère de l'Instruction publique. *La Commission des subsistances de l'An II: Procès verbaux et actes.* Ed. Pierre Caron. Paris, 1924.

Miromesnil, Armand-Thomas Hue de. *Correspondance politique et administrative de Miromesnil, premier président du Parlement de Normandie.* Ed. P. LeVerdier. 5 vols. Rouen and Paris, 1899–1903.

Napoléon, *Correspondance officielle.* Paris, 1970.

Napoléon sténographié au Conseil d'Etat. Ed. Alfred Marquiset. Paris, 1913.

Necker, Jacques. *Sur la législation et le commerce des grains,* 2 vols. Paris, 1775.

[Ouvrard, G-J], *Mémoires de G-J Ouvrard sur sa vie et ses diverses opérations financières.* 3 vols. Paris, 1826–1827.

Pasquier, Etienne-Denis. *A History of My Time: Memoires of Chancellor Pasquier.* 3 vols. Ed. Duc d'Audiffret Pasquier. Trans. Charles E. Roche. London, 1893–1894.

——. *The Memoires of Chancellor Pasquier, 1767–1815.* Ed. Robert Lacour- Gayet. Trans. Douglas Garman. Rutherford, Madison and Teaneck, N.J., 1968.

Procès-verbaux des comités d'agriculture et du commerce de la Constituante, de la Législative et de la Convention. 4 vols. Ed. Fernand Gerbaux and Charles Schmidt. Paris, 1906.

Quesnay, François. *Oeuvres économiques et philosophiques de F. Quesnay, fondateur du système physiocratique.* Frankfurt and Paris, 1888.

Recueil général des anciennes lois françaises. Ed. F.-A. Isambert, et al. 29 vols. Paris, 1822–33.

Rémusat, Charles de. *Mémoires de ma vie.* 5 vols. Ed. Charles-Henri Pouthas. Paris, 1958–1967.

Say, Jean-Baptiste. *An Economist in Troubled Times: Writings.* Ed. and trans. R. R. Palmer. Princeton, 1997.

Tableau des boulangers de Paris, pour l'éxercice de l'an 1831. Paris, 1831.

Tillet. *Expériences et observations sur le poids du pain....* Paris, 1781.

Turgot, [Anne Robert Jacques]. *Ecrits économiques.* Ed. Christian Schmidt. n.p., 1970.

——. *Oeuvres de Turgot.* Ed. Eugène Daire. 2 vols. Paris, 1844.

——. *Oeuvres de Turgot et documents le concernant....* Ed. Gustave Schelle. 4 vols. Paris, 1922.

Ville de Rouen. *Analyses des délibérations de l'Assemblée municipale et électorale....* Rouen, 1905.

Ville de Rouen. Conseil municipal. *Analyse des Procès verbaux des séances du 22 décembre 1800 au 20 novembre 1874.* 2 vols. Rouen, 1982.

Secondary Sources

A travers la Haute-Normandie en Révolution 1789–1800, n.p., 1992.

Abel, Wilhelm. *Agricultural Fluctuations in Europe from the Thirteenth to the Twentieth Centuries*. Trans. Olive Ordish. New York, 1980.

Adas, Michael. "From Avoidance to Confrontation: Peasant Protest in Pre-Colonial & Colonial Southeast Asia", *Comparative Studies in Society and History* 23 (April 1981): 217–247.

Adelman, Irma, and Morris, Cynthia Taft. "Patterns of Market Expansion in the Nineteenth Century: A Quantitative Study." *Research in Economic Anthropology* 1 (1978): 231–324.

Administration et contrôle de l'économie, 1800–1914. Ed. Michel Bruguière, Jean Clinquart, et al. Geneva, 1985.

Afanassiev, Georges. *Le Commerce des céréales en France au XVIIIe siecle*. Paris, 1894.

Aftalion, Florin. *The French Revolution: An Economic Interpretation*. Trans. Martin Thom. Cambridge, Mass, 1990.

Agriculture and National Development: Views on the Nineteenth Century. Ed. Lou Ferleger. Ames, Iowa, 1990.

Airiau, Jean. *L'Opposition aux Physiocrates à la fin de l'ancien régime: aspects économiques et politiques d'un libéralisme éclectique*. Paris, 1965.

Albaum, Martin. "The Moral Defenses of the Physiocrats' laissez-faire." *Journal of the History of Ideas* 16 (April 1955): 179–197.

Aminzade, Ronald. *Ballots and Barricades: Class Formation and Republican Politics in France, 1830–1871*. Princeton, 1993.

Andrews, Richard Mowery. *Law, Magistracy and Crime in Old Regime Paris, 1735–1789*. Cambridge, 1994.

Anon. "Salaires et revenus dans la généralité de Rouen, au XVIIIe siècle." *Bulletin de la société libre d'émulation de Rouen* (1885–6).

Appleby, Joyce Oldham. *Economic Thought and Ideology in Seventeenth-Century England*. Princeton, 1978.

Arbellot, Guy. "La grande mutation des routes de France au XVIIIe siècle." *AESC* 28 (May–June 1973): 765–791.

Ashley, Percy. *Modern Tariff History: Germany, United States, France*. New York, 1970.

Aspects de l'économie politique en France au XVIIIe siècle. Economies et Sociétés. Ed. G. Facarello. Paris and Grenoble, 1984.

Aymard, Maurice. *Venise, Raguse et le commerce du blé pendant la seconde moitié du XVIe siècle, Ports-Routes-Trafics*. 20. Paris, 1966.

Baker, Keith Michael. *Condorcet: From Natural Philosophy to Social Mathematics*. Chicago & London, 1975.

——. *Inventing the French Revolution: Essays on French Political Culture in the Eighteenth Century*. Cambridge, England, 1990.

Bamford, Paul W. "Entrepreneurship in 17th Century and 18th Century France: Some General Conditions and a Case Study." *Explorations in Entrepreneurial History* 9 (April 1958): 204–213.

Bardet, Jean-Pierre. *Rouen aux XVIIe et XVIIIe siècles: les mutations d'un espace social.* Paris, 1983.

—— and Ricque, Marie Paule. *Rouen vers 1770.* Caen, 1972.

Barksdale, Dudley. "Liberal Politics and Nascent Social Science in France: The Academy of Moral and Political Sciences, 1803–1852." 2 vols. Ph.D. diss., University of North Carolina, Chapel Hill, 1986.

Baudot, Marcel. *Le problème du pain à Evreux en 1788–1789.* n.p., 1939.

Baudrillart, Henri Joseph Léon. *Les populations agricoles de la France: la Normandie (passé et présent), enquête faite au nom de l'Académie des sciences morales et politiques.* Paris, 1880.

Baulant, Micheline. "Le Prix des grains à Paris de 1437 à 1788." *AESC* 23 (May–June 1968): 520–540.

Bechhofer, Frank, and Elliot, Brain. *The Petite-Bourgeoisie: Comparative Studies of the Uneasy Stratum.* New York, 1981.

Bell, David. *Lawyers and Citizens: The Making of a Political Elite in Old Regime France.* Oxford and New York, 1994.

——. "Lawyers into Demagogues: Chancellor Maupeou and the Transformation of Legal Practice in France, 1771–1789." *Past and Present* 130 (1991): 107–41.

Bercé, Yves Marie. *Histoire des croquants: étude des soulevements populaires au XVIIe siècle dans le sud-ouest de la France.* Geneva, 1974.

Bercé, Yves Marie. *Fête et révolte: des mentalités populaires du XVIe au XVIIIe siècle.* Paris, 1976.

Bergeron, Louis. *Banquiers, négociants et manufacturiers parisiens du Directoire à l'Empire.* Paris, 1985.

——. *France under Napoleon.* Trans. R. R. Palmer. Princeton, 1981.

——. *Les Capitalistes en France, 1780–1914.* Paris, 1978.

Bertaud, Jean-Paul. *Histoire du Consulat et de l'Empire: Chronologie commentée, 1799–1815.* Paris, 1992.

Bickart, Roger. *Les Parlements et la notion de souveraineté nationale au XVIIIe siècle.* Paris, 1932.

Binet, Pierre. *La Réglementation du marché du blé en France au XVIIIe siècle et à l'époque contemporaine.* Paris, 1939.

Biollay, Léon. *Etudes économiques sur le XVIIIe siècle; Le Pacte de Famine.* Paris, 1885.

Bisson de Barthélemy, Paul. *L'Activité d'un procureur général au parlement de Paris à la fin de l'ancien régime: les Joly de Fleury.* Paris, 1964.

Bloch, Camille. *Le commerce des grains dans la généralité d'Orléans d'aprés la correspondance inédite de l'intendant Cypierre.* Orléans, 1898.

Bloch, Camille, Pierre Caron, et al. *Le commerce des céréales: instructions, recueil de textes et notes.* Paris, 1907.

Bloch, Marc. *French rural history: an essay on its basic characteristics.* Trans. Janet Sondheimer. Berkeley, 1966.

Bohstedt, John. "Gender, Household, and Community Politics: Women in English Food Riots, 1790–1810." *Past and Present* 120 (August 1988): 88–122.

——. "The Politics of Riot and Food Relief in England during the Industrial Revolution." *Proceedings of the Consortium on Revolutionary Europe, 1750–1850* 1 (1980): 82–90.

——. "The Myth of the Feminine Food Riot: Women as Proto-Citizens in English Community Politics, 1790–1810." In *Women and Politics in the Age of Democratic Revolutions*. Ed. Harriet B. Applewhite and Darline G. Levy. Ann Arbor, 1990.

——. *Riots and Community Politics in England and Wales 1790–1810*. Cambridge, England, 1983.

Bois, Guy. "Le prix du froment à Rouen au XVe siecle." *AESC* (Nov.–Dec. 1968): 1262–1283.

Bord, Gustave. *Histoire du blé en France. Le Pacte de famine*. Paris, 1887.

Bordes, Maurice. *L'Administration provinciale et municipale en France au dix-huitième siècle*. Paris, 1972.

——. "Les intendants éclairés de la fin de l'ancien régime." *Revue d'histoire économique et sociale* 49 (1961): 57–83.

Bosher, John. *The French Revolution*. New York and London, 1988.

——. *Franch Finances, 1770–1795*. Cambridge, England, 1970.

——. "L'Opinion des physiocrats sur l'Angleterre." Mémoire de D.E.S., Université de Paris, 1954.

——. *The Single Duty Project: A Study of the Movement for a French Customs Union in the Eighteenth Century*. London, 1964.

Bossenga, Gail. *The Politics of Privilege: Old Regime and Revolution in Lille*. Cambridge, England, 1991.

Bouderon, H. "La lutte contre la vie chère dans la généralité de Languedoc au XVIIIe siècle." *Annales du Midi* 66 (1954): 155–170.

Bouloiseau, Marc. *Le séquestre et la vente des biens des émigrés dans le district de Rouen (1792-an X)*. Paris, 1937.

Bourde, André J. *Agronomie et agronomes en France au XVIIIe siècle*. 3 vols. Paris, 1967.

——. *The Influence of England on the French Agronomes, 1750–1789*. Cambridge, England, 1953.

Bourguet, Marie-Noëlle. *Déchiffrer la France: La statistique départementale à l'époque napoléonienne*. Paris, 1988.

Bourne, Henry E. "Food Control and Price Fixing in Revolutionary France. I," *Journal of Political Economy* 27 (1919): 73–94, 188–209.

Bouton, Cynthia. "Gendered Behavior in Subsistence Riots: The French Flour War of 1775." *Journal of Social History* 23 (Summer 1990): 735–754.

——. *The Flour War: Gender, Class, and Community in Late Ancien Régime French Society*. University Park, PA., 1993.

Bouvet, Jean. *La Question des subsistances en Maconnais à la fin de l'ancien régime et au début de la Révolution (1788–1790)*. Macon, 1945.

Bouvet, Michel, and Bourdin, Pierre-Marc. *A travers la Normandie des XVIIe et XVIIIe siècles*. Caen, 1968.

Bouvier, Jean. "A propos de la crise dite 'de 1805.' Les crises économiques sous l'Empire." *AHRF* 199 (January–March 1970): 100–109.

Brenner, Robert, "Agrarian Class Structure and Economic Development in Pre-Industrial Europe." *Past and Present* 70 (February 1976): 30–75.

Bricourt, Michel; Lachiver, Marcel; and Queruel, Julien. "La crise de subsistance des années 1740s dans le ressort du Parlement de Paris." *Annales de Démographie Historique* (1974).

Brinley, Thomas. "The Rhythm of Growth in the Atlantic Economy of the Eighteenth Century." In *Research in Economic History* vol. 3, pp. 1–46. Ed. Paul Uselding. Greenwich, Conn., 1978.

British Economic Fluctuations 1790–1939. Ed. Derek H. Aldcroft and Peter Fearon. London, 1972.

Brochin, Maurice. *Les réglements sur les marchés des blés de Paris sous l'Ancien Régime*. Paris, 1917.

Brown, Howard G. "Politics, Professionalism, and the Fate of the Army Generals after Thermidor." *FHS*, 19 (Spring 1995): 133–52.

——. *War, Revolution, and the Bureaucratic State: Politics and the Army Administration in France, 1791–1799*. Oxford, 1995.

Bruguière, Michel. "Finance et noblesse: L'entrée des financiers dans la noblesse de l'Empire." *AHRF* 199 (January–March 1907): 161–170.

——. *Gestionnaires et profiteurs de la Révolution: L'administration des finances françaises de Louis XVI à Bonaparte*. Paris, 1986.

Buller, A. H. Reginald. *Essays on Wheat*. New York, 1919.

Butel, Paul. "Crise et mutation de l'activité économique à Bordeaux sous le Consulat et l'Empire." *AHRF* 199 (January–March 1970): 110–134.

——. *L'économie française au XVIIIe siècle*. Paris, 1993.

——. "Guerre et commerce: l'activité du port de Bordeaux sous le régime des licenses 1708–1815." *RHMC* 19 (1972): 128–149.

Cahen, Léon. "Le prétendu pacte de famine. Quelques précisions nouvelles." *Revue Historique* 176 (September–October 1935): 173–216.

Caire, Guy. "Bertin, ministre physiocrate." *Revue d'histoire économique et sociale* 38 (1960): 257–284.

Cameron, Ian A. *Crime and Repression in the Auvergne and the Guyenne, 1720–1790*. Cambridge, England, 1981.

Capitalism in Context: Essays on Economic Development and Cultural Change in Honor of R. M. Hartwell. Eds. John A. James and Mark Thomas. Chicago, 1994.

Carré, Antonio. *Necker et la question des grains à la fin du XVIIIe siècle*. Reprint. New York, 1979.

Carrière, Charles. *Négociants marseillais au 18e siècle*. 3 vols. Marseilles, 1973.

Caty, R. "Une ascension sociale au début du XIXe siècle: Jean-Louis Bethfort et le commerce des blés à Marseille de 1802 à 1820." *Provence historique* 92 (April–June 1973): 164–216.

Censer, Jack, and Jeremy Popkin. *Press and Politics in Pre-Revolutionary France*. Berkeley, 1987.

Chaline, Olivier. "Le juge et le pain: Parlement et politique d'approvisionnement en 1788–1789, d'après les papiers du Procureur général de Rouen." 39 *Annales de Normandie* (March 1989): 21–35.

Chapuisat, Edouard. *Necker (1732–1804)*. Paris, 1938.

Chassagne, Serge. *Oberkampf, un entrepreneur au siècle des lumières*. Paris, 1980.

Chaussinand, G. "Capital et structure sociale sous l'Ancien Régime." *AESC* (March–April 1970): 436–476.

Chevet, Jean-Michel. "Le marquisat d'Ormesson (1700–1840). Essai d'analyse économique." Thèse de troisième cycle, EHESS, 1982.

Church, Clive F. *Revolution and Red Tape: The French Ministerial Bureaucracy, 1770–1850*. Oxford, 1981.

Clérembray, Félix. *La Terreur à Rouen 1793–1794–1795, d'après des documents inédits*. Rouen and Paris, 1901.

Cobb, Richard. *The People's Armies: The armées révolutionnaires: Instrument of the Terror in the Departments April 1793 to Floréal Year II*. Trans. Marianne Elliot. New Haven and London, 1987.

——. *The Police and the People; French Popular Protest 1789–1820*. Oxford, 1970.

——. *Terreur et subsistances 1793–1795; Etudes d'histoire révolutionnaire*. Paris, [1964].

Cole, Charles Woolsey. *Colbert and a Century of French Mercantilism*. 2 vols. Hamden, Conn., 1964.

——. *French Mercantilism, 1683–1700*. Reprint. New York, 1971.

Collingham, H.A.C. *The July Monarchy: A Political History of France, 1830–1848*. With R.S. Alexander. London and New York, 1988.

Collins, James B. "The Role of Atlantic France in the Baltic Trade: Dutch Traders and Polish Grain at Nantes, 1625–1675." *Journal of European Economic History* 13 (Fall 1984): 239–290.

Comité des travaux historiques et scientifiques. *La Révolution française et le monde rural, Actes du colloque tenu en Sorbonne les 23, 24 et 25 octobre 1987*. Paris, 1989.

Comité Régional d'Histoire de la Révolution Française (Haute-Normandie) et al. *A travers la Haute-Normandie en révolution 1789–1800: études et recherches*, n. p., 1992.

Conan, Jules. "Les débuts de l'école physiocratique. Un faux départ: l'échec de la réforme fiscale." *Revue d'histoire économique et sociale* 36 (1958): 45–63.

Crafts, N.F.R. "Economic Growth in France and Britain 1830–1910: A Review of the Evidence." *JEH* 44 (March 1984): 49–67.

Cubells, Monique. "Les mouvements populaires du printemps 1789 en Provence." *Provence historique* 36 (July–September 1986): 309–23.

Curmond, Henri. *Le commerce des grains et l'école physiocratique*. Paris, 1990.

Dakin, Douglas, *Turgot and the Ancien Régime in France*. Reprint. New York, 1965.

Dallas, George. *The Imperfect Peasant Economy: The Loire Country, 1800–1914*. Cambridge, England, 1982.

Dalton, George. *Economic Anthropology and Development; Essays on Tribal and Peasant Economies*. New York and London, [1971].

——. "The Impact of Colonization on Aboriginal Economies in Stateless Societies." In *Research in Economic Anthropology*, pp. 131–184. Ed. George Dalton. Greenwich, Conn., 1978.

Dammane, D. "L'économie politique sous le consulat et l'empire. Misére de l'économie, sciences des richesses." *Economies et sociétés* 20 (October 1986): 49–62.

Dardel, Pierre. *Commerce, industrie et navigation à Rouen et au Havre au XVIIIe siècle, rivalité croissante entre ces deux pors, la conjoncture*. Rouen, 1966.

——. *Importateurs et exportateurs rouennais au XVIIIe siècle. Antoine Guymonneau et ses opérations commerciales (1715–1741).* Dieppe, 1945.

——. *Navires et marchandises dans les ports de Rouen et du Havre au XVIIIe siècle.* Paris, 1963.

——. *Le trafic maritime de Rouen aux XVIIe et XVIIIe siècles, essai statistique.* Rouen, 1946.

——. *Statuts des boulangers et Barème du Prix du Pain à Lillebonne vers 1461; comparaison avec le Barème actuel.* Rouen, 1934.

Darnton, Robert, "Le Lieutenant de police J.P. Lenoir, la Guerre des farines et l'approvisionnement de Paris à la veille de la révolution." *RHMC* 17 (1969): 611–624.

Dejean, Etienne. *Un préfet du Consulat: Jacques-Claude Beugnot.* Paris, 1907.

Dejoint, Georges. *La politique économique du Directoire.* Paris, 1951.

Desert, Gabriel. *La Révolution française en Normandie.* Toulouse, 1989.

Desmarest, Charles. *Le commerce des grains dans la généralité de Rouen à la fin de l'ancien régime.* Paris, 1926.

Desportes, Françoise. *Le pain au Moyen Age.* Paris, 1987.

Dewald, Jonathan. *The Formation of a Provincial Nobility: The Magistrates of the Parlement of Rouen, 1499–1715.* Princeton: Princeton, 1980.

Dimet, Jacques. *1789: Evreux, La Révolution.* Paris, 1988.

Dorfman, Robert. *Prices and Markets.* 2nd ed. Englewood Cliffs, N. J., 1972.

Doyle, William. *The Oxford History of the French Revolution.* Oxford and New York, 1990.

——. *The Parlement of Bordeaux and the End of the Old Regime 1771–1790.* London and Tonbridge, 1974.

Dupâquier, J., M. Lachiver and J. Meuvret. *Mercuriales du Pays de France et du Vexin français (1640–1792). Monnaie, Prix, Conjuncture.* Vol. 7. Paris, 1968.

Dupâquier, Jacques. "La non-révolution agricole du XVIIIe siècle," *AESC* 27 (January–February 1972): 80–2.

Echeverria, Durand. *The Maupeou Revolution: A Study in the History of Libertarianism in France, 1770–1774.* Baton Rouge and London, 1985.

Economic Development and Social Change: The Modernization of Village Communities. American Museum Sourcebooks in Anthropology. Ed. George Dalton. Garden City, N.Y., 1971.

Economics and the Historian. Ed. Tom G. Rawski, et al. Berkeley, 1996.

L'Economie politique en France au XIXe *siècle.* Eds. Yves Breton and Michel Lutfalla. Paris, 1991.

Ehrard, Jean. *L'idée de nature en France dans la première moitié du XVIIIe siècle.* 2 vols. Paris, 1963.

Egret, Jean. *Louis XV et l'opposition parlementaire 1715–1774.* Paris, 1970.

——. *Necker, Ministre de Louis XVI.* Paris, 1975.

Essays in French Economic History. Ed. Rondo Cameron. Homewood, IL, 1970.

Etner, François. *Histoire du calcul économique en France.* Paris, 1987.

Evrard, F. "Les subsistances en céréales dans le département de l'Eure de 1788 à l'an V." *Bulletin d'histoire économique de la Révolution* (1909): 18–22.

Faccarello, G[ilbert]. *Aux origines de l'économie politique libérale: Pierre de Boisguilbert.* Paris, 1986.

Farge Arlette. *La vie fragile: violences, pouvoirs et solidarités à Paris au XVIIIe siècle*. Paris, 1986.

——. *Déliquence et criminalité: le vol d'aliments à Paris au XVIIIe siècle.* Civilisation et mentalités. Paris, 1974.

Farr, James R. "Artisans, Magistrates and the Moral Economy in 16th-Century Dijon." American Historical Association Annual Meeting, Chicago, 28 December 1984.

Faucon, G.H. *La juridiction consulaire de Rouen, 1556–1905, d'après les documents authentiques et avec l'agrément du Tribune de Commerce de Rouen.* Evreux, 1905.

Faure, Edgar. *La disgrâce de Turgot. 12 mai 1776*. Paris, 1961.

Festy, Octave. *L'agriculture pendant la Révolution française. Les conditions de production et de récolte des céréales. Etude d'histoire économique*. Paris, 1967.

——. *Les délits ruraux et leur répression sous la Révolution et le Consulat*. Paris, 1956.

Floquet, Amable-Pierre. *Histoire du parlement de Normandie*. 7 vols. Rouen, 1840–1842.

The Formation of Nation States in Western Europe. Ed. Charles Tilly. Princeton, 1975.

Forrest, Alan. *The French Revolution and the Poor*. New York, 1981.

Forster, Robert. *Merchants, Landlords, Magistrates. The Depont Family in Eighteenth-Century France*. Baltimore and London, 1980.

——. *The Nobility of Toulouse in the Eighteenth Century: A Social and Economic Study*. Baltimore, 1960.

——. "Obstacles to Agricultural Growth in Eighteenth-Century France." *AHR* 75 (October 1970): 1600–1615.

Fox-Genovese, Elizabeth. *The Origins of Physiocracy: Economic Revolution and Social Order in Eighteenth-Century France*. Ithaca, 1976.

——. "The Physiocratic Model and the Transition from Feudalism to Capitalism." *Journal of European Economic History* 4 (Winter 1975): 725–737.

François Quesnay et la physiocratie. 2 vols. Paris, 1958.

Frêche, Georges. "Etudes statistiques sur le commerce céréalier de la France meridionale au XVIIIe siècle." *Revue d'histoire économique et sociale* 49 (1971): 5–43, 180–223.

——. *Histoire des prix des céréales à Toulouse (1650–1715)*. Paris, 1964.

Frêche, Georges, and [Frêche], Geneviève. *Les prix dex grains, de vins et des légumes à Toulouse (1489–1868)*. Paris, 1967.

The French Revolution and the Creation of Modern Political Culture. Eds. Keith Baker, Colin Lucas and François Furet. Oxford, 1987–94.

Frondeville, Henri and Odette de. *Les conseillers du Parlement de Normandie*. Vol. 1, *Les conseillers de Normandie de 1641 à 1715*. Rouen, 1970.

Furet, François. *Interpreting the French Revolution*, trans. Elborg Forster. Cambridge (England) and Paris, 1985.

Garnier, Bernard. "La mise en herbe dans le Pays d'Auge aux XVIIe et XVIIIe siècles." *Annales de Normandie* 25 (March 1975): 157–77.

——. "Pays herbagers, pays céréaliers et pays "ouverts" en Normandie (XVIe-début XIXe siècle)." *Revue d'histoire économique et sociale* 53 (1975): 493–525.

Gay, Peter. *The Enlightenment: An Interpretation*. 2 vols. New York, 1966, 1969.

Geiger, Reed G. *Planning the French Canals: Bureaucracy, Politics and Entreprise under the Restoration*. Newark, London and Toronto, 1994.

Gille, Bertrand. *La Banque en France au XIX*[e] *siècle. Travaux de droit, d'économie, de sociologie et de sciences politiques* 81. Geneva, 1970.

Girard, Colette C. "La disette de 1816–1817 dans la Meuthe," *Annales de l'Est* 6 (1995): 33–362.

Godechot, Jacques. *Les Institutions de la France sous la Révolution et l'Empire*. 2nd ed. Paris, 1968.

Goldsmith, James L. "The agrarian history of preindustrial France. Where do we go from here?" *Journal of European Economic History* 13 (Spring 1984): 175–199.

Gomel, Charles. *Les causes financières de la Révolution française*. 2 vols. Paris, 1982.

Gosselin, E. *Journal des principaux épisodes de l'époque révolutionnaire à Rouen et dans les environs. . . .* Rouen, 1867.

Grab, Alexander I. "The Politics of Subsistence: The Liberalization of Grain Commerce in Austrian Lombardy under Enlightened Despotism." *JMH* 57 (1985): 185–210.

Grange, Henri. *Les idées de Necker*. Paris, 1974.

Grantham, George, "Agricultural Supply During the Industrial Revolution." *JEH* 49 (March 1989): 43–72.

Gruder, Vivian. *The Royal Provincial Intendants*. Ithaca, 1968.

Guéneau, Louis. "La disette de 1816–1817 dans une région productrice de blé, la Brie." *Revue d'histoire moderne* 9 (Jan-Feb 1929): 18–46.

Guerreau, Alain. "Mesures du blé et du pain à Macon (XIVe–XVIIIe siècle)." *Histoire et Mesure* 3 (1988): 163–219.

La Guerre du blé au XVIIIe siècle: La critique populaire contre le libéralisme économique au XVIIIe siècle. Eds. Florence Gauthier and Guy-Robert Ikni. Montreuil, 1988.

Gullickson, Gay L. "Agriculture and Cottage Industry: Redefining the Causes of Proto-Industrialization." *JEH* 43 (December 1983): 831–850.

——. *Spinners and Weavers of Auffay: Rural Industry and the Sexual Division of Labor in a French Village, 1750–1850*. Cambridge, England, 1986.

Habermas, Jürgen. *The Structural Transformation of the Public Sphere: An Inquiry into a Category of Bourgeois Society*, trans. Thomas Burger. Cambridge, England, 1989.

Hall, John R. *The Bourbon Restoration*. London, 1909.

Hardman, John. *French Politics 1774–1789: From the Accession of Louis XVI to the Fall of the Bastille*. London and New York, 1994.

Hardy, James D., Jr. *Judicial Politics in the Old Regime: The Parlement of Paris during the Regency*. Baton Rouge, 1967.

Harsin, Paul. *Les doctrines monétaires et financières en France du XVIe au XVIIIe siècle*. Paris, 1928.

Heckscher, Eli. *Mercantilism*, 2 vols., rev. 2nd ed. Ed. E.F. Soderland. Trans. Mendel Shapiro. London, 1955.

Henriot, Marcel. "Les subsistances en Côte d'Or (l'application de la loi du 16 septembre 1792 dans le département). In *Commission de recherche et de publication des documents relatifs à la vie économique de la Révolution*. Paris, 1945.

Hicks, John. *Classics and Moderns. Collected Essays on Economic Theory*. Vol. 3. Cambridge, Mass., 1983.

Higgs, David. *Nobles in Nineteenth-Century France: The Practice of Inegalitarianism*. Baltimore and London, 1987.

Higonnet, Patrice. *Class, Ideology and the Rights of Nobles during the French Revolution*. Oxford, 1981.

Hincker, François. *La Révolution française et l'économie: Décollage ou catastrophe?* Paris, 1989.

Hirsch, Jean-Pierre. *Les deux rêves du Commerce: entreprise et institution dans la région lilloise* (1780–1860). Paris, 1991.

Hirschman, Albert O. *The Passions and the Interests: Political Arguments for Capitalism before Its Triumph*. Princeton, 1974.

Histoire de la Normandie. Univers de la France. Ed. Michel de Bouard. Toulouse, 1970.

Histoire du Havre et de l'estuaire de la Seine. Ed. André Corvisier. Toulouse, 1987.

Histoire, économies, sociétés, Journées d'études en honneur de Pierre Léon (6–7 mai 1977). Lyon, 1978.

Hoffman, Philip. *Growth in a Traditional Society: The French Countryside 1450–1815*. Princeton, 1996.

——. "Land Rents and Agricultural Productivity: The Paris Basin, 1450–1789," *JEH* 51 (December 1991): 771–805.

Hoffman, Philip T., Gilles Postel-Vinay, and Jean-Laurent Posenthal, "Private Credit Markets in Paris, 1690–1840." *JEH* 52 (June 1992): 293–306.

Hohenberg, P. "Change in Rural France in the Period of Industrialization." *JEH* 32 (1972): 219–240.

——. *A Primer on the Economic History of Europe*. New York, 1968.

Hoock, Johan. "Le phénomène Savary et l'innovation en matière commerciale en France aux XVIIe et XVIIIe siècles." In *Innovations et renouveaux techniques de l'antiquité à nos jours*. Ed. Jean-Pierre Kintz. Association Interuniversitaire de l'Est, Vol. 24. Strasbourg, n.d, pp. 113–23.

Hufton, Olwen. *The Poor of Eighteenth Century France, 1750–1789*. Oxford, 1974.

——. "Social Conflict and the Grain Supply in Eighteenth-Century France." *JIH* 14 (Autumn 1983): 303–331.

Hunt, Lynn Avery. "Committees and Communes: Local Politics and National Revolution in 1789," *Comparative Studies in Society and History* 18 (July, 1976): 312–46.

——. *Revolution and Urban Politics in Provincial France: Troyes and Reims, 1786–90*. Stanford, 1978.

——. *Politics, Culture and Class in the French Revolution*. Berkeley, California, 1984.

Institut National d'Etudes Démographiques. *François Quesnay et la Physiocratie*. 2 vols. Paris, 1958.

Jones, D.J.V. "The Corn Riots in Wales, 1793–1801." *The Welsh History Review* 2 (1965): 323–350.

Joynes, D. Carroll. "Jansenists and Ideologues: Opposition Theory in the Parlement of Paris, 1750–1775," Ph.D. diss., University of Chicago, 1981.

Lebovics, Herman. *The Alliance of Iron and Wheat in the Third French Republic, 1860–1914: The Origins of the New Conservatism*. Baton Rouge and London, 1988.

Kaiser, Thomas E. "Money, Despotism, and Public Opinion in Early Eighteenth-Century France: John Law and the Debate on Royal Credit." *JMH* 63 (March 1991): 1–28.

——. "Public Credit: John Law's Scheme and the Question of *Confiance*." *Proceedings of the Annual Meeting of the Western Society for French History* 16 (1989): 72–81.

——. "The Abbé de Saint-Pierre, Public Opinion, and the Reconstitution of the French Monarchy," *JMH* 55 (December 1983): 618–43.

Kaplan, Steven Laurence. *The Bakers of Paris and the Bread Question, 1700–1775*. Durham and London, 1996.

——. *Bread, Politics and Political Economy in the Reign of Louis XV*. 2 vols. The Hague, 1976.

——. *The Famine Plot Persuasion in Eighteenth-Century France. Transactions of the American Philosophical Society*. 72, part 3 (1982).

——. "Lean Years, Fat Years: The 'Community' Granary System and the Search for Abundance in Eighteenth-Century Paris." *FHS* 10 (Fall 1977): 197–230.

——. "Luxury Guilds in Paris in the Eighteenth Century." *Francia* 9 (1982): 257–298.

——. "The Paris Bread Riot of 1725." *FHS* 14 (Spring 1985): 23–56.

——. "Réflexions sur la police du monde du travail, 1700–1815," *Revue historique* 261, no. 1 (January–March 1979): 17–77.

——. "Religion, Subsistence and Social Control: The Uses of Saint-Genevieve." *Eighteenth-Century Studies* 12 (Winter 1979–80): 142–168.

——. *Provisioning Paris: Merchants and Millers in the Grain and Flour Trade during the Eighteenth Century*. Ithaca, 1984.

Kaplow, Jeffrey. *Elbeuf during the Revolutionary Period: History and Social Structure*. Baltimore, 1964.

Kemp, Tom. *Economic Forces in French History*. London, 1971.

Keohane, Nannerl. *Philosophy and the State in France: The Renaissance to the Enlightenment*. Princeton, 1980.

Kroen, Sheryl Tracy. "The Cultural Politics of Revolution and Counterrevolution in France, 1815–1830." Ph.D. diss., University of California, Berkeley, 1992.

Kula, Witold. *Measures and Men*. Trans. R. Setzer. Princeton, 1986.

Labrousse, Ernest. *La crise de l'économie française à la fin de l'Ancien Régime et au début de la Révolution*. Paris, 1943.

——. *Esquisse du mouvement des prix et des revenues en France au XVIIIe siècle*. 2 vols. Paris, 1933.

Labrousse, Ernest et al. *Histoire économique et sociale de la France*. Vol. 2: *Des derniers temps de l'âge seigneurial aux préludes de l'âge industriel 1660–1789*. Paris, 1975.

Lachiver, Marcel. *Les années de misère: La famine au temps du grand roi, 1680–1720*. Paris, 1991.

Lachmann, Ludwig M. *Capital, Expectations, and the Market Process: Essays on the Theory of the Market Economy*. Kansas City, 1977.

Lantier, Maurice. "La crise des subsistances en 1784, à Saint-Lô." *Annales de Normandie* 25 (March 1975): 13–31.

——. *Crise des subsistances à Saint-Lô au printemps 1789*. Caen, 1974.

Latham, A.J.H., and Neal, Larry. "The International Market in Rice and Wheat, 1868–1914." *EHR*, 2nd series, 36 (May 1983): 260–280.

Laugier, Lucien. *Un Ministère reformateur sous Louis XV. Le Triumvirat (1770–1774)*. Paris, 1975.

Laurent [first name N/A]. "La réglementation municipale de la distribution des pains et de la boulangerie à Dijon sous le régime du Maximum et pendant la disette de l'an II." In *Commission de Recherche et de publication des documents relatifs à la vie économique de la Révolution*, pp. 177–185, Paris, 1945.

Lefebvre, Georges. *The Coming of the French Revolution*. Trans. R.R. Palmer. Princeton, 1971.

——. *The Great Fear of 1789: Rural Ranic in Revolutionary France*. Trans. Joan White. Princeton, 1982.

——. *Napoleon*. 2 vols. Trans. Henry F. Stockhold. New York, 1969.

——. *Les Paysans du Nord pendant la Révolution française*. Lille, 1924.

——. *Questions agraires au temps de la Terreur*. Strasbourg, 1932.

Le Goff, T.J.A. *Vannes and Its Region: A Study of Town and Country in Eighteenth-Century France*. Oxford, 1981.

Lemarchand, Guy. *La fin du féodalisme dans le Pays de Caux: conjoncture économique et démographique et structure sociale dans une région de grande culture de la crise du XVIIe siècle à la stabilisation de la Révolution (1640–1795)*. Commission d'Histoire de la Révolution française, Mémoires et Documents, XLV. Paris, 1989.

——. "Les troubles de subsistances dans la généralité de Rouen." *AHRF* 35 (October–December 1963): 401–27.

Leroy, G. *La Famine à Melun en l'An III*. Melun, 1902.

Le Roy Ladurie, Emmanuel. "Au palmarès des pataquès." *Histoire, économie et société* 4 (1985): 433–8.

Letaconnoux, J. *Les subsistances et le commerce des grains en Bretagne au XVIIIe siècle*. Rennes, 1909.

L'Huiller, "Une crise de subsistances dans le Bas-Rhin (1810–1812). Origines, aspects principaux, évolution." *AHRF* 14 (1937): 518–36.

Livesey, Gerard James Christopher. "An agent of enlightenment in the French Revolution: Nicolas Louis François de Neufchâteau, 1750–1828." Ph.D. diss., Harvard University, 1994.

Ljublinski, Vladimir Sergueevitch. *La guerre des farines. Contribution à l'histoire de la lutte des classes en France, à la veille de la Révolution*. Trans. Françoise Adiba and Jacques Radiguet. Grenoble, 1979.

Lucas, Colin. "The First Directory and the Rule of Law." *FHS* 10 (Fall 1977): 231–260.

Lyons, Martyn. *Napoleon Bonaparte and the Legacy of the French Revolution*. London, 1994.

Mansel, Philip. *Louis XVIII*. London, 1981.

Margairaz, Dominique. *Foires et marchés dans la France préindustrielle. Recherches d'Histoire et de Sciences Sociales/Studies in History and the Social Sciences*, 33. Paris, 1988.

Marion, Marcel. *Machault d'Arnouville: Etude sur l'histoire du contrôle général des finances de 1749 à 1754*. Paris, 1891.

Marjolin, Robert. "Troubles provoqués en France par la disette de 1816–1817." *Revue d'histoire moderne*. New ser. 10 (November–December 1933): 425–448.

Markets in History: Economic Studies of the Past. Ed. David W. Galenson. Cambridge, England, 1989.

Marlin, Roger. *La crise des subsistances dans le Doubs*. Besançon, 1960.

Martin de Saint-Léon, Etienne. *Histoire des corporations de métiers*. Reprint. New York, 1975.

Martin-Fugier, Anne. *La vie élégante ou la formation du Tout-Paris, 1815–1848*. Paris, 1990.

Martineau, Jean. *Les Halles de Paris des origines à 1789: Evolution matérielle, juridique et économique*. Paris, 1960.

Mathiez, Albert. *La vie chère et le mouvement social sous la Terreur*. Paris, 1927.

Maza, Sarah. *Private Lives and Public Affairs: The Causes Célèbres of Prerevolutionary France*. Berkeley, 1993.

Mazauric, Claude. *Babeuf et la conspiration pour l'égalité* (Paris, 1962).

——. *La Révolution à Rouen: Aspects politiques, sociaux et institutionels (1789-An III)*. Rouen, 1967.

Meek, Reginald. *Studies in the Labour Theory of Value*. 2nd ed. London, 1973.

Meuvret, Jean. *Etudes d'histoire économique; recueil d'articles*. Cahiers des Annales 32. Paris, 1971.

——. *Le problème des subsistances à l'époque Louis XIV*. 3 vols. Paris, 1977–88.

Meyer, Jean. *Les capitalismes*. Paris, 1981.

Meyssonnier, Simone. *La balance et l'horloge: La génèse de la pensée libérale en France au XVIIIe siècle*. Paris, 1989.

Mille, Jerome. *Un physiocrate oublié: G.F. Le Trosne (1728–1780); Etude économique, fiscale et politique*. Paris, 1905.

Miller, Judith A. "Politics and Urban Provisioning Crises: Bakers, Police, and Parlements in France, 1750–1793." *JMH* 64 (June 1992): 227–62.

Minard, Philippe. "L'inspection des manufactures en France, de Colbert à la Révolution." Thèse de doctorat, Université de Paris-I, 1994.

Ministère de l'Economie, des Finances et du Budget. *Etat, finances et économie pendant la Révolution française, colloque tenu à Bercy les 12, 13, 14 octobre 1989. . . .* Paris, 1991.

Ministère de l'Instruction Publique. Bibliothèque de l'Ecole des Hautes Etudes. *Sciences Philologiques et Historiques*. 55e fasc. *Les Etablissements de Rouen*. Paris, 1883.

Monahan, W. Gregory. *The Year of Sorrows: The Great Famine of 1709 in Lyon*. Columbus, OH, 1993.

Moriceau, Jean-Marc. "Au rendez-vous de la 'Révolution agricole' dans la France du XVIIIe siècle: A propos de la grande culture." *Annales HSS* 49 (January–February 1994): 27–63.

——. *Les fermiers de l'Ile-de-France. Ascension d'un groupe social (XVe–XVIIIe siècles)* (Paris, 1994).
—— and Postel-Vinay, Gilles. *Ferme, entreprise, famille. Grande exploitation et changements agricoles. Les Chartier. XVIIe–XIXe siècles, Les Hommes et la Terre* 21 (Paris, 1992).
Morineau, Michel. *Les faux-semblants d'un démarrage économique: agriculture et démographie en France au XVIIIe siècle. Cahiers des Annales* 30. Paris, 1971.
——. "Prix et 'révolution agricole'." *AESC* 24 (March-April 1969): 403–23.
——. "Révolution agricole, révolution alimentaire, révolution démographique." In *Annales de Démographie historique 1974*. Paris and The Hague, 1974, pp. 335–71.
——. "Y a-t-il eu une révolution agricole en France au XVIIIe siècle?" *Revue historique* 486 (April–June 1968): 299–326.
Murphy, Antoine. "Le développement des idées économiques en France (1750–1756)." *RHMC* 33 (Oct–Dec 1986): 521–541.
——. *Richard Cantillon: Entrepreneur and Economist*. Oxford, 1986.
Naissance des libertés économiques: liberté du travail et liberté d'entreprendre: le décret d'Allarde et la loi Le Chapelier, leurs conséquences, 1791-fin XIXe siècle. Ed. Alain Plessis. Paris, [1993].
Newell, William. "The Agricultural Revolution in Nineteenth-Century France." 33 *JEH* (December 1973): 697–731.
O'Brien, D.P. *The Classical Economists*. Oxford, 1975.
O'Brien, Patrick, and Keyder, Caglar. *Economic Growth in Britain and France, 1780–1914*. London, 1978.
——. "Les voies de passage vers la société industrielle en Grande-Bretagne et en France (1780–1914)." *AESC* 34 (1979): 1284–1304.
Oury, Bernard. *A Production Model for Wheat and Feedgrains in France (1946–1961)*. Amsterdam, 1966.
Patrick, Alison. *The Men of the First French Republic: Political Alignments in the National Convention*. Baltimore, 1972.
La pensée économique pendant la Révolution française, Actes du colloque international de Vizille (6–8 septembre 1989). Ed. G. Faccarello and Ph. Steiner. Grenoble, 1990.
——. *Genèse d'une ville moderne; Caen au XVIIIe siècle*. 2 vols. Lille, 1974.
——. *Une histoire intellectuelle de l'économie politique, XVIIe–XVIII siècle. Civilisations et Sociétés* 85. Paris, 1992.
Persson, Karl Gunnar. "On Corn, Turgot, and Elasticities: The Case for Deregulation of Grain Markets in Mid-Eighteenth Century France." *Scandinavian Economic History Review* 41 (1993): 37–50.
Peterson, Christian. *Bread and the British Economy, c.1770–1870*. Ed. Andrew Jenkins. Hants, UK, 1994.
Peuchet, J[acques]. *Mémoires tirés des Archives de la Police de Paris, depuis Louis XIV jusqu'à nos jours*. 5 vols. Paris, 1838.
Petersen, Susanne. "L'approvisionnement de Paris en farine et en pain pendant la Convention." *AHRF* 56 (July–September 1984): 366–85.
Pigout, Jean. *La Révolution en Seine-Maritime: Bolbec 1789–1794*. Luneray, France, n.d.

Pinkney, David H. *Decisive Years in France, 1840–1847*. Princeton, 1986.
Poisson, Charles. *Les fournisseurs aux armées sous la Révolution française: Le Directoire des achats (1792–1793)*. Paris, 1932.
Poitrineau, Abel. *La vie rurale en Basse-Auvergne au XVIIIe siècle (1726–1789)*. Paris, 1965.
Polanyi, Karl. *The Great Transformation*. New York: Rinehart, 1944.
Post, John D. *The Last Great Subsistence Crisis in the Western World*. Baltimore and London, 1977.
Procacci, Giovanna. *Gouverner la misère: La question sociale en France (1789–1848)*. Paris, 1993.
Ragan, Bryant Timmons, Jr. "Rural Political Culture in the Department of the Somme during the French Revolution." Ph.D. diss., University of California, Berkeley, 1988.
Ramsay, Clay. *The Ideology of the Great Fear: The Soissonais in 1789*. Baltimore and London, 1992.
Raynaud, B. "Les discussions sur l'ordre naturel au XVIIIe siècle." *Revue d'économie politique* 19 (1905): 231–248.
Recktenwald, Horst Claus. *Political Economy: A Historical Perspective*. London, 1973.
Reddy, William M. *The Rise of Market Culture: The Textile Trade and French Society, 1750–1900*. Cambridge, England, 1984.
Reinhardt, Joan Wesselmann. "A French Town under the Old Regime: Aumale in the Seventeenth and Eighteenth Centuries." Ph.D. diss., University of Wisconsin-Madison, 1983.
Reshaping France: Town, Country and Region during the Revolution. Eds. Alan Forrest and Peter Jones. Manchester and New York, 1991.
Revisions in Mercantilism. Ed. D.C. Coleman. London, 1969.
La Révolution française et le développement du capitalisme. Actes du colloque de Lille, 19–21 novembre 1987. Revue du Nord, special volume. Eds. Gérard Gayot and Jean-Pierre Hirsch. No. 5, 1989.
Riley, James C. "French Finances, 1727–1768." *JMH* 59 (June 1987): 209–43.
——. *The Seven Years' War and the Old Regime in France: The Economic and Financial Toll*. Princeton, 1986.
Roehl, Richard. "French Industrialization: A Reconsideration." *Explorations in Economic History* 13 (July 1976): 233–81.
Roehner, Bertrand, et al. *Un siècle de commerce du blé en France, 1825–1913: Les fluctuations du champs des prix*. Paris, 1991.
Rogers, John W., Jr. "The Opposition to the Physiocrats: A Study of Economic Thought and Policy in the Ancien Régime, 1750–1780." Ph.D. diss., Johns Hopkins University, 1971.
Rosanvallon, Pierre. *La monarchie impossible: Les Chartes de 1814 et de 1830*. Paris, 1994.
Rose, R.B. "Eighteenth Century Price Riots, the French Revolution, and the Jacobin Price Maximums." *International Review of Social History* 4 (1959): 432–445.
——. "The French Revolution and the Grain Supply." *Bulletin of the John Rylands Library* 39, 1 (1956–1957): 171–187.

Rosenthal, Jean-Laurent. "Rural Credit Markets and Aggregate Shocks: The Experience of Nuits St. Georges, 1756–1776." *JEH* 54 (June 1994): 288–306.
Rothkrug, Lionel. *Opposition to Louis XIV: The Political and Social Origins of the French Enlightenment*. Princeton, 1965.
Roussier, Michel. *Le Conseil général de la Seine sous le Consulat*. Paris, 1960.
Rudé, Georges. *The Crowd in History: A Study of Popular Disturbance in France and England 1738–1840*. New York, 1964.
——. "La taxation populaire de mai 1775 à Paris et dans la région parisienne." *AHRF* 28 (1956): 139–179.
——. "La taxation populaire de mai 1775 en Picardie, en Normandie et dans le Beauvaisis." *AHRF* 35 (1961): 305–326.
Ryan, W.J.L., and Pearce, D.W. *Price Theory*. Revised ed. New York, 1977.
Samuelson, Paul A. *Economics: An Introductory Analysis*. 3rd ed. New York, 1955.
Schelle, Gustave. *Vincent de Gournay*. Paris, 1897.
Schumpeter, Joseph. *Economic Doctrine and Method: An Historical Sketch*. Trans. R. Aris. London, 1954.
——. *History of Economic Analysis*. Ed. Elizabeth Boody Schumpeter. New York, 1954.
Schwartz, Robert M. *Policing the Poor in Eighteenth-Century France*. Chapel Hill, N.C., 1988.
Sewell, William. *Work and Revolution: The Language of Labor from the Old Regime to 1848*. Cambridge, England, 1980.
Sibalis, Michael David. "Corporatism after the Corporations: The Debate on Restoring the Guilds under Napoleon I and the Restoration." *FHS* 15 (1988): 718–730.
Sion, Jules. *Les paysans de Normandie Orientale; Pays de Caux, Bray, Vexin Normand, Vallée de la Seine. Etude géographique*. Rouen, 1946.
Smith, David Kammerling. "'*Au bien du commerce*:' Economic Discourse and Visions of Society in France." Ph.D. diss. University of Pennsylvania, 1995.
Smith, Michael Stephen. *Tariff Reform in France, 1860–1900: The Politics of Economic Interest*. Ithaca and London, 1980.
Soboul, Albert. *The Sans Culottes of the Year II: The Popular Movement and Revolutionary Government, 1793–1794*. Trans. Remy Inglis Hall. Princeton, 1980.
Sonenscher, Michael. *Work and Wages: Natural Law, Politics and the Eighteenth-Century French Trades*. Cambridge, England, 1989.
Soublin, Léopold. *Le premier vote des normands (1789)*. Fécamp, 1981.
Stone, Bailey. *The Parlement of Paris, 1774–1789*. Chapel Hill, N.C., 1981.
Stouff, Louis. *Ravitaillement et alimentation en Provence aux XIVe et XVe siècles, Civilizations et Sociétés* 20. Paris and La Haye, 1970.
Sutherland, D.M.G. *France 1789–1815: Revolution and Counterrevolution*. New York, 1986.
Sydenham, M.J. *The First French Republic, 1792–1804*. Berkeley and Los Angeles, 1973.
Sylvestre, A.-J. *Histoire des professions dans Paris et ses environs*. Paris, 1853.
Szostak, Rick. *The Role of Transportation in the Industrial Revolution: A Comparison of England and France*. Montreal, 1991.

Themes in the Historical Geography of France. Ed. Hugh D. Clout. London, 1977.
Thompson, E.P. "The Moral Economy of the English Crowd in the Eighteenth Century." *Past and Present* 50 (1971): 76–136.
Tilly, Louise. "The Food Riot as a Form of Political Conflict in France," *JIH* 1 (1971): 23–57.
Tocqueville, Alexis de. *The Old Regime and the French Revolution.* Trans. Stuart Gilbert. New York, 1955.
Tombs, Robert. *France: 1814–1914.* London and New York, 1996.
Toutain, J.C. *Le produit de l'agriculture française de 1700 à 1958*, 2 vols. Paris, 1961.
Tulard, Jean. *Le Consulat et l'Empire (1800–1815).* Paris, 1970.
An Ungovernable People: The English and Their Law in the Seventeenth and Eighteenth Centuries. Eds. John Brewer and John Styles. New Brunswick. N.J., 1980.
Usher, Abbot Payson. *The History of the Grain Trade in France*. Cambridge, Mass., 1913.
Van Kley, Dale. "Church, State, and Revolution: The Debate over the General Assembly of the Clergy in 1765," *JMH* 51 (1979): 29–66.
Vardi, Liana. *The Land and the Loom: Peasants and Profit in Northern France, 1680–1800.* Durham and London, 1993.
Velde, François R. "The Financial Market and Government Debt Policy in France, 1746–1793." *JEH* 52 (March 1992): 1–39.
Vergnaud, Maurice. "Agitation politique et crise de subsistances à Lyon de septembre 1816 à juin 1817." *Cahiers d'histoire* 2 (1957): 163–178.
Viard, Pierre-Paul. "La disette de 1816–1817, particulièrement en Côte d'Or." *Revue historique* 159 (Sept–Oct 1928): 95–117.
Vidalenc, J. "La crise des subsistances et les troubles de 1812 dans le Calvados." In *Actes du 84e congrès national des sociétés savantes. Dijon, 1959. Section d'histoire moderne et contemporaine*. Paris, 1960, pp. 321–64.
Vivre en Normandie sous la Révolution. Ed. Olivier Chaline and Gérard Hurpin. Rouen, 1989.
Wallon, Henri. *La Chambre de commerce de la Province de Normandie (1703–1791).* Rouen, 1903.
Ware, Norman J. "The Physiocrats: A Study in Economic Rationalization." *American Economic Review* 21 (1931): 607–619.
Watson, Donald Steven. *Price Theory and Its Uses.* 2nd ed. Boston, 1968.
Weir, David R. "Crises économiques et les origines de la Révolution française." *AESC* 46 (July–August 1991): 917–47.
——. "Tontines, Public Finance, and the Revolution in France and England, 1688–1789," *JEH* 49 (March 1989): 95–124.
Welch, Cheryl B. *Liberty and Utility: The French Idéologues and the Transformation of Liberalism.* New York, 1984.
Weulersse, Georges. *Le mouvement physiocratique en France (de 1756 à 1770).* 2 vols. Paris, 1910.
——. *La physiocratie à l'aube de la Révolution 1781–1792.* Ed. Corinne Beutler. Paris, 1984.
——. *La physiocratie à la fin du régne de Louis XV (1770–1774).* Paris, 1959.

——. *La physiocratie sous les ministres de Turgot et de Necker (1774–1781)*. Paris, 1950.

——. *Les physiocrats*. Paris, 1931.

White, Eugene N. "Was There a Solution to the *Ancien Régime*'s Financial Dilemma?" JEH 49 (September 1989): 545–68.

Williams, Dale Edward. "Midland Hunger Riots in 1766." *Midland History* 3 (1976): 256–297.

——. "Morals, Markets and the English Crowd in 1766." *Past and Present* 104 (1984): 56–73.

Woloch, Isser. *The New Regime: Transformations of the French Civic Order, 1789–1820s*. New York and London, 1994.

Work in France: Representations, Meaning, Practice and Organization. Eds. Steven Laurence Kaplan and Cynthia Koepp. Ithaca, 1985.

Woronoff, Denis. *The Thermidorian Regime and the Directory, 1794–1799*. Trans. Julian Jackson. Cambridge and Paris, 1984.

Zupko, Ronald Edward. *French Weights and Measures Before the Revolution: A Dictionary of Provincial and Local Units*. Bloomington: Indiana, 1978.

Index